Rural Geography

Rural Geography

Processes, Responses and Experiences in Rural Restructuring

Michael Woods

Los Angeles • London • New Delhi • Singapore • Washington DC

SAGE Publications Ltd
1 Oliver's Yard
55 City Road
London EC1Y 1SP

SAGE Publications Inc.
2455 Teller Road
Thousand Oaks, California 91320

SAGE Publications India Pvt Ltd
B 1/I 1 Mohan Cooperative Industrial Area
Mathura Road, New Delhi 110 044
India

SAGE Publications Asia-Pacific Pte Ltd
33 Pekin Street # 02-01
Far East Square
Singapore 048763

British Library Cataloguing in Publication data

A catalogue record for this book is available
from the British Library

ISBN-13 978-0-7619-4760-8
ISBN-13 978-0-7619-4761-5

Library of Congress Control Number 2004095884

Typeset by C&M Digitals (P) Ltd., Chennai, India
Printed in Great Britain by TJ International, Padstow, Cornwall

FSC
Mixed Sources
Product group from well-managed
forests and other controlled sources
Cert no. SGS-COC-2482
www.fsc.org
© 1996 Forest Stewardship Council

Summary of Contents

Contents

Contents

Foreword

Although the history of Geography is replete with references to and emphases on regions, land and communities which might be considered as 'rural', the emergence of Rural Geography as a specific line of geographical enquiry spans only the past 30 years or so and might only be thought to have 'taken off' during the 1980s. In this relatively short period of time there has been a significant assembling and considering of material relating to the changing nature of rurality and rural areas, and there have also been noticeable attempts to bring wider theoretical frameworks and insights into the rural domain. Such attempts, sometimes confident, sometime faltering, have facilitated emphases on space, society, politics, economics, culture and nature (and hybrids of these) in our understanding of the rural, and rather than representing a series of episodic paradigm shifts they have deposited a rather palimpsestual landscape of theoretical enquiry.

Many of us would now argue that it is a good time to take stock of these approaches and their achievements. What have we learned about rurality as an object of desire, a focus of processes, a social construction, and how capable are we of understanding the different ways in which rurality is restructured and recomposed? To what extent has rural geography been content to establish itself and its concerns as a legitimate category of enquiry, and to what degree have rural geographers been persuasive in attributing significance to rural phenomena? Has rural geography been duped by the romantic and nostalgic appeal of idyllistic rurality, or has it been successful in lifting the cultural covers of idyll in order to expose the more problematic underbelly of social marginalization, poverty and homelessness in rural settings?

Mike Woods's book is an excellent and timely contribution to these necessary processes of stocktaking. He presents a clear, lively, informative and engaging account of rural restructuring, in terms of both the processes and practices that underpin rural change, and of multifaceted political-economic and social-cultural responses to that rural change. Mike is himself a leading rural scholar who has been engaged in agenda-setting research into rural politics and governance, so he is excellently placed to offer a summary of the state of the art of rural geography.

But more than that, he directs important attention to issues that are likely to make rural areas into even more significantly contested areas in the future. Traditional practices of food production will be cross-cut by political and ethical issues around food, health and landscape. The rural idyll has already been punctured by the dystopic imagery of death, destruction and emptiness following the outbreak of foot and mouth disease. Now urban-based consumers are insisting on having a greater say about the nature of food, farming, landscape and even about the ethics of traditional rural practices such as hunting. Seemingly urban-based government is insisting on having its say about both the look of the countryside and its place as a repository for housing and economic growth. Brussels-based policy-making continues to CAP the headless politic of the rural economy. By contrast, rural identity politics are making an increasingly noisy and visible contribution, infusing these debates with the supposedly united views of country people.

There seems little doubt that 'rural' and 'countryside' will continue as distinct discursive areas in everyday life, yet it is equally clear that there are many different countrysides and many different geographies of the rural. Mike Woods's book provides a scholarly framework from which to launch out into your own understandings of how rurality is being reconstructed. His challenge is that such understandings should be politically aware and relevant, yet sensitive and open to difference. I hope that you will accept that challenge critically and radically, for in your response, and in those like yours, lies the health and impact of future rural geographies.

Paul Cloke
University of Bristol

Acknowledgements

The production of a textbook such as this is a process of detection, exploration, examination, selection, collation, synthesis, editing and re-presentation. It by necessity draws on the work and ideas of a vast array of geographers, sociologists and other rural researchers, as credited in the text. In addition to the published papers and books that are referenced in the bibliography, I have been guided, informed and inspired by a great many conference papers, seminar presentations, discussions and informal conversations that have given me new insights, suggested fresh ways of approaching topics, led me to different readings, theories and case studies, and taught me about rural studies outside of the UK. I am grateful to my friends and colleagues in the rural studies community for these inadvertent contributions, which are not easy to cite formally.

I have also drawn inspiration, and gained insights, from my colleagues and students in the Institute of Geography and Earth Sciences at the University of Wales, Aberystwyth, the lively, dynamic and convivial atmosphere of which both facilitated and, occasionally, distracted from the writing of this book. In particular, I would like to acknowledge the support of those colleagues and research students with whom I have had the pleasure of collaborating on rural research: Bill Edwards, Mark Goodwin, Jon Anderson, Graham Gardner, Rachel Hughes, Simon Pemberton, Catherine Walkley, Eldin Fahmy, Owain Hammonds and Suzie Watkin.

I am grateful, too, to Robert Rojek and David Mainwaring at Sage for their careful stewardship of this project, as well as to the reviewers of earlier drafts of the manuscript for their generous comments and suggestions.

Much of the artwork in the book has been produced by Ian Gulley in the Institute's drawing office at Aberystwyth, with his characteristic skill and attention to detail.

The author and publishers wish to thank the following for permission to reproduce material:

Figure 1.2,   reproduced from P. Cloke (1977) 'An index of rurality for England and Wales', *Regional Studies*, 11, figure 2, p. 44, and P. Cloke and G. Edwards (1986) 'Rurality in England and Wales 1981: a replication of the 1971 index', *Regional Studies*, 20, figure 2, p. 293, by kind permission of Taylor and Francis Ltd. (www.tandf.co.uk/journals)

Figure 7.1, reproduced from R. Liepins (2000) 'New energies for old ideas', *Journal of Rural Studies*, 16, pp. 25–35, figure 1, with permission from Elsevier. Copyright 2000.

Figure 8.3, reproduced from Tranquil Area maps published by the Campaign to Protect Rural England, by kind permission of the Campaign to Protect Rural England. Copyright, Campaign to Protect Rural England and Countryside Agency, October 1995.

Figure 11.1, reproduced from B. Edwards, M. Goodwin, S. Pemberton and M. Woods (2000) *Partnership Working in Rural Regeneration*, figure 1, p. 7, by kind permission of the Policy Press.

Every effort has been made to trace all the copyright holders, but if any have been overlooked, or if any additional information can be given, the publishers will be pleased to make the necessary amendments at the first opportunity.

Part 1

INTRODUCING RURAL GEOGRAPHY

1

Defining the Rural

Introduction

Clear your mind and think of the word 'rural'. What image do you see? Maybe
you see the rolling green downland of southern England, or the wide open spaces
of the American prairie? Perhaps it's the golden woodlands of the New England
fall, or the forests of Scandinavia? The Rocky Mountains or the sun-baked outback
of Australia? Are there any people in your rural picture? If so, what are they doing?
Are they working? Or maybe they are tourists? What age are they? What colour
are they? Are they men or women? Rich or poor? Do you see any buildings in
your rural scene? Perhaps a quaint thatched cottage, or a white-washed farmstead?
Maybe a ranch, or a simple log cabin? Or do you see a run-down dilapidated
home, barely fit for human habitation; or an estate of modern, identikit, housing?
Is there any evidence of economic activity? Farming, probably, but then do you see
a farmyard of free-range animals, as the children's storybooks would have us
believe, or do you see battery hen sheds, or endless fields of industrially produced
corn? Maybe you see quarrying or mining or forestry. But what about factories, or
hi-tech laboratories or office complexes? Are there any shops, or banks, or schools –
or have they been converted into holiday homes? Are there any roads or traffic in
your image? Is there any crime, or any sign of police on patrol? Do you see any
problems of ill-health, or alcoholism, or drug abuse? Who owns the land that you
are picturing? Who has access to it?

Do you still have a clear picture of what 'rural' means to you, or are you
beginning to think that defining the rural is more complicated than you thought?
There is, alas, no simple, standard, definition. Whatever picture of the 'rural' you
have conjured up, it will probably be different from that imagined by the person
sitting nearest to you as you read this book. This is not to say that we all have an
entirely individual understanding of rurality. Our perceptions will be shaped by a
wide range of influences that we will share with other people: where we live,
where we holiday, which films we watch, which books we read. Local and national

cultural traditions are also important, as is what we learn at school, what we read in the newspapers and the political propaganda that we receive from pressure groups. In some countries, 'rural' is not a widely used concept at all but visitors to those countries will recognize spaces that look to them to be 'rural'. Thus, if our understanding of what 'rural' means is not individually specific, it is at least culturally specific. Someone living in the crowded countryside of south-east England will probably have a different idea about rurality from someone living in deepest North Dakota. A farming family in rural New Zealand will have a different idea from a city-dwelling tourist from Amsterdam. And so on … .

Yet, if 'rural' is such a vague and ambiguous term, in what sense can we talk about 'rural studies', or 'rural geography' or 'rural sociology'? This chapter introduces the different ways in which academics have attempted to produce a definition of rural, setting out the pros and cons of each approach, before eventually describing how the concept of rurality will be treated in this book.

Why Bother with Rural?

So, if 'rural' is such a difficult concept to define, why bother with it at all? For a start, distinctions between urban and rural, city and country, have a long historical pedigree and great cultural significance. Raymond Williams, one of the leading chroniclers of English language and literature, has observed that,

> 'Country' and 'city' are very powerful words, and this is not surprising when we remember how much they seem to stand for in the experience of human communities … On the actual settlements, which in the real history have been astonishingly varied, powerful feelings have gathered and have been generalised. On the country has gathered the idea of a natural way of life: of peace, innocence and simple virtue. On the city has gathered the idea of an achieved centre: of learning, communication, light. Powerful hostile associations have also developed: on the city as a place of noise, worldliness and ambition; on the country as a place of backwardness, ignorance, limitation. A contrast between country and city, as fundamental ways of life, reaches back into classical times. (Williams, 1973, p. 1)

So deep is this cultural tradition that differentiating between town and countryside is one of the instinctive ways in which we place order on the world around us. In academic usage, however, the term is more recent. Sociologist Marc Mormont, for example, has suggested that the use of 'rural' as an academic concept evolved during the 1920s and 1930s – a time when the countryside was undergoing major social and economic transformations – in an attempt to define the essential features of 'rural' society in the face of rapid urbanization and industrialization (Mormont, 1990). Very often, the definitions of rural society produced reflected a particular moral geography, with the 'rural' associated with values such as harmony, stability and moderation. These more judgemental ideas about the urban–rural dichotomy have been removed over time from academic thought, but the distinction remains a useful one for researchers for at least two reasons.

First, many governments officially distinguish between urban and rural areas and govern them through different institutions with different policies. For England, for example, the government published two separate policy papers in November 2000, one for 'urban policy' and one for 'rural policy', and much of the latter will be administered by the Department of the Environment, Food and *Rural* Affairs and implemented through the government's *Countryside* Agency.

Secondly, many people living in rural areas identify themselves as 'rural people' following a 'rural way of life'. So strong is this sense of identity that when they are faced with problems such as unemployment, the decline of staple industry (such as agriculture) or the loss of local services, they do not build links of solidarity with people experiencing the same problems in urban areas, but rather assert their rural solidarity as a basis for resistance to a perceived 'urban threat'. An example of this can again be seen in the UK, where over 400,000 people joined a march in London in September 2002 organized by the Countryside Alliance to protest at the perceived neglect of rural areas and rural interests by the central government (there is more on this in Chapter 14).

These two factors mean that although researchers may be able to identify the same social and economic processes at work in rural areas as in urban areas, they also know that the processes are operating in a different political environment and that the reactions of people affected may be different. The analysis of these differences, however, brings us back to the problem of what we mean by 'rural'. Halfacree (1993) identified four broad approaches that had been taken to defining the rural by rural researchers. These are (i) descriptive definitions; (ii) socio-cultural definitions; (iii) the rural as locality; and (iv) the rural as social representation. Each of these approaches will now be introduced and critiqued in turn.

Descriptive Definitions

Descriptive definitions of rurality are based on the assumption that a clear geographical distinction can be made between rural areas and urban areas on the basis of their socio-spatial characteristics, as measured through various statistical indicators. The simplest way of doing this is by population and this is the approach adopted in most official definitions of rural areas. After all, it appears to be fairly logical – we all know that towns and cities have larger populations than villages and dispersed rural communities. But, at precisely what population does a rural area become urban? As Table 1.1 shows, there is considerable variation in the maximum population size of a rural settlement permissible under the official definitions of rural and urban areas used in different countries.

There are other problems too. First, the population recorded depends on the boundaries of the area concerned. For example, if the population of the town in which I live, Aberystwyth in West Wales, is measured on its official community boundaries, then it comes in at just under 10,000 – sufficient to qualify as rural on some definitions. Yet the community boundary cuts right across the university campus. If the total population for the actual built-up urban area is counted, the real tally is nearer 20,000. Similarly, there are many rural counties in the United States that have larger populations than many incorporated urban areas, simply because they cover a much more extensive territory.

Secondly, simple population figures reveal nothing about the function of a settlement, or about the settlement's relation to its surrounding local area. A town of 1,000 people in Nebraska may be a definite urban centre for a dispersed rural population, but a village of 1,000 people in Massachusetts may be perceived to be rural in its regional context. Thirdly, distinctions based solely on population are arbitrary and artificial. Why should a settlement with 999 residents be classified as rural, and one with 1,000 residents be classified

Table 1.1 Official population-based definitions of rural settlements

Definition used by	Maximum population of a rural settlement	Notes
Iceland	300	Minimum population of an urban administrative unit
Canada	1,000	(+ population density less than 400 per km²) Census definition
France	2,000	
United States	2,500	Census definition
England	10,000	Countryside Agency definition
United Nations	20,000	
Japan	30,000	Minimum population of an urban administrative unit

as urban? What difference does that one extra person make?

Some official definitions of rurality have addressed these problems by developing more sophisticated models that also include reference to population density, land use and proximity to urban centres. In many countries a mix of different definitions is employed by different government agencies. For example, the website of the Rural Policy Research Institute (www.rupri.org) discusses nine different definitions used by parts of the United States government; whilst in the UK it has been recently estimated that there are over 30 different definitions of rural areas in use by different government agencies (ODPM, 2002). Many of these are actually 'negative' definitions in that they set out the characteristics of urban areas and designate anywhere that does not qualify as 'rural'. Three examples of this approach can be seen in the definitions used for the US and UK censuses and by the US Office of Budget and Management:

- The **US census** uses population to define urban areas as comprising all territory, population and housing units in places of 2,500 or more persons incorporated as cities, villages, boroughs (except in Alaska and New York), and towns (except in the six New England states, New York and Wisconsin). Everywhere else is classified as 'rural'.

- The **UK census** uses land use to define urban areas as any area with more than twenty continuous hectares of 'urban land uses' – including permanent structures, transport corridors (roads, railways and canals), transport features (car parks, airports, service stations, etc.), quarries and mineral works, and any open area completely enclosed by built-up sites. Everywhere else is classified as 'rural'.

- The **US Office of Budget and Management** defines metropolitan areas as at least one central county with a population of more than 50,000, plus any neighbouring county which has 'close economic and social relationships with the central county' – defined in terms of commuting patterns, population density and population growth. Anywhere outside a metropolitan area is classified as a 'non-metropolitan county' (Figure 1.1). Non-metropolitan counties are the most commonly used definition of a rural area in research and policy analysis in the United States.

All three of the above definitions, however, can be critiqued on the same grounds. First, they are dichotomous, in that they set up rural

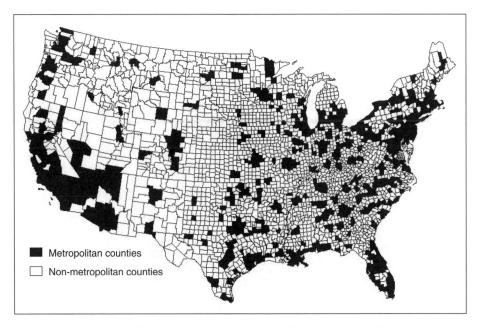

Figure 1.1   The US Office of Budget and Management's classification of metropolitan and non-metropolitan countries in the United States
Source: United States Department of Agriculture, Economic Research Service

areas in opposition to urban areas and recognize no in-between. Secondly, they are based on a very narrow set of indicators that reveal little about the social and economic processes that shape urban and rural localities. Thirdly, because rural areas are a residual category they are treated as homogeneous with no acknowledgement of the diversity of rural areas.

Indices of rurality

In an attempt to recognize some of the differences between degrees of rurality, and to overcome the problems that resulted from defining a rural area using just one or two indicators, Cloke (1977) and Cloke and Edwards (1986) constructed an 'index of rurality' for local government districts in England and Wales using a range of statistics from the 1971 and 1981 censuses. Significantly, the indicators used related not just to population (including population density, change, in-migration and out-migration and the age profile), but also household

amenities (percentage of households with hot water, fixed baths and inside WCs), occupational structure (percentage of workforce employed in agriculture), commuting patterns and the distance to urban centres. These indicators were fed into a formula that placed districts into one of five categories – extreme rural, intermediate rural, intermediate non-rural, extreme non-rural and urban (Figure 1.2).

Although the indices of rurality did mark an improvement on simple dichotomous definitions, it still provokes a number of critical questions. First, why choose the indicators that were used? What, for example, does the percentage of households with a fixed bath tell us about rurality? Secondly, how was the weighting between different indicators determined? Is agricultural employment more or less important than population density in determining rurality? Thirdly, how are the boundaries between the five different categories decided? At what point on the

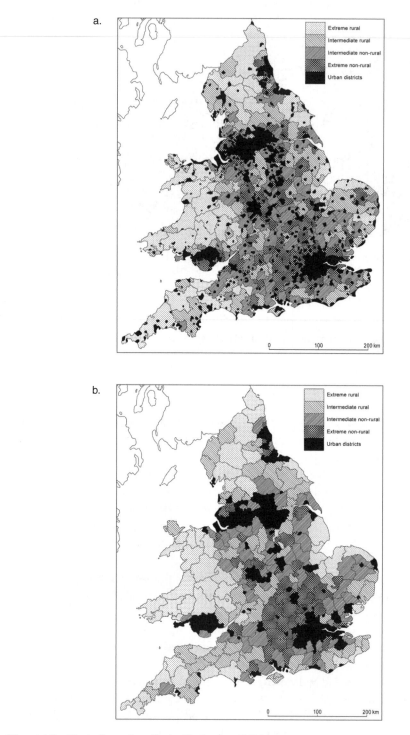

Figure 1.2 The indices of rurality for England and Wales, as calculated from the
1971 census (*a*) and the 1981 census (*b*)
Source: Cloke, 1977; Cloke and Edwards 1986

Table 1.2 Some urban/rural dichotomies employed in socio-cultural definitions

Author	Urban	Non-urban or rural
Becker	Secular	Sacred
Durkheim	Organic solidarity	Mechanical solidarity
Maine	Contact	Status
Redfield	Urban	Folk
Spencer	Industrial	Military
Tönnies	*Gesellschaft*	*Gemeinschaft*
Weber	Rational	Traditional

Source: Based on Phillips and Williams, 1984; Reissman, 1964

artificial scale produced by the formula does an 'intermediate rural' district become an 'intermediate non-rural' district?

More problematic still is the effect of using local government districts as the basis of the classification. Look at the two maps in Figure 1.2. On the 1971 map there are many isolated dots of black urban areas scattered across England and Wales. Yet, on the 1981 map they have disappeared. Did Britain suddenly become more rural during those ten years? No, local government had been reorganized in 1974, amalgamating the many small urban districts with their surrounding rural districts to create new, larger, districts – most of which came out as 'rural' when put through the formula for 1981. All that had happened was that the scale at which the index was calculated had changed.

Methodological flaws can be found with all the descriptive approaches employed to define rurality, but the real fundamental problem is identified by Halfacree (1993): 'Descriptive methods only describe the rural, they do not define it themselves' (p. 24). The descriptive definitions simply reflect preconceptions about what rural areas should be like, but offer no explanation as to why they are like that (or not).

Socio-cultural Definitions

Just as descriptive definitions have attempted to identify rural *territories*, so socio-cultural definitions have been used to try to identify

rural *societies*. In these approaches, distinctions are made between 'urban' and 'rural' society on the basis of residents' values and behaviours and on the social and cultural characteristics of communities. Two of the best-known examples are the models developed by Ferdinand Tönnies and by Louis Wirth. Tönnies based his distinction on the social ties found within rural and urban areas by contrasting the *Gemeinschaft*, or community, of the rural, with the *Gesellschaft*, or society, of the urban (see Tönnies, 1963). Wirth (1938), meanwhile, suggested that urban life was dynamic, unstable and impersonal, with an urban resident having different contacts through work, home and leisure, whereas rural life was stable, integrated and stratified, with the same people coming into contact with each other in different contexts. Other writers produced similar dichotomies (Table 1.2).

Dichotomies of this type over-emphasized the contrast between urban and rural societies. In response, the concept of a rural–urban continuum was devised, suggesting that communities could be identified as displaying different degrees of urban and rural characteristics. However, Pahl (1968) criticized the rural–urban continuum for continuing to oversimplify the dynamics of social and spatial milieux, arguing that 'some people are of the city but are not in it, whereas others are in the city but are not of it' (Phillips and Williams,

1984, p. 13). Pahl's own work identified so-called urban societies in rural Hertfordshire, whilst Young and Wilmott (1957) identified the supposed characteristics of rural communities in the East End of London.

The Rural as a Locality

The third approach to defining rural areas differed from the above two by focusing on the processes that might create distinctive rural localities. This approach was influenced by a wider debate within geography in the late 1980s that had explored how far local structures could shape the outcomes of social and economic processes. If, as some writers claimed, a 'locality effect' could be identified, might not it also be possible to distinguish between urban and rural localities? The challenge was therefore to identify the structural features that might allow this to be done: as Halfacree (1993) noted, 'rural *localities*, if they are to be recognised and studied as categories in their own right, must be carefully defined according to that which makes them *rural*' (p. 28).

Halfacree (1993) records that three main ways of doing this were attempted. First, it was suggested that rural space had to be associated with primary production (such as agriculture), or with 'the competitive sector'. Yet, as Halfacree notes, 'many urban localities could be similarly classified' (p. 28). Secondly, it was proposed that low population densities created distinctive connections between the rural and issues of collective consumption. Yet, again, Halfacree notes that the assertion is debatable, especially 'given the decline in the importance of friction of distance' (p. 28). Thirdly, rural localities were identified with a particular role in *consumption*, including the collective consumption of tourist sites and the private consumption of in-migrant house-buyers. However, it is not clear how this differs from gentrifying urban neighbourhoods and urban heritage sites.

The rural as locality approach faltered, therefore, because none of the structural features claimed to be rural could be proven to be uniquely or intrinsically rural. Instead, they simply highlighted the way in which the same social and economic processes appeared to be at work in both so-called urban and rural areas. Thus, in 1990 Hoggart proposed that it was time to 'do away with rural', arguing that it was a confusing 'chaotic conception' that lacked explanatory power:

> The broad category 'rural' is obfuscatory, whether the aim is description or theoretical evaluation, since intra-rural differences can be enormous and rural–urban similarities can be sharp. (Hoggart, 1990, p. 245)

So why are we still talking about the 'rural' more than a decade later? Because, as noted earlier, whatever academics might say about the difficulty of defining rural areas, there are still millions of people who consider themselves to be 'rural', to live in 'rural areas', and to follow a 'rural way of life'. It is the investigation of these perceptions that provides the foundation of the fourth approach.

The Rural as Social Representation

'There is an alternative way of defining rurality,' writes Halfacree, 'which, initially, does not require us to abstract causal structures operating at the rural scale. This alternative comes about because "the rural" and its synonyms are *words and concepts understood and used by people in everyday talk*' (Halfacree, 1993, p. 29). Thus, instead of trying to identify particular social characteristics or economic structures that are uniquely distinctive to rural areas, the social representation approach begins by asking what symbols and signs and images people conjure up when they think about the rural. This actually produces a more robust and flexible way of defining rurality, which can, for example, accommodate the effects of social and economic change in rural environments. As Mormont (1990) has argued, social and economic change means that there is no longer a single 'rural space' that can be functionally defined. Rather there are many imagined social spaces occupying the same territory.

The question of defining rurality hence becomes one of 'how people construct themselves as being rural', understanding rurality as 'a state of mind'. To employ a more technical vocabulary, rurality is 'socially constructed' (see Box 1.1) and 'rural' 'becomes a world of social, moral and cultural values in which rural dwellers participate' (Cloke and Milbourne, 1992, p. 360).

Box 1.1   Key term

Social construction: The way in which people give themselves, a place, an object or an idea an identity by attributing it with particular social, cultural, aesthetic and ideological characteristics. A social construct exists only in as much as people imagine it to exist.

This approach shifts attention from the statistical features of rural areas to the people who live there or visit it. It suggests that an area does not become 'rural' because of its economy or population density or other structural characteristics – but because the people who live there or use it think of it as being 'rural'. People have preconceived ideas about what 'rurality' means – informed by television, film, literature, holidays, life experience, etc. – and use this 'knowledge' to identify certain areas, landscapes, lifestyles, activities, people and so on as

being 'rural' (see for example Box 1.2). This in turn has a causal effect. If people think that they live in a rural area, and have preconceived ideas about what rural life should be like, it can influence their attitudes and behaviour. Similarly, people may be motivated to protect their image of what the countryside should be like if they feel it to be threatened – for example by housing development. Thus, as the rural is socially constructed differently by different people, conflicts can arise about what exactly it means to be rural and what rural areas should look like.

Box 1.2   What is rural? Views from rural Britain

In early 2002 a British pressure group, the Countryside Alliance, which represents traditional, pro-hunting and pro-farming rural interests, asked its members what it meant to be 'rural' and how 'rural' should be defined. These are some of the responses to the question 'What is rural?':

- 'A sparsely populated area, i.e. villages, hamlets and small towns necessitating travel for amenities not supplied in locality, i.e. cinema, bank, supermarket.'
- 'Rural should be defined as areas in which the primary land use is of an agricultural nature. This should include equestrian activities. Tourist activities should also be included. Dormitory villages should be excluded (definition of dormitory village

---

(Continued)

should be one where more than half of the working population travel more than 15 miles to work).'

- '"Rural" is as much a state of mind as an actual place. It is an acceptance and understanding of people and things living in a mainly agricultural area, the practices and traditions.'
- 'Rural is seeing the stars on a clear night, being able to breath unpolluted air, seeing wildlife in its natural habitat, being able to sleep without the constant noise of traffic. The beauty of nature in landscape, woodlands, hedgerows, etc.'
- 'Living and working in the countryside – with roots in the countryside from childhood. An understanding of the countryside and an unsentimental attitude to the animals, both wild and domesticated.'
- '"Rural" is where strange cars are noted'.

For more contributions, see www.countryside-alliance.org/policy/whatis/index.html.

The different ways in which the rural is socially constructed can be described as different 'discourses of rurality'. 'Discourse' in this sense means a way of understanding the world (see Box 1.3), and therefore discourses of rurality are ways of understanding the rural. As Halfacree (1993) comments, 'our attempts at defining the rural can be termed "academic discourses" because they are the constructs of academics attempting to understand, explain and manipulate the social world' (p. 31). But academics are not the only people to produce discourses. Frouws (1998) describes some of the *policy discourses* that have informed the government of rural areas in the Netherlands. These include the *agri-ruralist discourse*, in which the interests of agriculture are prioritized and 'farmers are considered as the principal creators and carriers of the rural as social, economic and cultural space' (Frouws, 1998, p. 59); the *utilitarian discourse*, in which the problems of rural areas are seen as the product of underdevelopment, and rural development initiatives are required to integrate rural areas into modern markets and socio-economic structures; and the *hedonist discourse*, in which the countryside is represented as a space of leisure and recreation and the 'ideal countryside' is perceived in terms of natural beauty and attractiveness.

Discourse: There are many different definitions of precisely what 'discourse' is, and the term is often used quite loosely. Put simply, however, discourses structure the way we see things. They are collections of ideas, beliefs and understandings that inform the way in which we act. Often we are influenced by particular discourses promoted through the media, through education, or through what we call 'common sense'. Derek Gregory, writing in *The Dictionary of Human Geography*, identifies three important aspects of discourse. (1) Discourses are not

Box 1.3   (Continued)

independent, abstract, ideas, but are materially embedded in everyday life. They inform what we do and are reproduced through our actions. (2) Discourses produce our 'taken for granted world'. They naturalize a particular view of the world and position ourselves and others in it. (3) Discourses always produce partial, situated, knowledge, reflecting our own circumstances. They are characterized by relations of power and knowledge and are always open to contestation and negotiation.

Just as important are the *lay discourses of rurality* produced and reproduced by ordinary people in their everyday lives, and the *popular discourses of rurality* that are disseminated through cultural media including art, literature, television and film. These two types of discourse are closely related as lay discourses will inevitably be influenced by popular discourses, and to some extent the opposite is also true. One of the most important popular discourses of rurality is that of the rural idyll (Bunce, 2003). This presents an aspirational picture of an idealized rurality, often emphasizing the pastoral landscape and the perceived 'peace and quiet', as Little and Austin (1996) and Short (1991) both describe:

> Rural life is associated with an uncomplicated, innocent, more genuine society in which traditional values persist and lives are more real. Pastimes, friendships, family relations and even employment are seen as somehow more honest and authentic, unencumbered with the false and insincere trappings of city life or with their associated dubious values. (Little and Austin, 1996, p. 102)

> [the countryside] is pictured as a less-hurried lifestyle where people follow the seasons rather than the stock market, where they have more time for one another and exist in more organic community where people have a place and an authentic role. The countryside has become the refuge from modernity. (Short, 1991, p. 34)

Whilst the 'rural idyll' is a myth, it has been influential in encouraging people to visit the countryside as tourists, and to move there as in-migrants. For many such people, elements of the rural idyll are entangled with lived experience to produce lay discourses that are never entirely matched in reality. Other lay discourses are more grounded in everyday life and can be cynical of, even negative towards, rural life.

Thinking about Rurality in Two English Villages: a Case Study

Examples of lay discourses of rurality can be found in the reports of two ethnographical studies of communities in rural southern England in the early 1990s – one by Michael Bell (1994) in the village of 'Childerley' (a pseudonym) in Hampshire, and the other by Owain Jones (1995) in an unnamed village in Somerset. The two villages are similar in that they both are within commuting distance of larger towns, and both have populations mixed between long-term, locally born residents and more recent in-migrants.

In Childerley, Bell found a number of in-migrants who described the rural nature of the village by drawing comparisons with the towns or cities that they had moved from. Usually, such comparisons emphasized the different pace of life:

In the towns, people are in a rush. That's the difference! In the towns, you get in your car [for everything]. I had a neighbor, lived there thirteen years. But I never spoke to her because she'd come out of her door, get in her car, go off, come back, and go indoors ... Here, the pace is that much slower. (In-migrant, quoted by Bell, 1994, pp. 91–92)

Life is like it was in the past here. You feel like you should lock it up every night. Coming home at night when we first moved here we used to think we should be closing a gate behind us at the bottom of the hill. (In-migrant, quoted by Bell, 1994, p. 93)

The influence of the 'rural idyll' can be seen in both these observations, yet Bell notes that even those who spoke most enthusiastically about the countryside ideal often qualified their statements. Furthermore, the perception of the countryside as a slower pace of life was shared by many longer-term residents. Bell quotes an 18-year-old farmer's son who comments that rural 'means a quieter lifestyle to start with. I don't know. You could call it an escape from the rat-race' (p. 91).

There are indeed a number of common elements that recur in the descriptions of rurality recorded by both Bell and Jones from all sections of the communities studied – and which reflect both geographical and social factors. First, the geographical context is important. Jones records a villager who suggested that rural meant 'a lack of industry, traffic, shops, offices, dense man-made environment' (p. 43), and another who commented that the village was rural because it was 'void of urban facilities, i.e., industry, street lighting' (p. 43). The presence of farming is also significant for many. Jones again reported the comments of one resident that, 'we are fortunate to have several local farms, animals graze the fields. Tractors track up and down the road. Not always a blessing!' (p. 42).

Secondly, rural life was associated with a close-knit sense of community, with people drawing on examples from their own experience:

the small size of the community has encouraged me to get involved in part so that I can meet other villagers and also in order to support village amenities such as the hall, church, pub and assorted events. (Villager quoted by Jones, 1995, p. 44)

People have got time, time for living, time to talk, which I think is smashing. I mean, even in our little country shop, they've got time to serve somebody rather than expect them to rush around and get it all themselves and get 'em out as quick as possible. (Villager quoted by Bell, 1994, p. 91)

Thirdly, Bell observes that many villagers felt that rural life was closer to nature than urban life. The presence of animals was one symbol of this. Bell quotes one resident who said that the word 'country' made him think of 'woods, fields, the plowed fields, the sheep, the cows, the walks I go on, the dells, the badger holes, the fox holes, the rabbits, the lot of woodpeckers you see, the deer' (p. 90); whilst Jones quotes one comment that the village was rural because 'we regularly get stuck behind cows on their way back from milking. We hear sheep, birds, tractors etc.' (p. 42). For some, however, rural life was not just about *seeing nature*, but also about *understanding nature*. Knowledge about the seasons, botany, hunting and traditional culinary methods was used to distinguish true rural people. As one recent in-migrant to Childerley – albeit with a rural family background – told Bell: 'My aunt always told me that I can't be a country girl until I learn to eat jugged rabbit' (p. 104).

Yet, both Bell and Jones also found people who felt that their village was no longer rural, or at least that it had lost some of its rural identity. This was often because of the decline of agriculture. One villager told Jones that 'very few of [the village] people work in agriculture so it is not as rural as it was 20 or 30 years ago' (p. 42), whilst Bell repeats a comment that Childerley 'is not really a rural area … It's not so farmery here' (p. 96).

Summary

'Rural' is one of those curious words which everyone thinks they know what it means, but which is actually very difficult to define precisely. Attempts by academics to define and delimit rural areas and rural societies have always run into problems, sometimes because the distinctions they have drawn have been rather arbitrary, sometimes because they have over-emphasized the differences between city and country, and sometimes because they have under-emphasized the diversity of the countryside. It is not surprising that by the late 1980s some geographers were suggesting that 'rural' be abandoned altogether as a category of analysis.

Yet, the concept of rurality is still important in the way that people think about their identity and their everyday life. As such, the dominant approach in rural studies today is to see 'rurality' as a 'social construct'. This means that geographers no longer try to draw precise boundaries around rural areas and sociologists no longer try to identify the essential characteristics of rural society. Rather, rural researchers now try to understand how particular places, objects, traditions, practices and people come to be identified as 'rural' and the difference that this makes to how people live their everyday lives.

This is the approach that is taken in this book. It is not a book about the geography of territorially delimited 'rural areas', neither is it about distinctively rural social processes. Indeed, many of the processes that will be discussed are at work in urban areas and urban society as well. Rather, the book is concerned with examining the processes that shape people's experiences and perceptions of contemporary rurality – and the responses that are adopted by individuals and institutions in order to protect or promote particular ideas about rurality. As such, the book is structured into four parts. After this opening, introductory, section, the second part examines the processes that are shaping the contemporary countryside, including processes of economic, social, demographic and environmental change. The third section explores responses to these processes, including political responses and strategies for rural development and conservation. Finally, the fourth part investigates how rural change is experienced in people's lives.

Further Reading

To read more about the different approaches to defining rurality, and about how rurality is 'socially constructed' by individuals, see two key papers by Keith Halfacree: 'Locality and social representation: space, discourses and alternative definitions of the rural', in *Journal of Rural Studies*, volume 9, pages 23–37 (1993) and 'Talking about rurality: social representations of the rural as expressed by residents of six English parishes', in

Journal of Rural Studies, volume 11, pages 1–20 (1995). For more on the case studies, see Michael Bell's book *Childerley: Nature and Morality in a Country Village* (University of Chicago Press, 1994), and Owain Jones's paper 'Lay discourses of the rural: developments and implications for rural studies', in *Journal of Rural Studies*, volume 11, pages 35–49 (1995). The concept of the 'rural idyll' is discussed in detail by Michael Bunce, 'Reproducing rural idylls', in Paul Cloke (ed.), *Country Visions* (Pearson, 2003).

Websites

The various definitions of rurality used in the United States are described and discussed by the Rural Policy Research Institute at www.rupri.org. For more contributions to the Countryside Alliance's debate on 'What is Rural?' see www.countryside-alliance. org/policy/whatis/index.html.

2

Understanding the Rural

Introduction

In the previous chapter we discovered how difficult it is simply to describe the 'rural'; yet as rural social scientists we need not just to be able to describe the processes shaping the 'rural' and their effects – we need also to try to *understand* these processes, and to propose and critique explanations as to why particular processes operate in particular ways in particular places and have particular outcomes. To do this we need to use theory. Using 'theory' may sound like a daunting prospect because it conjures up thoughts of heavy-weight philosophy, but in fact we all routinely use theories in our everyday life. We all implicitly use scientific theories whenever we switch on a light or open a door – and we also produce our own theories, for example when we speculate about plot developments in our favourite TV soap opera, or analyse the performance of our sports team.

Some theories are *empirical*, in that they are produced solely from evidence observed in a specific context. For example, I may produce a theory about the closure of a village shop based on observations of the number of people using the shop, examination of its accounts, and a survey of local residents about where they shop, that proposes that the shop has closed because residents are increasingly doing their shopping in a nearby town. Other theories are *conceptual*, in that they employ models and concepts that have been developed at a generalized or abstract level. For example, in explaining the closure of my village shop I might draw on Marxist theory to argue that the need for capitalist corporations to maximize profits has encouraged the expansion of supermarkets which undercut the prices of local shops, thus drawing away customers.

Traditionally, a lot of research in rural studies has been empirical in nature, but over the past 25 years a more *critical rural social science* has developed which has employed a range of conceptual theories in its analysis, including political-economic concepts (which are derived from Marxist theory), feminist theory and post-structuralism. The approach adopted by different researchers will be influenced by their disciplinary background and training. Contemporary rural studies is a very inter-disciplinary field, with similar types of research being conducted by geographers, sociologists, anthropologists, agricultural economists, planners and political scientists. However, the sub-disciplines were once more distinctive, focusing on different objects of enquiry, and drawing on different concepts, models and social theories. Accordingly, this chapter begins by describing the features of the three main traditions – geography, sociology and anthropology – and the contribution that they have made to the evolution of contemporary rural studies; and then proceeds to discuss two conceptual approaches that have been influential *across* rural studies in the past 25 years – the political-economy approach and the cultural approach.

The Geographical Tradition

Rural geography emerged as a distinct sub-discipline in the 1950s, as the dominance of regional geography faded. Before the 1950s much of what human geography did had been *de facto* rural geography as regional geography's central concern with how people interacted with their natural environment meant many human geography studies were conducted within rural areas. However, as the study of urban areas became fashionable within a new process-focused geography, rural geography was created by default from the remnants of the old approach, and was marginalized within geography until it was revitalized in the early 1970s with a new integrated perspective. The key concerns of rural geography as practised during this period (*c*.1960–1980) fell into three main areas (see also Table 2.1):

- *The geography of agriculture*. This interest reflected the economic importance of agriculture in the post-war era and policy concerns to modernize farming. The Rural Geography Study Group of the Institute of British Geographers had been known as the Agricultural Geography Study Group until 1974, and even in the late 1970s, over 40 per cent of research in British rural geography was concerned with agriculture (Clark, 1979). Topics of research included structural change in farming, agricultural land use patterns,

farm systems and the social geography of agriculture.

- *The organization and impact of human activity over rural space*. This included research on population distribution and migration, as well as transport networks and rural settlement patterns. In the immediate post-war period, analysis focused on the classification of rural settlement forms, for example in Sharpe's classic 1946 text on *The Anatomy of a Village*. Later, a more applied approach shifted its attentions to problems of rural settlement planning.

- *The rural landscape and land use*. This approach combined elements of the above two in its concern with describing and explaining the evolution of the rural landscape. Research in this field was especially significant in North American geography, where it has been particularly associated with the work of John Fraser Hart (see Hart, 1975; Hart, 1998), and with the Contemporary Agriculture and Rural Land Use Speciality Group of the Association of American Geographers (known as CARLU). It was not until 2002 that CARLU merged with the Rural Development Speciality Group to form a new Rural Geography Speciality Group.

These traditional approaches to rural geography tended to be very empirical, with little

Table 2.1 Chapter headings from selected rural geography textbooks

Clout (1972) *Rural Geography*	Rural depopulation; People in the countryside; Urbanization of the countryside; Land-use planning; Structural changes in agriculture; Forestry as a user of rural land; Landscape evaluation; Settlement rationalization in rural areas; Manufacturing in the countryside; Passenger transportation in Rural Britain; Integrated management of the countryside
Hart (1975) *The Look of the Land*	The plant cover; Some basic concepts; Land division in Britain; Land division in America; Farm size and farm tenure; Farm employment and farm management; Factors influencing farmers' decisions; Farm buildings; Agricultural regions and farmstead; House type and villages; Mining, forestry and recreation; The changing American countryside
Phillips and Williams (1984) *Rural Britain:* *A Social Geography*	The rural economy I: living off the land; The rural economy II: non-agricultural employment; Population and social change; Housing; Transport and accessibility; Planning; Services and retailing; Recreation and leisure; Deprivation; Policy issues and the future
Gilg (1985) *An Introduction to* *Rural Geography*	Agricultural geography; Forestry, mining and land use competition; Rural settlement and housing; Rural population and employment; Rural transport, service provision and deprivation; Rural recreation and tourism; Land use and landscape; Rural planning and land management

engagement with conceptual ideas. As Cloke (1989a) commented, 'when faced with the need to underpin research with a conceptual framework many rural geographers have preferred to concentrate on their principal interest, that is empirical investigation of rural issues' (p. 164); or as John Fraser Hart describes his approach: 'I try to understand what I see as I ramble through rural areas, and I supplement my observations with census data and maps that are based on these data' (University of Minnesota website). To the extent that theory was used, this was often restricted to the application of spatial models, such as Von Thunen's model of land use and Christaller's central place theory. These models were essentially generalized cartographic representations of empirical observations, and not only did they often fail to work when taken out of their original context, they also revealed nothing about the social, economic and political processes that produced the phenomena concerned.

Overall, the contribution of the geographical tradition to contemporary rural studies has been three-fold. First, it has left a sensitivity to space and spatial difference; secondly, it has left a concern for landscape; and thirdly, it has left an interest in human–environment interactions, which is now being reworked in new ways.

The Sociological Tradition

The beginnings of rural sociology can be traced back to the turn of the nineteenth and twentieth centuries. The first North American university course in rural sociology was taught at the University of Chicago in 1894, followed by a second at the University of Michigan in 1902. It was not until after the First World War, however, that rural sociology really took off, with rapid expansion in both Europe and North America, symbolized by the founding of a dedicated journal, *Rural Sociology*, in 1936. Significantly, the popularity of rural sociology in the interwar period was encouraged by the pressure for change encountered by rural societies in the face of rampant urbanization and industralization. Indeed, early rural sociology had a strong moral agenda, with close connections to churches in both Europe and North America, as well as to political movements such as the Commission on Country Life established by President Roosevelt in 1908. As Mormont (1990) notes, this moral agenda had two, often contradictory, elements: 'On the one hand there was an (agricultural) modernization movement attempting to transform the structures of the rural world in order to integrate it technically and economically into the modern industrialized world. On the other hand there was a (more ideological) movement of reaction against the social and political tensions of the age' (p. 23).

Reflecting these twin pressures, rural sociology developed a number of research foci, of which four stand out (see also Table 2.2):

- *Rural society versus urban society.* As discussed in the previous chapter, identifying the differences between rural and urban society has been a major concern of the sociological tradition.
- *Social relations within rural areas.* Sociologists explored the social structure of rural communities, including the role of kinship networks, the systems of hierarchy, and the importance of institutions such as the church.
- *The sociology of agriculture.* This differed from agricultural geography in two key ways: first, through a concern with the farm household as a social unit; and secondly, through a concern with the labour relations between farms and farm workers.
- *Change in rural society.* A particularly common theme running through much rural sociology was the impact of modernization and change. For some researchers the role of sociology was to assist rural modernization, for others it was about studying those distinctive aspects of rural societies that might be lost.

Although there was always a strong practical dimension to much rural sociological

Table 2.2 Chapter headings from selected rural sociology textbooks

Gillette (1913) *Constructive Rural Sociology*	Distinction between rural and urban community; Types of community as results of the differentiating effects of environment; Rural and urban increase; The social nature of the rural problem; Advantages and disadvantages of farm life; Improvement of agricultural production; Improvement of the business side of farming; Improvement of transportation and communication; Social aspects of land and labor in the United States; Rural health and sanitation; Making farm life more attractive; Socialization of country life; Rural social institutions and their improvement; Rural charity and corrections; Rural social surveys
Sorokin and Zimmerman (1929) *Principles of Rural–Urban Sociology*	The rural world and the position of the farmer-peasant class in the 'great society'; Bodily and vital traits of the rural–urban population; Rural–urban intelligence, experience, and psychological processes; A cross-section of rural–urban behavior, institutions and culture; Rural–urban migration
Jones (1973) *Rural Life*	What is rural?; A conceptual framework; Rural ways of life in Britain; Rural social structure and organisation I: family and neighbourhood; Rural social structure and organisation II: the rural community; Change in contemporary rural society; Rural–urban interaction and rural change

Table 2.3 Some rural community studies in the British Isles

Arensberg (1937); Arensberg and Kimball (1948)	Luogh and Rynamona, Co. Clare, Republic of Ireland
Rees (1950)	Llanfihangel-yng-Ngwynfa, Wales
Williams (1956)	Gosforth, Cumberland
Frankenberg (1957)	Glynceiriog, Wales
Dennis, Henriques and Slaughter (1957)	'Ashton', Yorkshire
Stacey (1960)	Banbury, Oxfordshire
Littlejohn (1964)	Westrigg, Northumberland
Williams (1963)	Ashworthy, Devon
Strathern (1981) (study undertaken in the 1960s)	Elmdon, Essex

work, rural sociology as a whole was more engaged with conceptual theories than rural geography. The socio-cultural approach to defining rurality, discussed in the previous chapter, was largely developed within rural sociology. As well as testing some of the dichotomies of rural and urban society empirically, rural sociologists also drew in this context on the social theories of leading thinkers such as Ferdinand Tönnies, Max Weber and Emile Durkheim, who had constructed ideas about rural and urban societies as part of their conceptual work on modern society. From the 1950s to the 1970s the concept of the rural–urban continuum became the major item of debate within rural sociology.

Indeed, one of the main contributions of the sociological tradition to contemporary rural studies is the understanding of how perceptions about rural–urban differences persist in lay discourses of rurality. Other key contributions include interests in social relations and social structures; the continuing importance of the household as a unit of analysis; and concerns with the provision of welfare services in rural areas, such as health, education and housing.

The Anthropological Tradition

There is significant overlap between the anthropological tradition and the sociological tradition, not least because much rural anthropological research has concerned itself with social structures and processes. The difference, however, has been methodological, with anthropology employing the technique of ethnography that usually involved researchers living within rural communities. The most notable products of the anthropological tradition are the numerous 'rural community studies' conducted in Britain and Ireland in the 1940s and 1950s (see Table 2.3). These studies were comprehensive investigations into individual communities that attempted to integrate research on social structures, economic activities, families and households, religion, politics and cultural activities. Although the community studies were essentially intensive empirical exercises, some researchers did draw on conceptual theories to try to understand the communities they studied. Many sought to identify the characteristics of rural society proposed by socio-cultural theories; whilst Frankenberg (1966) used community studies to position nine communities along a rural–urban continuum. Other writers imported concepts developed in anthropological work in the developing world, such as Erving Goffman's (1959) notions of front- and back-regions, to explain aspects of social interaction in rural communities.

In institutional terms, anthropology is not as strong in contemporary rural studies as geography or sociology (except in Australia where much rural research is conducted through anthropology departments), but the legacy of the anthropological tradition remains significant in three ways. First, 'the rural community' continues to be a major focus of research in rural studies; secondly, the anthropological tradition has contributed an enduring interest in rural identity; and thirdly, there has been a revival in contemporary rural studies of the ethnographic community study as a methodological approach, as demonstrated, for example, by Michael Bell's study of 'Childerley', discussed in the previous chapter (Bell, 1994).

The Political-economic Approach

If the three traditions outlined above point us to the beginnings of rural studies, the origins of contemporary rural social science as we know it today can be traced to a paradox that confronted rural research in the 1970s. Although the volume and range of empirical rural research remained buoyant, critics accused rural studies of losing its way theoretically, failing to engage with new developments in social theory and having little to say of relevance beyond the specific circumstances in which research was conducted (Buttel and Newby, 1980; Cloke, 1989a).

Much rural research was being conducted under contract for government agencies or large corporations, and tended to uncritically follow the agenda set by these powerful institutions. In contrast, the 1970s saw a new critical edge emerge elsewhere in the social sciences through engagement with neo-Marxist political-economy theories of the operation of capitalism (see Box 2.1). These propose that the social, economic and political structures that order the modern world are all shaped by the central need of the capitalist mode of production to create profit. Capitalism, it is argued, requires the polarization of society into different classes, of the bourgeoisie and the proletariat; capitalism requires that economic policies, institutions and geographies are organized to assist production at the lowest cost; capitalism requires that mass consumption is encouraged to create a demand for goods; and capitalism requires and creates uneven geographies of wealth and opportunity. The application of these ideas had transformed other fields, such as urban studies, and some younger researchers began to consider whether the same theoretical ideas could be introduced into rural studies.

Box 2.1   Key term

Political-economy: The study of the relations of production, distribution and capital accumulation, the efficacy of political arrangements for the regulation of the economy, and the impact of economically determined relations on social, economic and geographical formations. In contemporary geography the term 'political economy' is applied to studies that are influenced by Marxist theories, particularly those with an emphasis on the social characteristics of capitalist societies, including social inequalities and the imperative of capital accumulation.

The political-economic approach was hence pioneered by a few projects within rural sociology, including, notably, work by

Howard Newby and colleagues on agricultural labour relations and rural power structures in the East Anglia region of England (Newby, 1977;

Newby et al., 1978). The influence of these first studies soon spread within and beyond rural sociology, and by the early 1980s, organizations such as the Rural Economy and Society Study Group – established in the UK in 1978 – had created an inter-disciplinary space for political-economic research in rural studies.

As Buttel and Newby (1980) observed, the introduction of the political-economy approach led not only to new ways of thinking within rural studies, but also to new fields of enquiry. In particular, four key areas of concern can be identified with the political-economy approach in rural studies:

- *Agriculture as a capitalist enterprise.* The political-economy approach asserted that agriculture operates in the same way as any other form of capitalist production – by seeking to maximize profit. From this perspective the re-structuring of agriculture in the post-war period (see Chapter 4) was driven by the interests of capital accumulation, and the relationship between farmers and farm labourers is recast as an exploitative relationship.
- *Class.* Traditional rural studies tended to emphasize community solidarity over class differences, but the political-economy approach reversed that by investigating class conflict and oppression. 'Class' also became a basis for analysis of population change within rural areas, with later studies in the 1980s and 1990s examining the role of a new group, the 'service class', in migration to rural areas and the effect of middle class in-migrants displacing working class residents – or 'gentrification' (both of these issues are discussed in Chapter 6).
- *Change in the rural economy.* The political-economy approach connected rural

economic change to wider transformations in the capitalist economy. An urban to rural shift in manufacturing, for example, was explained by the relocation of production in lower-cost environments. Similarly, the Marxist concept of the 'commodity' has been employed to propose that rural landscapes and lifestyles have been 'packaged' to be sold and consumed through tourism and recreation (see Chapter 12).
- *The state.* The political-economy approach sees the state not as a neutral administration, but rather as complicit in creating favourable conditions for capitalism. As such, rural researchers have analysed the role of the state in areas such as agricultural policy and planning.

The approach based on theories of political economy had a major impact on rural research by providing a framework through which the study of rural economies and societies could be connected to wider social and economic processes. This helped to highlight that rural areas do not exist as isolated, discrete territories but rather are shaped and influenced by actors and events outside rural space. The political-economy approach also permitted the development of a more radical rural studies which sought to use research to expose social and economic inequalities in the countryside and to challenge established structures of power. However, the approach has limitations. From a political-economy perspective, rural areas cannot be identified as having sufficient common, distinctive characteristics that would allow for the positioning of the 'rural' as a discrete object of enquiry. Rather, the logic of the political-economy approach leads to the treatment of 'rural' localities just as other localities, in other words focusing on the 'local' as opposed to the 'rural'. The emphasis in political economy on economic structures and on collective identities such as classes also meant that individual agency and personal experiences tended to get marginalized in analysis. Thus, in the 1990s the emphasis in rural studies shifted again to a move to bring people back in through the enculturing of political-economy approaches.

Rural Studies and the Cultural Turn

At the end of the 1980s, human geography and the social sciences in general entered into what has been subsequently labelled the 'cultural turn'. This promoted a new understanding of culture as the product of discourses through which people signify their identity and experiences and which are constantly contested and re-negotiated, and cultural geographers started to explore spatial relations and the meaning of place through issues of identity, representation and consumption. As Cloke (1997a) observes, the cultural turn supported a resurgence of rural studies, lending both respectability and excitement to engagements with rurality. Rural geographers, for example, drew upon ideas of identity and representation to examine the ways in which rurality is discursively constructed – as discussed in the previous chapter. Additionally, several of the key concerns that were developed in cultural geography more broadly, including the spatiality of nature, landscape and otherness (see Chapter 15), all led to constructive engagement with rural spaces and environments.

Cloke (1997a) lists four areas of 'excitement' in rural studies in the mid-1990s that reflect the influence of the cultural turn:

- *Nature–society relations.* Rural researchers have explored the significance of nature in the constitution of rurality and the ways in which rural space becomes incorporated in human–nature engagements. These include work on the geographies of animals and flora, on non-human agencies and hybrid forms, and on perceptions of natural environments and landscapes (some of which are explored in Chapters 8 and 13).

- *Discourses of rural experience and imaginations.* As well as work on the social construction of rurality (see Chapter 1), a wide range of studies have been developed exploring the different rural lifestyles and experiences, with a particular focus on previously neglected 'other' rural groups (see Chapter 15 and following chapters).

- *Symbolic texts of rural cultures.* The cultural turn also focused attention on the ways in which rurality is represented in various media, and how such representations contribute to the reproduction of discourses of rurality. Research has, for example, focused on the history and heritage of rural symbolisms (such as pastoral art) that are reproduced in contemporary consumption, and on the representation of rural space, landscapes and life in contemporary popular media (see Chapter 11).

- *Movements.* Finally, research has begun to explore aspects of mobility in rural space, including, for example, work on tourism and travel as well on alternative rural lifestyles that embrace forms of nomadism and tribalism (see Chapter 21).

More latterly, new strands of research have been developed that can be added to the above, including in particular work on the geographies of food that has explored connections between production, consumption and representation (Goodman, 2001), on farming cultures (Morris and Evans, 2004), and on the body in rural space and embodied experiences of rurality (Little and Leyshon, 2003).

However, Cloke (1997a) also raises five concerns about the implications of the cultural turn. The first three relate to a perceived blunting of the radical rural studies that had developed with the political economy approach. Cloke asks whether the emphasis on identity in the cultural approach turned 'a commitment to emancipatory social practices and politics into a commitment to the political empowering of pleasure' (p. 373), replacing a politics of conviction with a politics of identity. Similarly, he also asks whether the openness of cultural approaches to different moral positions had promoted a moral thinking that is above and free from social interest. Drawing these concerns together, Cloke thirdly questions the capacity of cultural research to have a practical output, particularly given that policy-makers are frequently suspicious about the ability to draw general conclusions from qualitative research. The fourth concern is that research on representations of the rural has focused on the more 'seductive' high-culture texts such as art and literature at the expense of other texts that are more closely related to the everyday lives of most people. Finally, Cloke warns against 'research tourism' in work on rural 'otherness', undertaking partial studies of marginalized groups in the rural but 'without requisite attention to the importance of sustained, empathetic and contextualized research which is conducted under clear and acceptable ethical conditions' (p. 374). Although some of these concerns have been addressed by subsequent research, for example on popular forms of cultural representation of rurality such as television, others remain unresolved.

Summary

The revitalization of rural studies in recent years has owed much to the creativity generated by the fusion of ideas from different disciplinary traditions and the introduction of new theoretical perspectives from political economy and feminism to post-modernism and post-structuralism. The story, however, has not been a linear narrative of one dominant theory replacing another. As Cloke (1997a) observed, 'rural studies have witnessed a series of different conceptual fascinations, the result often being an interesting hybridization between them rather than any clear paradigmatic shift one to the other' (p. 369). Rural researchers have become adept in understanding how particular theoretical ideas can help to throw light on particular aspects of the rural economy and society. Whilst care must be taken not to combine incompatible world-views, this eclectic approach is followed in this book. The analysis of the processes of social, economic and political restructuring that are reshaping rural areas, for example, will be conducted in a political-economic framework; whilst the discussion of people's experiences of rural life owes much to the cultural turn. These theoretical references will, however, be implicit rather than explicit in the individual chapters, and will mainly be pointed to through the explanation and application of various key concepts. In this way I hope to demonstrate that a theoretically informed rural studies need not be difficult or challenging, but creates vast opportunities for understanding the changing countryside.

Further Reading

Three articles by Paul Cloke provide further discussion of the development of different theoretical influences in rural studies. First, his chapter on 'Rural geography and political economy', in volume 1 of Richard Peet and Nigel Thrift (eds), *New Models in Geography: The Political Economy Perspective* (Unwin Hyman, 1989), details the emergence of the political-economic approach in rural geography and discusses issues in its application. Second, his editorial 'Country backwater to virtual village? Rural studies and "the cultural turn"', in the *Journal of Rural Studies*, volume 13, pages 367–375 (1997), reflects critically on the influence of the cultural turn in rural studies. Finally, the chapter on '(En)culturing political economy: a life in the day of a "rural geographer"', in P. Cloke, M. Doel, D. Matless, M. Phillips and N. Thrift, *Writing the Rural* (Paul Chapman, 1994) provides a personalized account of how different theoretical developments (along with a range of other factors) have influenced Cloke's own work.

Part 2

PROCESSES OF RURAL RESTRUCTURING

3

Globalization, Modernity and the Rural World

Introduction

One of the key themes of this book is that of the *changing countryside*. It is, like the notion very landscape of rural areas in the developed world appears to testify to the changes wrought upon the countryside over the past 50 years or so – the sprawling extensions of rural settlements, new roads and power lines, redesigned field patterns, new forms of agricultural and industrial buildings, afforestation and deforestation, and the plethora of signs that help us to reach and 'interpret' 'protected' rural landscapes and sites. Nor is it just the physical appearance of rural space that has changed. Oral histories by people who have lived in rural communities throughout the second half of the twentieth century frequently comment on the changes they have experienced, including many changes to those intangible qualities that we saw in Chapter 1 are so often at the heart of lay definitions of rurality – things like the sense of community, solidarity, social order, tranquillity. Equally, it is easy to find statistics that quantify the degree of change in the social and economic characteristics of rural areas – the decline of agricultural employment, the in-migration of new residents, the closure of village services, and so on.

The perception of change is reinforced by the campaigning of political groups that regard change as a threat to rurality, and therefore seek to resist further change and to protect those aspects of the rural world that they see as being 'lost'. When countryside protesters marched from four peripheral regions of Britain to join 125,000 more demonstrators at the Countryside Rally in London's Hyde Park in July 1997 – an event organized by the Countryside Alliance pressure group primarily to protest at attempts to ban the hunting of wild mammals with hounds – one marcher was quoted in The *Guardian* newspaper claiming that 'rural people' were a distinctive culture, as threatened as any indigenous tribe in the rainforest (Woods, 2003a).

Such warnings convey a sense of urgency, but is contemporary rural change really anything new? In April 2000, 250 'rural leaders' gathered in Kansas City to discuss the policy challenges facing rural America. As one participant told the conference: 'At the dawn of the 21st century rural America faces unprecedented change', yet he then continued: 'for at least the last half century many rural communities have been on a demographic and economic roller coaster' (Johnson, 2000, p. 7). A historian would have probably stretched the timescale even further. The point being made is that the problem with much of the contemporary rhetoric about rural change is that it suggests a false dichotomy between a dynamic and threatening rural present and a stable, romanticized rural past. More accurately, the rural can be recognized as a continuous space of change – sometimes on a far greater and more disruptive scale than that experienced today. Are the changes experienced in recent decades by rural regions of North America, Australia and New Zealand really more significant that those that followed the arrival of European settlers from the sixteenth century onwards? Is contemporary rural change in Europe really as extensive as that experienced during the first agricultural revolution in the eighteenth and nineteenth centuries, or during the great period of industrialization and urbanization at the turn of the twentieth century?

Contemporary rural change is, however, distinguished by two characteristics. The first is the *pace and persistence* of change. Rural economies and societies are not just changing, but changing constantly and rapidly, affected by successive trends and innovations that roll in like the waves of an incoming tide. This vigorous pace of change is driven by the rate of technological innovation and social reform in late modernity. The second characteristic is the *totality and interconnectivity* of change. Many historical instances of rural change, such as the enclosure of farmland in Britain in the eighteenth century, were revolutionary for those directly affected but spatially limited. In contrast, today's processes of rural change resound around the globe. Rural areas, it seems, are tightly interconnected by global social and economic processes that cut across rural and urban space in a condition of advanced globalization.

This chapter examines these characteristics in more detail, seeking to identify some of the key processes of change and to illustrate some of the consequences, highlighting themes that will be developed further in later chapters. Drawing together the effects of modernity and globalization, the conclusion argues that it is the cumulative impacts of the processes operating under these short-hand concepts that enable us to talk about *rural restructuring*.

Modernity, Technology and Social Change

'It seems often to be assumed', writes David Matless, 'that the English village lies on the side of tradition against modernity, with those two terms in opposition' (1994, p. 79). In the same way that discourses of rural change have frequently reproduced a false dichotomy between a dynamic rural present and an unchanging rural past, so they have also promoted the equally problematic dualism of the modern city and the traditional countryside. As Matless discusses, the distinction is both unhelpful and misleading, but it has been a convenient fiction for the advocates and opponents of rural change alike. For preservationist movements, 'tradition' describes the order and endurance of rural society, contrasted with the moral disorder and uncertainty of modernity. For reformers, however, modernization has been the key to stimulating rural economies and raising the living standards of rural people, reducing inequalities with urban areas. In this sense, modernization has often described programmes of infrastructure development, such as electrification, road-building or the renovation of rural housing. Such projects left a significant mark on the rural landscape, but their greater significance is in the opportunities that they created for rural populations to participate in the new consumer society and purchase technological innovations that would change their lives.

The list of technological innovations that have changed aspects of rural social and economic life is extensive, but three examples can be flagged up here as illustrations. First, consider refrigeration. The development of refrigeration technologies, both for commercial storage and for domestic use, has had a revolutionary impact on our relationship to food in the developed world. Food can now be transported vast distances from the site of production to the site of consumption, and no longer needs to be consumed in season. Refrigeration created new food-processing industries and corporations, and enabled the development of supermarkets. These developments in turn helped to make agriculture into a global trade, encouraged specialization by farmers and strengthened the power of the food-processing and retail companies against that of the farmers. At a domestic level, refrigeration changed the shopping habits of rural consumers, reducing their dependence on local suppliers and allowing more irregular shopping trips to supermarkets in towns, thus contributing to the closure of rural shops and services.

Similarly, the development of motor vehicles has changed practices of both production and consumption in the countryside. Commercial farm vehicles, such as tractors and combine harvesters, changed the nature of farming and reduced the demand for farm labour, contributing to the decline of agriculture as a source of employment in rural areas. The growth of private car ownership, meanwhile, increased the mobility of rural people and loosened ties to rural communities. Commuting became possible, prompting counterurbanization and breaking the link between residence and employment. Mass tourism, too, was facilitated, rejuvenating the economies of some rural regions but also bringing environmental consequences.

Thirdly, the development of telecommunications technology has alleviated some of the problems of distance and peripherality experienced by many rural areas. At one level this has meant that for some newer, 'foot-loose', industries such as biotechnology and telematics, rural locations are no longer disadvantaged, permitting, as Howard Newby noted, 'rural areas to compete on an equal basis with towns and cities for employment' for the first time since the industrial revolution (quoted in Marsden et al., 1993, p. 2). At another level, rural people are now consumers via television, radio and the Internet, of the same cultural commodities and experience as urban residents, and the attraction of localized rural traditions, events and cultural practices has declined, even in spite of recent grassroots efforts to revitalize such activities.

Moreover, the impact of modernization on rural areas has not been restricted to technological innovation. Social change too has had an effect, with similar trends operating in rural societies as in urban societies. The decline of organized religion (more pronounced in Europe, Australia and New Zealand than in the United States), for example, has eroded the prominence and power of churches and chapels as one of the traditional tenets of rural communities. Mass participation in secondary and higher education in the developed world has meanwhile altered the life courses of rural young people, taking many out of their communities to colleges and universities and restricting their opportunity to return due to a shortage of graduate-level jobs.

Collectively these processes match lay understandings of the term 'modernization', but they also constitute a transformation in rural societies that reflects a more philosophical conceptualization of modernity. This holds that one of the fundamental features of modernity is the separation of the natural and the human. Modernization has arguably engineered this separation in rural society by reducing employment in those occupations that involve direct contact with the natural world (for example, agriculture, forestry); by introducing technologies into farming that are designed to intervene between the human worker and nature, or to manipulate or resist nature; by developing technologies to overcome the vulnerability of rural societies to natural phenomena, such as difficult terrain or harsh weather; and by diminishing the cultural connection of rural people with nature through, for example, festivals celebrating seasons of the year. Modern agriculture and food marketing distances food consumers from the place and process of production (such that surveys frequently show that children have little knowledge of where their food comes from), and nature itself has become packaged and delimited in the countryside in nature reserves and national parks.

As the twentieth century drew to a close it was suggested that we were moving from the era of modernity into a condition of postmodernity, in which the order, structure and normative ideals of modernity would be dissolved into a world characterized more by flux, fluidity and multiplicity. Postmodernity does not suggest any reversal of (or even an end to) the physical modernization of rural space described above, but it does suggest a change in the attitude and perceptions of those who live in and shape rural space, and of the academics who attempt to research it. The postmodern rural is less precisely defined and delimited than modern countryside – the blurring of the rural and the urban is recognized, as is the existence of many different rurals occupying the same space but socially constructed differently by people from different standpoints (see Chapter 1). The postmodern countryside is perhaps expressed too in the rejection of some of the idealistic orthodoxies of modernization, including growing scepticism towards science in the wake of food-related disease scares and resistance towards GM agriculture, as well as attempts by colonizers seeking to get 'back to nature' to deconstruct the modernist separation of the natural and the human. These issues will be picked up again in later chapters (see Chapters 4, 15 and 21).

Globalization and the Rural

The rural areas of the developed world have been subject to the influences of global trade and migration since the first European explorers introduced new crops to their home countries from the nascent colonies, and the first European colonists began to cultivate the wildernesses of America, Australia and New Zealand. However, in identifying globalization as one of the pre-eminent forces of our time, globalization is conceptualized not as the movement of goods, people and capital around the world, but as the advanced interconnection and interdependence of localities across the world (see Box 3.1).

Box 3.1   Key term

Globalization: The advanced interconnection and interdependence of localities around the world, reflecting the compression of time and space. It is defined by Held et al. (1999) as 'the widening, deepening and speeding up of worldwide interconnectedness in all aspects of contemporary social life, from the cultural to the criminal, the financial to the spiritual' (p. 2). The sense of an inevitable binding together is conveyed even more strongly by Albrow (1990), for whom globalization is 'all those processes by which the peoples of the world are incorporated into a single world society, global society' (p. 9).

Globalization is therefore, in essence, about power – about the lack of power of rural regions to control their own futures, and about the increasing subjection of rural regions to networks and processes of power that are produced, reproduced and executed on a global scale. The power of global capitalism, and, by extension, global corporations, is one clear example of this and is as significant in traditional rural economic sectors such as agriculture as in any industry. But globalization is about more than just trade or corporate ownership. Indeed, Pieterse (1996) argues that globalization should not be seen as a monolith, but that there are many globalizations, sometimes contradictory, always fluid and often open-ended. As Gray and Lawrence (2001) demonstrate in an examination of rural Australia in the context of globalization, Pieterse's argument presents a means of understanding the multiple ways in which globalization of different forms impacts upon rural areas, and the opportunities that exist for rural actors to determine their response.

This section discusses three forms of globalization that have a particular relevance for contemporary rural societies – economic globalization, the globalization of mobility and the globalization of values – and explores their roles in driving rural change and the consequences for rural societies.

Economic globalization

The term 'global economy' most likely conjures up an image of Manhattan skyscrapers or the trading floor of a stock exchange. Yet, the most immediate contact with the global economy that most of us have on a regular basis is in the aisles of our local supermarket. There on the shelves are row after row of food products that have been sourced from all over the world, processed and sold by global corporations, aimed at a global market and often promoted through a multinational advertising campaign. As Table 3.1 shows, the food that you eat in a single meal is likely to have travelled further than you will in a whole year. The location of the supermarket will make little difference; all of the products listed in Table 3.1 for Iowa are grown in the state itself, yet supermarkets buying from agri-food corporations or through large wholesale markets (Figure 3.1), will go for the cheapest, best-selling or most convenient option, wherever sourced. Even if local produce is sold, it may have come via a circuitous route. An investigation for British television found that beef from cattle reared in South Wales was transported nearly 500 miles to the slaughterhouse, processing and packaging plant and distribution centre before being sold in a supermarket close to the initial farm (*Guardian*, 10 May 2003).

Table 3.1 Approximate 'food miles' travelled by typical food products from source to place of consumption in Iowa and London

Cedar Falls, Iowa			London, England		
Product	**Source**	**Miles (km)**	**Product**	**Source**	**Miles (km)**
Chicken	Colorado	675 (1085)	Chicken	Thailand	6643 (10689)
Potatoes	Idaho	1300 (2100)	Potatoes	Israel	2187 (3519)
Carrots	California	1700 (2735)	Carrots	South Africa	5979 (9620)
Tomatoes	California	1700 (2735)	Tomatoes	Saudi Arabia	3086 (4936)
Mushrooms	Pennsylvania	800 (1290)	Prawns	Indonesia	7278 (11710)
Lettuce	California	1700 (2735)	Lettuce	Spain	958 (1541)
Apples	Washington	1425 (2300)	Apples	USA	10133 (16303)
Radishes	Florida	1200 (1930)	Peas	South Africa	5979 (9620)

Sources: Pirog et al., 2001; *Guardian*, Food supplement, 10 May 2003

Figure 3.1 Rungis wholesale market, Paris. Centres like this are the major nodes in the global agricultural economy
Source: Woods, private collection

The globalization of trade is one of three major features of economic globalization that impact on rural economies and societies, and is an intensifying trend. Bruinsma (2003) suggests that agriculture experienced a first wave of globalization in the late nineteenth century following the introduction of rail travel and steamships, cutting transport costs across the Atlantic and reducing price differentials. Following the First World War levels of global trade slumped, with exports from the United States falling by 40 per cent between 1929 and 1933, and imports by 30 per cent over the same period. After the Second World War, however,

Table 3.2 World exports of selected livestock produce as a percentage of total world consumption

	1964–66	1974–76	1984–86	1997–99
Bovine products	9.4	10.3	12.2	16.4
Pig meat	5.7	6.0	7.9	9.6
Poultry meat	4.0	4.7	6.3	13.9
All meat	7.4	7.9	9.4	12.7
Milk and dairy products	6.0	7.6	11.1	12.8

Source: Bruinsma, 2003

global trade steadily increased and began to account for a significant share of agricultural produce. As Table 3.2 shows, the proportion of milk and dairy goods produced for export more than doubled between 1964–66 and 1997–99, and the proportion of poultry meat exported more than trebled. Other sectors of the rural economy have been similarly incorporated into global trade flows. For example, forestry is increasingly part of a global industry, with exports accounting for 30 per cent of world production of sawnwood, 30 per cent of wood-based panels and 7 per cent of industrial roundwood (Bruinsma, 2003).

Adaptation to the new global economy has resulted in a number of significant changes to the practice of agriculture in the developed world, with knock-on effects for the wider rural community. Farms have become more specialized as the need to provide a range of produce to local markets has evaporated and greater profitability can be achieved by maximizing sales of single products to food processing companies and supermarkets; the ties between farmers and local rural communities have been weakened as the sales transaction has disappeared; and agriculture has become more vulnerable to global economic factors (Figure 3.2). When British agriculture slumped into depression in the late 1990s, including one year when average farm income fell by 46 per cent, the crisis was precipitated by the strength of sterling reducing income

Figure 3.2 The 'McFarmer' burger advertised by this fast food outlet in Switzerland hints at an attempt to respond to local tastes but ultimately represents the homogenization and corporatization of food consumption
Source: Woods, private collection

from exports compounding the effects of an earlier ban on the export of British beef imposed by the European Union due to an epidemic of BSE (mad cow disease).

The second feature of economic globalization impacting on rural areas is the rise of global corporations. Again this is most explicitly pronounced in agriculture. For example, the global seed market is dominated by just four corporations – Monsanto, Syngenta, DuPont and Aventis. Over 80 per cent of maize exports from the United States, and over 65 per cent of soybean exports, are controlled by three firms (Bruinsma, 2003). Three companies control over 75 per cent of the retail food distribution system in Australia (Bruinsma, 2003). Moreover, many of the individual companies that dominate in these particular sectors are connected through joint ventures and strategic alliances into three 'food chain clusters' headed by the corporations, Cargill and Monsanto, ConAgra, and Novartis and Archer Daniels Midland (Hendrickson and Heffernan, 2002). As Box 3.2 illustrates, these 'food chain clusters' operate on a truly global scale, vertically and horizontally integrating the various components of the food production process such that they have control, in the slogan of ConAgra, from 'seed to shelf'. The power of the 'food chain clusters' is immense. As well as being major landowners and employers in many rural areas, their dominance of food processing gives them considerable influence in determining prices paid to farmers, and their involvement in research and development could enable them to shape the future direction of agriculture. It is no coincidence that Monsanto and Novartis have been at the forefront of developing GM technologies, as will be discussed in Chapter 4.

Box 3.2   The Novartis/ADM food chain cluster

Novartis was created from a merger of CIBA–Geigy and Sandoz to form the world's largest agrochemical company, with 15 per cent of the global agrochemical market in 1997. It subsequently merged its seed and chemical business with AstraZeneca to create Syngenta, one of the five dominant global seed companies. Novartis formed a joint venture, Wilson Seeds, with Land o' Lakes, a farmer cooperative that also has joint ventures with Archer Daniels Midland (ADM), a leading grain collection and food processing company. ADM's stakes in farmer cooperatives including Growmark, Countrymark, United Grain Growers and Farmland Industries, gives it access to substantial parts of North American agriculture, including 75 per cent of the Canadian corn and soybean market region and 50 per cent of the US corn and soybean market region. ADM owns 50 per cent of A.C. Toepfer, a German corporation that is one of the world's largest grain trading companies, and has joint ventures with the Chinese government. It has interests in processing firms for wet and dry corn, rice, peanuts, animal feed, wheat, oilseed and malting, including investments in Mexico, the Netherlands, France, Britain, Bolivia, Brazil and Paraguay. ADM owns Haldane Foods in the UK and produces the Harvest Burger vegetarian alternative in the United States, whilst Novartis owns Gerber baby food. As such the cluster has a web of interests stretching around the globe and from 'seed to shelf'. The formation of the cluster through joint ventures enabled Novartis to access food processing and gave ADM a direct link to farmers.

For more see Mary Hendrickson and William Heffernan (2002) Opening spaces through relocalization: locating potential resistance in the weaknesses of the global food system. Sociologia Ruralis, 42, 347–369.

Corporate concentration is no less pronounced in the food retailing sector. Over 40 per cent of food retail sales in the United States are accounted for by five supermarket chains – Kroger, Albertsons, Wal-Mart, Safeway and Ahold USA – some of which are beginning to expand on a global scale. Wal-Mart now operates in the UK, Germany, Argentina, Brazil, Canada and Mexico, and has joint ventures in China and Korea. Ahold has interests in the Netherlands, Latin America, Portugal, Spain, Poland, the Czech Republic, Scandinavia and the Far East. The French supermarket chain Carrefour, meanwhile, is also the largest retailer in Brazil, Argentina, Spain, Portugal, Greece, Belgium and Taiwan (Hendrickson and Heffernan, 2002). Supermarkets have a two-fold influence in rural areas. As large-scale purchasers from farmers and farm cooperatives, they exercise considerable power over farmgate prices. But as large-scale retailers, with the capacity to undercut smaller shops, supermarkets have also been accused of contributing to the closure of independent rural stores and specialist butchers, bakers and greengrocers in small towns and villages (see Chapter 7).

The third feature of economic globalization to impact on rural areas is the growing significance of global regulatory frameworks. As rural economies become integrated into global trade networks, so the capacity of national governments to regulate the economic life of rural regions is diminished, with power shifted upwards to bodies such as the World Trade Organization (WTO). Agriculture is one of the most contentious political flashpoints in the negotiations that set the WTO's policies, as the organization's underlying agenda of trade liberalization (which is supported by the agri-food conglomerates and by a number of net-agricultural-exporter nations) clashes with domestic political pressures in Europe and the United States to protect internal agricultural markets (see Chapter 9 for more). The resolution of this impasse will reverberate down to the level of individual farms and rural communities, with the potential that a pro-free-trade outcome would remove the subsidies and price support mechanisms that have effectively bankrolled agriculture in some peripheral rural regions for decades (Chapters 4 and 9).

The globalization of mobility

It is not just the mobility of commodities and capital that has been liberalized by globalization, but also the mobility of people. Technological developments mean that we are able to travel across the globe in relatively short periods of time for relatively little cost. For travellers from most developed world nations bureaucratic requirements for visas and permits have been gradually relaxed, and many of us have an opportunity to participate in an effectively global labour market if we so choose. Mass migration has, of course, long been a significant factor in the evolution of rural societies (see Chapter 6), but the movement of people in and out of rural areas today is different insofar as it must be positioned within this context of heightened global mobility. For example, migration flows are no longer predominantly uni-directional forces. Many rural areas may be experiencing net in-migration through counterurbanization, but this tends to disguise a fluid situation in which there is also significant out-migration, and in which people may move in and out of rural areas (as well as within rural areas) several times during the course of their life. One consequence of this is that people have become less tied to particular places and therefore that the coherence and stability that once characterized rural communities have been eroded. These issues are discussed further in Chapter 6, whilst further implications for rural housing are considered in Chapter 16.

The majority of migration into rural areas is still of a domestic origin, but there are also flows of immigration directly into rural regions. Notably, this reflects the mobility of both the haves and the have-nots under globalization. On the one hand, it includes the purchase of holiday homes and second homes by wealthy foreigners, as well as more permanent moves by individuals seeking a new start in life. For example, over 20,000 Britons purchase property in rural France each year (Hoggart and Buller, 1995). On the other hand, immigration reflects the dependency of many labour-intensive forms of agriculture on migrant workers, particularly in the United States. An estimated 69 per cent of all seasonal farm workers in the United States are foreign-born, including more than 90 per cent of the seasonal workforce in California (Bruinsma, 2003). The majority have come from Mexico, and as such they are part of a long tradition that extends back over most of the twentieth century (Mitchell, 1996), and which is a vital component in the story of American agricultural capitalism. However, as will be discussed further in Chapter 18, migrant workers have often been subjected to extreme exploitation and poor pay and working conditions. Moreover, immigration of any form can provoke ethnic and cultural tensions in rural communities, particularly where the new arrivals are perceived to 'threaten' nationalistic notions of rurality, or local cultural traditions and languages. As such, racism is increasingly acknowledged as a problem in many rural areas (see Chapter 20).

On a more temporary basis, global mobility also encompasses the rise of global tourism, with some 692 million people taking holidays outside their country of residence in 2001. Long-haul tourism has played a major role in regenerating rural economies, with New Zealand in particular acquiring a global reputation as a centre for rural adventure tourism (see Chapter 12). However, the growth of tourism also brings social and environmental challenges for rural areas, including demands for structural changes in their local economies and, like other forms of globalization, involves a loss of power by rural communities as the way in which their rurality is represented and promoted is reconfigured to appeal to the preconceptions of international tourists (Cater and Smith, 2003).

Cultural globalization

A third dimension of globalization is the rise of the global media and emergence of a global mass culture, founded on the common consumption of the same films, television, literature, music and so on. In this global culture, much of our perception of and knowledge about the countryside is derived from films, books and television programmes in which a stylized representation of rural life is portrayed and in which the regional distinctions between, for example, a farmyard in England and a farmyard in Pennsylvania, are ignored. In particular, our knowledge of nature is frequently based on children's literature, Disney films and natural history programmes – all of which tend to humanize animals – rather than on actual interaction with nature in the countryside. The result, so rural campaigners claim, is that there is lack of understanding of rural life and rural traditions, leading to conflicts over practices such as hunting and some methods of farming.

A promotional article for the pro-hunting British pressure group the Countryside Alliance, for instance, remarked that 'a generation brought up on *The Animals of Farthing Wood*, Walt Disney films and visits to theme parks is easy meat for single-issue pressure groups who exploit this lack of understanding of the realities of the countryside to their own ends' (Hanbury-Tenison, 1997, p. 92), whilst a recent book celebrating hunting in America argued that 'to attempt to "manage" nature after such a kindergarten-cartoonish fashion as *Bambi* portrays and fosters … would soon spell ecological catastrophe … *Bambi* – that monstrously unnatural Hollywood propaganda beast – must die' (Petersen, 2000, p. 158).

The dissemination of such homogenized cultural references is one contributing factor in a wider process of the *globalization of values*, in which certain Western values and principles are encoded into international treaties and charters and enforced on a global scale. Examples of this include the European Convention on Human Rights and the International War Crimes Tribunal, but also the promotion of global environmental standards and of animal rights. These latter initiatives are commonly rooted in scientific and philosophical discourses and may therefore lead to different conclusions from the lay understandings of nature passed down by rural people. As such, conflicts can arise as they are put into practice. For example, the *Chasse, Pêche, Nature et Tradition* (hunting, fishing, nature and tradition) party polled 12 per cent of the vote in the 1999 European Parliamentary elections in France on a platform of opposition to an EU directive that would reduce the hunting season for migratory birds, which they presented as part of a wider assault on indigenous rural values.

Resisting globalization

Globalization is not all-powerful. As noted earlier, it is perhaps more accurate to think of there being multiple globalizations, some of which are contradictory, and which present numerous opportunities for resistance and contestation. In the contemporary countryside, instances of resistance to globalization can be observed when farmers blockade ports or distribution plants to protest at imports or the prices paid by supermarkets; when pro-hunting groups rally to protect their 'sport'; and when environmental campaigners fight oil corporations in rural Alaska or logging companies in the forests of the Pacific North West (see also Box 3.3).

Resistance to globalization need not be confrontational. Hendrickson and Heffernan (2002) suggest that the global agri-food complex, for example, has a number of vulnerable points that offer the potential for farmers, workers and consumers to develop alternative structures. They cite the example of the Kansas City Food Circle that brings together local producers and consumers in an arrangement that cuts out the corporate 'middle-man' and reconnects the community with local food sources. Other examples include the promotion of farmers' markets that enable producers to sell directly to local consumers (see Chapter 10; also Holloway and Kneafsey, 2000), and the Italian 'slow-food' movement that aims to resist the global spread of 'American' fast food and to promote the aesthetic qualities of traditional regional cuisine (Miele and Murdoch, 2002). Grassroots action has also been mobilized to respond to the withdrawal of services from rural areas by transnational corporations by establishing community shops, credit units and community transport schemes.

Box 3.3   *José Bové and anti-globalization protests*

In August 1999, a group of farmers from the *Confédération Paysanne* (smallholders' confederation) were arrested for 'dismantling' a new branch of McDonalds in the small French town of Millau. The protest was represented by its leader, José Bové, and his supporters, as part of an on-going struggle against globalization and its impact on French agriculture. The *Confédération Paysanne* had campaigned since the 1970s on behalf of small farmers, and its previous protests had included opposition to a large battery-chicken complex and to GM crop trials. Many of its members had also bene-fited from some aspects of globalization, and the Millau incident was precipitated by a typical tussle in the politics of global trade. In retaliation for a European Union ban on the import of hormone-treated beef, the United States had doubled the customs tariff on a number of European food products, including Roquefort cheese – the production of which employed over 1,300 people in the Millau region. Bové and his supporters, however, perceived that the real agenda behind the 'trade war' was the ambition of US-based agri-food complexes to dominate the European market by opening it up to modified foods such as hormone-treated beef, to the detriment of European farmers. Thus, they directed their response to the tariffs at McDonalds – a company that symbolized US-led globalization and the promotion of cheap, homogenized food, or '*malbouffe*'. At his trial in June 2000, Bové reinforced this analysis by calling environmental, land rights and anti-globalization campaigners from around the world as witnesses. A simultaneous festival outside the court, attended by over 20,000 anti-globalization protesters, further helped to subvert the legal process into what the French newspaper, *Libération* described as 'the trial of globalization'.

For more see José Bové and François Dufour (2001) The World Is Not For Sale: Farmers against Junk Food (Verso); Michael Woods (2004) Politics and protest in the contemporary countryside, in L. Holloway and M. Kneafsey (eds), Geographies of Rural Societies and Cultures (Ashgate).

Summary

Rural areas have always been spaces of change, shaped by economic cycles, trade fluctuations, new technologies, migration flows, political upheavals and environmental conditions. In the late twentieth century – and the early twenty-first – however, rural areas across the developed world have experienced a period of change distinguished by its intensity, persistence and totality. Driven by the twin forces of technological and social modernization and globalization, contemporary rural change has affected all areas of rural life – from the domestic routines of rural families to the investment decisions of global agri-food corporations; from the ownership of rural property to the management of the rural environment. It is in this way that the countryside can be described as undergoing 'restructuring'.

'Restructuring' is a widely used term in contemporary rural studies but its meaning can be quite loose. In some cases, 'restructuring' is used to imply nothing more than that change is taking place, whilst in other cases it has a more precise and theoretically grounded application. Hoggart and Paniagua (2001) contend that the concept has been devalued through over-use and misapplication and argue for a more careful usage:

> For us, when seen as a shift in society from one condition to another, 'restructuring' should embody major qualitative, and not just quantitative, change in social structures and practices. Unless we want to trivialize the concept, its use should be restricted to transformations that are both inter-related and multi-dimensional in character; otherwise we have descriptors that are more than adequate, like industrialization, local government reorganization, electoral dealignment or growth in consumerism. To clarify, in our view restructuring is not a change in one 'sector' that has multiplier effects on other sectors. Restructuring involves fundamental readjustments in a variety of spheres of life, where processes of change are causally linked. (Hoggart and Paniagua, 2001, p. 42)

From this perspective, sector-specific changes such as farm diversification or the closure of rural schools, cannot be considered to be 'restructuring' in their own right. Placed in a wider context, however, they can be interpreted as the local expressions of inter-connected processes of rural restructuring driven by globalization, technological innovation and social modernization. Rural restructuring as pitched at this scale has produced causally linked effects across a multiplicity of sectors with consequences that are qualitative as well as quantifiable.

This book follows the logic of the above analysis by next exploring how rural restructuring has been operationalized and expressed through changes in agriculture, the wider rural economy, the social composition of the rural population, the organization of rural communities and services, and the management of the rural environment. It then proceeds to examine the responses to rural restructuring that have been adopted both by those responsible for governing rural areas and by those living in rural areas, before finally investigating the experiences of change and the contemporary countryside of people from all parts of the rural population.

Further Reading

There is relatively little published work that explicitly examines the experience of rural areas under globalization. The best account, which is written from the perspective of rural Australia but contains extensive general material on globalization, is Ian Gray and Geoff Lawrence's *A Future for Regional Australia* (Cambridge University Press, 2001). For more on the globalization of agriculture, and particularly the role of global 'food chain clusters', see Mary Hendrickson and William Heffernan, 'Opening spaces through relocalization: locating potential resistance in the weaknesses of the global food system', in *Sociologia Ruralis*, volume 42, pages 347–369 (2002). For more on rural restructuring and the debates over the application of the concept, see Keith Hoggart and Angel Paniagua, 'What rural restructuring?', in *Journal of Rural Studies*, volume 17, pages 41–62 (2001).

4

Agricultural Change

Introduction

Agriculture is one of the most potent and enduring emblems of rurality. For centuries, agriculture was in most rural regions not only the overwhelmingly dominant source of employment, but also the driving force of the rural economy and a pervasive influence in the organization of rural society and culture. The legacy of this historical centrality of agriculture to the countryside is still evident today in many discourses of rurality, as discussed in Chapter 1. Yet, a major component of the restructuring of rural areas over the course of the past century has been the fundamental transformation of agriculture in the developed world which has seen farming move from the centre towards the periphery of everyday life as experienced by most residents of rural areas. In many developed countries, including the United States, Canada, the UK and France, less than a fifth of the rural population are now dependent on agriculture for their livelihood, substantially fewer than even twenty or thirty years ago (see Table 4.1). Other countries have seen similarly dramatic shifts – in Spain, for example, more than eight out of ten rural people were dependent on agriculture in 1970; by 2000 it was less than one in three. There are, of course, individual localities in which agriculture is still the major employer, but these are increasingly confined to the more remote rural regions and even within such localities farming tends to be significant rather than dominant in the local labour market.

The shifting position of agriculture within rural economies and societies is a product of reforms that have transformed virtually every aspect of farming in developed countries since the end of the Second World War. Over this period farms have become increasingly integrated into a modern capitalist economy. This is not to suggest that every individual farm is run as a capitalist enterprise, in the sense that there is a division between the owners and the workers, but that even the many farms that are still run on traditional, family-based, lines are obliged to participate in the capitalist marketplace in order to sell produce, and therefore are

Table 4.1 Agriculture-dependent population as a percentage of the total rural population for selected countries, 1950–2000

	1950	1960	1970	1980	1990	2000
Canada	54.1	45.6	34.5	29.6	15.6	12.0
Denmark	80.2	68.1	55.1	42.9	36.6	25.3
France	70.5	58.7	47.0	30.9	21.1	13.6
Germany	82.0	62.9	42.9	40.0	26.7	20.2
Hungary	90.7	71.7	53.8	47.5	44.8	33.9
Ireland	68.2	67.6	54.6	41.6	33.2	24.8
Italy	96.2	75.8	52.6	37.8	25.8	16.1
Japan	95.9	85.4	65.7	44.1	30.8	18.2
Spain	—	94.4	85.0	67.4	47.8	32.7
Sweden	66.8	44.1	49.1	40.8	29.3	21.1
UK	34.6	27.9	24.3	23.0	19.6	16.8
USA	36.4	23.5	17.3	14.1	12.2	9.7

Note: These statistics use each country's own definition of a rural area and therefore are not directly comparable. In all cases a small proportion of the agriculturally dependent population may live in areas classified as 'urban'.

Source: The Food & Agriculture Organization (FAO) www.fao.org

subject to the whims and demands of capitalism. This has a transformative effect because capitalism is a dynamic force that requires constant innovation to maximize profit margins and secure the reproduction of capital. The remainder of this chapter examines how the capitalist imperative produced changes in the organization and practice of agriculture and raises questions about the implications for the wider countryside.

California: the Laboratory of Capitalist Agriculture

California is one of the most important agricultural economies in the world, and the producer of one of the most extensive ranges of crops and farm produce. Traditionally, agricultural geographers have attributed its prolificacy to environmental factors, notably the diversity of micro-climates within the state. However, as Dick Walker has argued, this explanation underplays the extent to which Californian agriculture was *manufactured* over a relatively short period at the start of the twentieth century. Between 1905 and 1940, the cumulative output of Californian agriculture rose from around $5 billion per year to over $20 billion, and by the 1920s the state had become the largest agricultural producer in the United States. The booming agricultural economy was one significant factor in drawing thousands of migrants to California during the 1920s and 1930s, including farmers escaping the devastation of the Dust Bowl in the American Mid-West. These migrants, whose experience was vividly recounted by John Steinbeck in the novel *The Grapes of Wrath*, moved to California in search of wealth, pursuing an American dream that was wholly capitalist in its ideology. Hence Walker advances a political-economic analysis of Californian agriculture that reveals it as the laboratory of farming as a capitalist industry (Walker, 2001; see also Henderson, 1998).

Historians have traced the origins of agrarian capitalism to parts of northern Europe in the sixteenth and seventeenth centuries, but what California represented at the turn of the twentieth century was wholesale application of capitalist principles to agriculture in a manner unfettered by aristocratic landowners or a partially subsistence-based peasantry, as was the case in Europe. Moreover, the development of agriculture in California was fully integrated with the development of a wider 'resource capitalism' encompassing mining, oil and gas extraction, forestry, fishing and hydroelectric generation (Walker, 2001). As such, California became the laboratory of capitalist agriculture, innovating and developing strategies, techniques and technologies that have subsequently become fundamental features of modern agriculture across the developed world.

Walker argues that farm development was driven by petty bourgeois investment in agriculture, some of which came directly from migrants, some from business owners in the expanding urban centres, and more still from the profits of mining and mineral exploitation. A modern banking system was developed to assist the circulation of capital within California, and as savings banks were prohibited from investing in mining, investment flowed into farming (Henderson, 1998). The largest bank in the United States was created by the amalgamation of unit banks in farming towns, with a system of credit arrangements extended to farmers that 'did not just provide capital; it was also a brilliant device for overcoming space–time discontinuities in agricultural production and marketing' (Walker, 2001, p. 184). At the same time, the pressure for returns on investments led to innovations in agricultural organization and practice aimed at maximizing the value of outputs.

Arid areas were irrigated and swamplands reclaimed, both with state support, fertilizers developed and experiments undertaken with improving soils and slopes. Similar effort was put into improving the quality of the plants and stock that are the raw materials of agriculture. The mass importation of plant varieties from around the world in the late nineteenth century gave way to the development of extensive nursery and seed industries to supply industrial agriculture, and later to the creation of the biotechnology industry.

Factory-farming was introduced to increase the throughput of farms, with the mass-rearing of poultry and confined dairy farming both pioneered in California in the early twentieth century. Many 'factory farms' required industrial-scale inputs of labour, yet the high costs of capitalization in California demanded that labour must be cheap. As Mitchell (1996) observes, 'large-scale, capital-intensive farming simply could not rely only on family labour: crops would rot before they could all be picked. Nor could it rely exclusively on a local pool of labour for such temporary work. In that case local farmers would have to pay the *yearly* reproductive costs of their workers (and their families) out of *seasonal* profits' (p. 59). Hence, the need was filled by migrant workers, both from elsewhere in the United States and from Mexico and Asia, creating an agricultural workforce that had none of the paternalistic ties to the landowner associated with farm workers in Europe.

Capitalist agriculture as a system of capital accumulation is characterized not just by labour exploitation, but also, among other things, by the creation of higher value through the commodity chain. The booming cities of San Francisco, Los Angeles and San Diego formed an initial market for Californian agriculture, but capitalism demanded the creation of new and higher-order markets. Farmers hence established cooperatives to improve the processing and marketing of their produce. The development of railroad and shipping links helped to facilitate an export trade, but, significantly, investment was also put into food preservation, and by the end of the nineteenth century California had the world's largest canning industry. Frozen foods and dried milk were other inventions of the state. California was also the centre of the development of modern supermarkets during the 1920s and 1930s, most notably the Safeway chain, thus creating a new form of mass food retailing. Moreover, as part of the search for new markets, the Californian food processing industry led the development of new food products, such as the fruit cocktail, each producing new outputs for, but also new demands on, agriculture.

All these innovations have been translated in some form to other parts of North America, Europe and the developed world as key elements in the restructuring of agriculture. Only the initial availability of capital investment has been less easily reproduced elsewhere. Thus in many countries it fell to the state to provide investment for the capitalization of agriculture (that is, the purchase of machinery, seeds, fertilizer, chemicals, etc.), through government grants and subsidies.

State Intervention in Agriculture

The intervention of the state in agriculture reflects the dual purpose of agriculture in the capitalist economy. Agriculture is a means of capital reproduction in its own right, but it is also needed to provide raw materials for industry and food for workers and consumers. This latter purpose can be regarded as falling within the capitalist state's role of social regulation – in other words, governments have an interest in ensuring that agriculture produces enough to feed a nation's population at a cost that is generally affordable, whilst enabling

farming to continue functioning as a capitalist industry. Alongside this, governments have an interest in checking uneven economic development between regions (if only to maintain tax bases and to avoid unmanageable population shifts) and therefore to help rural economies to remain viable. Both these imperatives have led to the substantial involvement of the state in regulating and supporting agriculture, through a variety of methods.

One of the earliest examples of government action to support agriculture was the establishment in 1862 of the United States Department of Agriculture (USDA), with a remit of distributing seeds and plants to farmers, together with information about how to use them. This was followed by the foundation of government-sponsored 'land grant colleges' to teach agricultural sciences and to help to 'modernize' farming. By the early twentieth century, the growing political power of the American farmers' movement combined with occasional agricultural depressions and concern about the failure rate of new farms to stimulate a new strategy of direct state intervention in the agricultural market. In 1916 the Federal Farm Loan Act introduced direct financial assistance from the US government to producer cooperatives, in 1927 the McNary–Hangen Bill brought in the first fixed prices for agricultural products, and the 1930s saw the creation of marketing bodies and mechanisms for controlling production. Collectively, these American initiatives set the precedent for state intervention in agriculture of four types: through training; through price support, including the purchase of surplus products; through marketing; and through production controls.

Similar policies were adopted in other countries. Agriculture was one of the first departments of the Canadian federal government to be created in the 1860s, with a responsibility for agricultural research and training. From the 1930s, the Canadian government started to intervene in the agricultural markets, for example by establishing the Canadian Wheat Board in the late 1940s as the sole purchaser of wheat, oats and barley destined for export, and of domestic feed grain. The Australian government similarly introduced a Wheat Board in 1948 and intervention mechanisms to stabilize the wool sector in the 1960s.

In Europe, state intervention in agriculture was shaped by the effects of the two world wars. Not only had war disrupted (and in some parts of Europe, devastated) farm production, restrictions on trade had limited the supply of many goods, and the need to feed troops during the wars was replaced in their aftermath by an imperative to feed displaced and rapidly urbanizing populations. Price supports were first introduced in Britain during the First World War, but it was after the Second World War that these principles were most explicitly enshrined in the 1947 Agriculture Act, which established a system of guaranteed prices for farmers, as well as state involvement in marketing, training and the regulation of agricultural wages. Similar objectives were expressed in the section of the 1957 Treaty of Rome that formulated the Common Agricultural Policy (CAP) of the new European Economic Community (later to become the European Union) (see Box 4.1).

Box 4.1   The Common Agricultural Policy

The Common Agricultural Policy shall have as its objectives: (a) to increase agricultural productivity by promoting technical progress and by ensuring the rational development of agricultural production and the optimum utilisation of the factors of production, in particular, labour; (b) thus to ensure a fair standard of living for the agricultural community, in particular by increasing the individual earnings of persons engaged in agriculture; (c) to stabilise markets; (d) to assure the availability of supplies; (e) to ensure that supplies reach consumers at reasonable prices. (Article 39 of the Treaty of Rome (1957), quoted in Winter, 1996, p. 118)

The Common Agricultural Policy was a milestone in the development of capitalist agriculture in four ways. First, it was the first agreement to regulate agriculture at a transnational scale, thus marking a significant step towards a regulated global agricultural economy. Secondly, it created a common agricultural market in Europe equivalent to that of the United States and an agricultural exporting unit able to compete with the US (and other major exporters including Australia, Canada and New Zealand) in global trade. Thirdly, in seeking to ensure standards of living for the 'agricultural community', it tied farming to the wider rural community in a manner that reflected the fact that over half of the EEC's rural population was at the time dependent on agriculture, but which has subsequently complicated attempts at reform. Fourthly, it set as its first objective the unqualified increase of agricultural productivity, thus expressing clearly the imperative that was already the driving force of agricultural development in North America, Australia, New Zealand and the UK, and which is encapsulated in the term 'productivism' (see Box 4.2).

Box 4.2   Key term

Productivism: The dominant policy trend in agriculture from the 1940s to the mid-1980s. The central aim was to increase agricultural production. This involved the intensification and industrialization of agriculture, including the introduction of agri-chemicals, mechanization and the specialization of farms. The system was underpinned by state subsidies.

Table 4.2   Application of inorganic fertilizers (nitrogen, phosphate and potash) in four Western European countries

	Application (thousand tonnes)			
	1956	**1965**	**1975**	**1985**
West Germany	2114	2897	3300	3185
France	1924	3123	4850	5694
Netherlands	468	566	638	701
UK	—	1555	1800	2544

Source: After Ilbery and Bowler, 1998

Productivist Agriculture

The rise of productivist agriculture after the Second World War was characterized by change in three structural dimensions – intensification, concentration and specialization (Bowler, 1985; see also Ilbery and Bowler, 1998). *Intensification* involved the pursuit of higher productivity through the substantial capitalization of agriculture, including significant investment in machinery and the farm infrastructure, and increasing utilization of agri-chemicals and other biotechnologies. Evidence of this could be seen around the developed world. In Canada, for example, purchases of herbicides soared from C$53.3 million in 1973 to C$121.4 million in 1976, whilst the use of nitrogen fertilizer in the prairie provinces of Canada increased ten-fold from 50.4 thousand tonnes in 1948 to 569.9 thousand tonnes in 1979 (Wilson, 1981). Total use of inorganic fertilizers similarly increased in Europe, if at a less dramatic rate (Table 4.2).

The prairie states of the United States, meanwhile, witnessed the transformative impact of rapid advances in farm machinery during the 1960s and 1970s: 'Tractors doubled and then quadrupled in size and price in the space of a few years, as did the array of specialized machinery for dealing with individual crops. This allowed single operators to cover vast acreages in a day' (Manning, 1997, pp. 151–152). The attraction of large machinery was not restricted to the United States. Sales of large four-wheel-drive tractors in Wales rose from fewer than 100 in 1977 to 1,500 in 1992 (Harvey, 1998).

Concentration aimed to maximize cost-effectiveness by creating larger farm units. In 1951 the average farm size in Manitoba, Canada, was 137 hectares, by 1976 it was 240 (Wilson, 1981). Over the same period, average farm size in England and Wales increased from under 40 hectares to nearly 50 hectares, and again to over 60 hectares by 1983

Table 4.3 Size of agricultural holdings in seven Western
European countries, 1975 and 1987

	Under 10 ha (%)		10–50 ha (%)		Over 50 ha (%)	
	1975	**1987**	**1975**	**1987**	**1975**	**1987**
Denmark	32.5	19.0	59.9	64.0	7.6	17.0
Germany	54.3	49.6	42.8	44.6	2.9	5.8
France	41.4	35.0	48.0	48.2	10.6	16.8
Ireland	31.6	31.2	59.8	59.8	8.6	9.0
Italy	88.6	89.2	10.0	9.4	1.4	1.4
Netherlands	52.4	49.7	45.6	46.4	2.0	3.9
UK	26.2	30.8	44.3	38.1	29.5	31.1

Source: After Winter, 1996

(Marsden et al., 1993). A similar trend continued into the 1980s in many developed countries (Table 4.3). The corollary has been a decrease in the total number of farms. For example, the number of farms in Canada fell by 40 per cent between 1961 and 1986, and in Australia the number of farms fell by a quarter over 25 years (Gray and Lawrence, 2001; Wilson, 1981).

Efficiency was also promoted by concentration in the commodity chain. Farms moved to contracts with a single purchaser, either a government-sponsored marketing board, or food processing companies and retailers. In the early 1980s, 95 per cent of poultry and peas produced in the UK were farmed under contract for a food processor, as were 65 per cent of eggs, 50 per cent of pigs and 100 per cent of sugarbeet (Bowler, 1985).

Specialization also helped to enhance cost-effectiveness. Investment in expensive specialist machinery tailored to a single crop meant that diversity was discouraged, as did the ability to sell a single crop under contract to a single purchaser. The production of particular agricultural products hence became concentrated on fewer, larger farms. For example, an 81 per cent increase in the average area of cereal cultivation per farm in the UK between 1967 and 1981 was accompanied by a 27 per cent decrease in the number of farms growing cereals (Ilbery, 1985). In Canada, the top 5 per cent of poultry farms by sales

accounted for 75 per cent of all receipts by the late 1980s (Troughton, 1992).

Specialization occurs in other ways too. As agricultural employment has been restructured, generalist farmworkers contracted to a single employer have been supplanted by specialist agricultural contractors working for a number of farmers as required, for example as combine harvester operators. It is notable, for instance, that whilst employment in farming continued to fall in the United States during the 1990s, employment in agricultural services increased by 27 per cent between 1990 and 1996 (Rural Policy Research Institute, 2003).

These changes in agricultural practice and organization had a number of effects on the wider rural economy, society and environment. First, there was a physical impact on the landscape as field sizes were increased, hedgerows removed, grasslands ploughed and new crops introduced. Further, less visible but serious environmental consequences included pollution, soil erosion and the loss of habitats, as discussed in greater detail in Chapter 8. Secondly, significant social effects resulted from the dislocation of agriculture from the community. Mechanization meant that less labour was required in agriculture – it is estimated that the total amount of farmwork undertaken in the United States decreased by over a third between 1950 and 1970 (Coppock, 1984) – such that farming declined as a source of employment

Table 4.4 Corporate concentration in primary processing
in New Zealand

	Percentage of output produced by top three processing companies		
	1960	**1986**	**1992**
Dairying	42.0	—	75.0
Meat freezing	37.5	—	67.0
Wool scouring	34.2	50.0	—
Fruit and vegetable processing	78.5	80+	—

Source: After Le Heron, 1993

in rural communities. In France, for example, there were over 5 million people employed in agriculture in 1954, but only 3 million by 1968, and 2 million by 1975 (INSEE, 1993). Ties were also weakened as farmers started to sell more of their produce to food processing companies and supermarkets rather than through local shops and markets, and as more and more farmland passed into the ownership of corporations and absentee proprietors.

Thirdly, there has been a spatial effect as traditional agricultural geographies have been remoulded. The concentration of agricultural production included regional specialization in production sectors such as dairy and fruit farming; in other regions the entire balance of agriculture shifted with the targeting of government subsidies, large proportions of Illinois and Iowa, for example, being converted from grazing land to arable land in the 1970s and 1980s (Manning, 1997). Intensive, commercial agriculture was established in some peripheral regions, such as parts of Andalucia in Spain, for the first time, whilst in other less favoured or pressurized rural regions farming declined at an above average rate as individual farms found themselves unable to compete in the globalized agricultural market.

Fourthly, the 'industrialization' of agriculture has had a political and economic impact by shifting power from individual farmers to corporations engaged in different stages of the commodity chain. The growing corporate

presence in farming is a key feature of productivist, capitalist, agriculture. Corporate landowners have become increasingly significant in particular production sectors (for example, fruit, sugar) and particular regions (such as California and Florida). One company, for instance, owns 80 per cent of all land used for hop growing in Tasmania (Gray and Lawrence, 2001). Other specialist corporations have emerged as contract farming businesses, working for landowning clients. One of the largest such firms in the UK, Velcourt, farmed nearly 25,000 hectares (60,000 acres) in the mid-1990s on behalf of insurance companies, pension funds and private landowners (Harvey, 1998). However, corporate power has been most substantially advanced by the increasing dependence of independent and family farmers on a relatively small range of companies as suppliers and buyers. On the one hand, farmers rely on a limited number of companies to supply seed, agri-chemicals and machinery. On the other hand, they rely on an equally limited range of companies to buy their products. In New Zealand the top three processing companies accounted for over three-quarters of all dairy products in 1992, up from 42 per cent in 1960, with similar trends of concentration evident in other sectors (Table 4.4) (Le Heron, 1993). As noted in Chapter 3, many of the companies involved in the different stages of the process have been linked together through

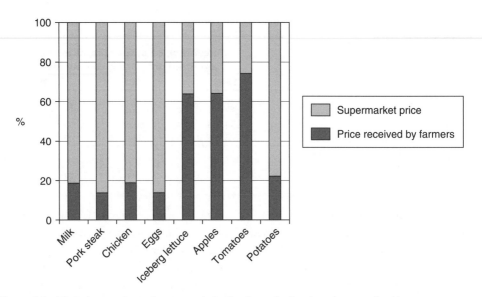

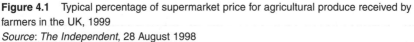

Figure 4.1 Typical percentage of supermarket price for agricultural produce received by farmers in the UK, 1999
Source: *The Independent*, 28 August 1998

shareholdings and strategic alliances in global 'food chain clusters' dominated by large transnational corporations including Monsanto, Cargill and ConAgra. Vertical integration of this kind is done to maximize returns on capital, and one way of achieving that is to squeeze the payments made to farmers, such that only a small proportion of the supermarket price of food finds its way back to the producer (Figure 4.1).

Box 4.3   Disease – an unanticipated consequence of productivism?

The use of biotechnology to eradicate or control plant and animal diseases was one of the means by which farmers attempted to increase productivity during the productivist era. Ironically, however, some of the techniques employed in productivist agriculture are now suspected of assisting the spread of some diseases and even of creating new livestock diseases. In 1986 the first case of bovine spongiform encephalopathy (BSE) (also known as 'mad cow disease') was officially confirmed in cattle in England. A brain disorder, BSE was new in cattle, but a similar disease, scrapie, has long affected sheep. It was soon established that the disease had probably originated in scrapie-infected sheep offal fed to cattle – part of a wider practice of feeding naturally herbivorous livestock, such as cattle, with cheap, industrially produced feed manufactured from the by-products of slaughtered animals, including chicken litter, pig offal and cattle remains (Macnaghten and Urry, 1998). Between 1986 and 1996 over 160,000 cases of BSE were confirmed in the UK, infecting at least 54 per cent of dairy herds and 34 per cent of breeding herds (Woods, 1998a). Following the introduction of a ban on the inclusion of sheepmeal in animal feed in 1988, incidences of the disease began to fall,

Box 4.3 (Continued)

but a more serious worry remained. If BSE had been transmitted to cattle from sheep by infected meat, could BSE be transmitted to humans through the consumption of infected beef? Could it, indeed, be the cause of a new variant of a similar human brain disorder, Creutzfeld–Jakob Disease (CJD) that had been recorded during the 1980s? When in March 1996 British government scientists reported that exposure to the BSE agent was 'the most plausible interpretation' of the cause of new variant CJD, the effect was dramatic. The European Union imposed an immediate ban on the export of British beef, and beef sales within the UK itself fell sharply. In an attempt to restore consumer confidence and resume exports, the government embarked on an eradication strategy involving the slaughter of over a million cattle and costing over £2.5 billion (Macnaghten and Urry, 1998). BSE has been controlled in the UK, but remains a threat. Outbreaks have occurred in Europe, particularly France, Switzerland and most notably Germany – where the scare led to the resignation of the agriculture minister and the appointment of a new minister from the Green party, committed to reforming productivist farming. Isolated incidences in Canada in May 2003 and the United States in December 2003 have raised fears that the disease may have spread to North America.

British farming had barely recovered from BSE when a second epidemic – this time of foot and mouth disease (FMD) (also known as 'hoof and mouth disease') – struck in 2001. Unlike BSE, foot and mouth is not a new disease. It is endemic in many parts of the developing world, but had been eradicated in most of the developed world, where it is considered as one of the most serious agricultural diseases. It is not usually fatal to infected animals, but does reduce productivity and is therefore feared as a serious economic threat. Moreover, it can spread between species and infect all hoofed livestock, including cattle, sheep and pigs. The 2001 outbreak in Britain was the world's worst ever epidemic of foot and mouth disease, and whilst modern agriculture cannot be blamed as the source of FMD, its practices did intensify the speed and scale of the epidemic. High stock densities on farms and, more particularly, the practice of transporting animals long distances across the country to centralized livestock markets and abattoirs helped the disease to spread rapidly across Britain. Again, the epidemic was controlled only through a large-scale cull of over 4 million at-risk animals, and the effective 'closure' of significant parts of the British countryside to public access, with a considerable knock-on impact on other parts of the rural economy, especially tourism.

For more details see the websites of the UK government's Inquiries into the BSE and foot and mouth epidemics: www.bse.org.uk and www.defra.gov.uk/footandmouth/. For more on BSE see P. Macnaghten and J. Urry (1998) Contested Natures (Sage), ch. 8; Michael Woods (1998) Mad cows and hounded deer: political representations of animals in the British countryside. Environment and Planning A, 30, 1219–1234.

The Farm Crisis

The productivist regime in agriculture has had profound and far-reaching effects on the rural economies, societies and environments of the developed world. Some of these may be judged (depending on your perspective) as positive, some as negative; some have been intentional targets, others unintended consequences (see Box 4.3). However, in terms of its central objective of increasing agricultural production, productivism was an undoubted success. Between 1961 and 1990, agricultural production in the developed world increased by around 62 per cent, so successful, indeed, that the developed world is today producing more agricultural goods than can be

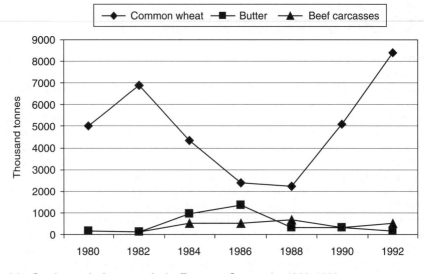

Figure 4.2 Surplus stocks in storage in the European Community, 1980–1992
Source: After Winter, 1996

sold at profit in the marketplace (this is not the same as supply outstripping demand on a domestic level – the UK, for example, was estimated to be only 79 per cent self-sufficient in indigenous food in 2000). Instead, over-production has been underpinned by price support mechanisms as governments have intervened to buy surplus produce at an agreed minimum price. As Figure 4.2 shows, in 1980 the European Community was stor-ing nearly 5 million tonnes of surplus wheat; by 1982 the total had increased to nearly 7 million tonnes. Although the so-called 'wheat mountain' was subsequently (temporarily) reduced, surplus stocks of butter, beef carcasses and other products increased. This system was a core element of productivist policy, intended to guarantee a stable income to farmers, but with overproduction it began to place a finan-cial burden on society as a whole. By 1984 the implementation of the Common Agricultural Policy (CAP) was consuming 70 per cent of the European Community's entire budget, and a quarter of that was being spent on storing surplus produce. The actual expenditure on

storage was nearly five times greater than it had been in 1973 (Winter, 1996).

In an attempt to relieve the pressure of over-production, all the major agricultural produ-cing nations began to search for new markets by increasing exports. The result was fierce competition, sporadic 'trade wars' between the major economic blocs, and a depression in world commodity prices. Large producers, who could compete effectively – and who were anyway favoured by the nature of government subsidies – gained financially during this period, but for smaller farmers, more exposed to price fluctuations and more vulnerable to the intrusion of imports into domestic markets, it was the precipitation of a farm crisis.

In the United States the problem of over-production was compounded by drought and, most significantly, rising interest rates. Ever since it was pioneered in turn-of-the-century California, credit had been the catalyst for agri-cultural modernization. During the 1960s and 1970s, in particular, farmers had been encour-aged to borrow money to invest in machinery and farm modification. Farmer debt in the US almost doubled between 1970 and 1980 (Le Heron, 1993). This was sustainable so long as interest rates remained low, commodity prices remained stable, and land values continued to rise (in Iowa, for example, farmland prices nearly quadrupled during the 1970s, see Stock, 1996). However, in the early 1980s the collapse of commodity prices coincided with fiscal pressures that pushed US interest rates into double figures. Over the next decade it is estimated that 200,000 to 300,000 farmers defaulted on their loans, many of them in the 'farm belt' of Iowa, Minnesota and Wisconsin (Dudley, 2000). At the height of the crisis in 1986–7, nearly one million people – farmers and their families – were forced out of agriculture over a 12-month period (Dyer, 1998). The farm crisis fundamentally changed American agriculture, reducing the commercial significance of small family farms, but it also entailed severe personal implications for the individuals and communities affected (see Box 4.4). Among the observed consequences have been problems with stress and increased suicide rates in rural communities (Dudley, 2000), and growing political alienation, some of which has fuelled support for extreme right-wing militia groups (Dyer, 1998; Stock, 1996).

Box 4.4 *Personal stories of the farm crisis*

The human side of the US farm crisis is revealed by interviews with farming families conducted by Kathryn Marie Dudley in Minnesota and Janet Fitchen in New York State. One farming couple interviewed by Dudley, Dick and Diane, described the spiral of circumstances that pushed their farm into crisis. They had purchased land in the 1970s at a low interest rate of 6 per cent and took out other loans for operating expenses. In 1982, however, poor crops and rising interest rates put them behind on repayments. In order to finance spring planting in 1984 they had to negotiate a new loan package, consolidating their debts with the mortgage on their land – but at a new interest rate of 11 per cent. As interest rates peaked at 19 per cent in 1985 their annual interest payment averaged $1,000 a week. To continue farming they were forced to borrow from the 'lender of last resort', the Farmers Home Administration, who gave them a second mortgage on their land and new operating loans at subsidized interest rates. With this package the farm could just 'break even' and the couple relied on the wife's salary from a teaching job for living expenses. The experience left the family still in debt and bitter at the way in which they had been treated and the opportunities that they felt had been denied to them but given to others.

Dick and Diane survived the farm crisis. Len and Yolanda, a farming couple interviewed by Fitchen, did not. For them the financial pressures of the 1980s coincided with the decision by their children to find jobs outside farming. The last straw was a surcharge by their milk haulier because they were the only farm along a back road. As Len explained, 'the only way to deal with all this would be to expand. Back twenty-five years ago, a family farm could survive with 25 cows, but nowadays you have to have at least 50 – and we simply couldn't do all the work ourselves' (Fitchen, 1991, p. 25). Len and Yolanda sold their cattle as part of a federal buyout, auctioned off the equipment and sold the farm to an in-migrant from the city.

For more on these and other accounts of the farm crisis see Kathryn Marie Dudley (2000) Debt and Dispossession: Farm Loss in America's Heartland (University of Chicago Press); Janet Fitchen (1991) Endangered Spaces, Enduring Places: Change, Identity and Survival in Rural America (Westview Press).

A reduction in interest rates combined with adjustments in the agricultural sector eventually served to defuse the US farm crisis, but the fundamental problem of over-production has persisted. The continuing efforts of policy-makers to agree substantial reforms to agricultural policy in Europe and the United States, the significance of agriculture in international trade negotiations, and the radical route pursued by New Zealand, are all discussed later in this book in Chapter 9. However, it would be wrong to suggest that productivism remains unchecked. Since the 1980s, numerous initiatives have been adopted to gradually reform agriculture by shifting government subsidies away from production. The implementation of these measures has been described as the 'post-productivist transition' (Box 4.5).

Box 4.5   Key term

Post-productivist transition: The general term employed to refer to changes within agricultural policy and practice that have shifted the emphasis away from production towards the creation of a more sustainable agriculture. The post-productivist transition (PPT) has been driven by a diverse set of initiatives aimed at promoting a range of social and economic objectives. As the term 'transition' implies, the concept suggests not an abrupt switch from productivist policy (*q.v.*) but rather a gradual process of reform and adaptation.

The Post-productivist Transition

When compared with the focused drive of productivism, the post-productivist transition is a far more ambiguous and multi-faceted concept. It is clear that it is a move away from productivism, but what it is a move towards is less certain. Some elements of post-productivist policy have emphasized environmental goals such as replanting woodland (see Chapter 13); others have emphasized social goals such as the protection of the family farm, yet underlying the policy shift as a whole is a concern with finding an economically viable model for agriculture without the disbenefits that have become associated with productivism. In broad terms, however, the post-productivist transition has been understood as involving four key components – extensification; farm diversification; an emphasis on countryside stewardship; and enhancing the value of agricultural products.

Extensification aims to reverse the intensification of agriculture, slowing production and reducing the amount of chemical and other artificial inputs used by farmers. This has been promoted in part by removing or restricting subsidies that supported intensive farming, but also through specific initiatives to actively encourage more extensive forms of agriculture, such as the temporary retirement of farmland from production. The most notable example of this is the European Union's set-aside scheme, launched on a voluntary basis in 1988. Under the scheme, farmers received compensation payments for retiring at least 20 per cent of their arable land from production for a minimum of five years. However, initial projections that 6 million hectares would be set aside proved over-optimistic, with less than 2 million hectares (or 2.6 per cent of arable land in the EU) included in the first phase of the scheme. Participation increased to substantial levels only following the introduction of a compulsory set-aside scheme for cereal farmers in 1992 (Table 4.5), and by 2001 involved 12.4 per cent of arable land in the EU.

Table 4.5 Land retired under the European Union's set-aside scheme

	Land retired (thousand hectares)		
	1988–92	**1993–4**	**2001–2**
Austria	—	—	103.9
Belgium	0.9	19	27.5
Denmark	12.8	208	217.7
Finland	—	—	198.0
France	235.5	1578	1575.8
Germany	479.3	1050	1156.2
Greece	0.7	15	45.7
Ireland	3.5	26	36.4
Italy	721.8	195	232.9
Luxembourg	0.1	2	2.1
Netherlands	15.4	8	22.6
Portugal	—	61	99.1
Spain	103.2	875	1610.6
Sweden	—	—	269.2
United Kingdom	152.7	568	847.9
European Union total	1725.8	4605	6445.6

Source: Ilbery and Bowler, 1998; European Union DGVI

Farm diversification seeks to reduce the dependency of farm households on agricultural production so that farms remain viable as an economic and social unit even as production is decreased. Technically, farm diversification refers only to 'the development of non-traditional (alternative) enterprises on the farm' (Ilbery, 1992, p. 102). However, together with income generated by farm household members through off-farm employment, diversification contributes to pluriactivity, described by Ilbery and Bowler (1998) as 'the generation by farm households of income from on-farm and/or off-farm sources in addition to the income obtained from primary agriculture' (p. 75). Farm diversification has been supported by direct grants, loans and training schemes. The type of diversified activity adopted by farms will vary depending on the farm's location and structure, the interests of the farm household and the potential market, but significant examples include the development of farm tourism, on-site farm shops, horse riding centres, on-site food

processing, pick-your-own fruit enterprises and craft shops, as well as diversification into new crops and livestock.

The significance of income from pluriactivity has increased as further volatility in commodity prices has pushed direct agricultural income down. The average farming family in the United States in 1997 earned over 88 per cent of their income from off-farm sources, over half of which (equivalent to an average of $25,000 per farm) came from off-farm employment (Johnson, 2000). Similarly, in England over a quarter of farms received income from pluriactivity in 1997–8, again largely through off-farm employment (Table 4.6). Across Europe, research indicated that around 58 per cent of farm households were pluriactive in the late 1980s, but also that levels of pluriactivity varied considerably between regions, ranging from 27 per cent in Picardie (France) and 33 per cent in Andalucia (Spain) to 72 per cent in West Bothnia (Sweden) and 81 per cent in Freyung-Grafenau (Germany) (Fuller, 1990;

Table 4.6   Non-agricultural income of farmers and spouses in England, 1997–1998

	% of farmers receiving	Average income (all farms)	Average income (on farms receiving)
On-farm non-agricultural income (e.g. tourism, farm shops)	4	£200	£5,600
Off-farm income	58	£4,800	£8,400
of which: Self-employment	8	£800	£9,900
Employment	14	£1,600	£11,100
Social payments	18	£200	£1,300
Investments, pensions, etc.	40	£2,200	£5,500
All non-agricultural income	58	£5,000	£8,600
All income from pluriactivity (excludes social payments, investments, pensions, etc.)	23	£2,600	£11,200

Source: Cabinet Office, 2000

Ilbery and Bowler, 1998). Indeed, farm engagement in pluriactivity reflects a range of factors, including the relative prosperity of the locally dominant agricultural sector, the opportunities for off-farm employment or on-farm diversification, and historic social and economic structures. Thus, Campagne et al. (1990) identified three different types of pluriactivity occurring in different regions of France. In the Languedoc they found a long history of pluriactivity, the income from which is invested in the farm. In the more marginal farming region of the Savoie, in contrast, involvement in a diverse range of activities off-farm was necessary for survival, whilst in the more prosperous, arable region of Picardie, pluriactivity tended to be more entrepreneurial, including farm-based enterprises.

The emphasis on *countryside stewardship* is arguably both a form of extensification and a contributor to diversification, but has a distinctive logic. It recognizes the role played by farming in creating and maintaining the rural landscape, but seeks to reward farmers directly for their stewardship of the countryside, rather than regarding this as a by-product of agricultural production. Under initiatives of this kind, farmers have been paid to restore

hedgerows, walls, ponds and orchards, to maintain stiles and gates that help public access, to implement management plans for sensitive habitats, and – in some parts of the United States – simply to keep farmland in agricultural use, regardless of the type or level of production. Politically, however, some farmers have objected that they are being forced to become 'park-keepers', whilst other critics have argued that the schemes reward the wrong farmers:

> To collect a grant for restoring a meadow or planting a new hedge you need first to have destroyed the originals. The farmers making most from environmental payments are those who did the greatest damage during the frenzied years of all-out production. (Harvey, 1998, pp. 60–61)

Finally, a fourth strategy has been to enable farms to lower production levels by *enhancing the value* of their outputs, particular by specializing in quality 'region-branded' produce that can be sold at a premium. Since 1992, specialist regional foods in the European Union have been awarded a 'protected designation of origin' (PDO) or a 'protected geographical indication' (PGI) to restrict the use of place-related branding. Examples include Parma ham, Belfort cheese and Jersey Royal potatoes. Even without protected status, the marketing of regionally branded food can evoke an implication of quality and thus increase retail prices. Kneafsey et al. (2001), for example, discuss the marketing of regional foodstuffs from Wales, including Welsh Black Beef, Saltmarsh Lamb, Llyn Beef and Llyn Rosé Veal.

The various initiatives that have been clustered together under the banner of the 'post-productivist transition' have, as indicated above, begun to change aspects of agricultural policy and practice. However, the extent to which they add up to a fundamental 'restructuring' of agriculture is questionable. Evans et al. (2002) critique the concept of 'post-productivism' on both empirical and theoretical grounds. Empirically, they argue that the evidence for post-productivism has been selectively presented. Some proclaimed features of the transition, such as the diversification of farms into new crops (such as evening primrose) and livestock (such as llamas) still reflect the logic of productivism; whilst other observed changes, such as extensification, can result from factors other than post-productivism – which has been pursued with differing degrees of enthusiasm by different governments. Moreover, there is considerable evidence for the continuing strength of productivism. As Evans et al. (2002) note, 'political emphasis on the need for farmers to be able to compete in a liberalized global market seems to place greater emphasis worldwide on the continuation of productivist principles' (p. 316), and this can be identified for instance in the deregulation of agriculture in New Zealand (see Chapter 9). Certainly, the UK government spent £2,636.8 million on subsidies and payments supporting agricultural production in 2000–01, and just £376.1 million on 'post-productivist' initiatives, including set-aside, farm diversification and countryside stewardship.

On theoretical grounds, Evans et al. (2002) argue that the notion of a 'post-productivist transition' sets up an overly simplistic dualism between a productivist era, that perhaps was never as straightforward as suggested, and a post-productivist era, evidence for which is contentious at best. The complexities of agricultural change during the closing years of the twentieth century have hence been glossed over by debates that have focused on the timing and categorization of the post-productivist transition, and which have failed to engage with behavioural and actor-orientated research on rural change (Wilson, 2001) and the evidence of farm-level dynamics (Argent, 2002). Wilson (2001) suggests that the concept might be modified by looking beyond agriculture to wider rural change and adopting new terms that better reflect the larger picture. Evans et al. (2002), however, are blunter, describing post-productivism as 'a distraction from developing theoretically informed perspectives on agriculture' (p. 325), and proposing that it be abandoned in favour of a more critical, varied, engagement with broader social and economic theory.

The Future of Farming?

Whatever the outcome of the ongoing struggle to reform agricultural policy, the future of farming is already being shaped by the transnational corporations that dominate the agribusiness and retailing sectors. More than ever, twenty-first-century agriculture is driven by the capitalist imperative to maximize returns on investment. Increasingly, however, this means improving the product, as opposed to maximizing production. Many of the strategies for achieving this build on techniques and methods developed in the productivist era, and of these the most controversial is the use of genetic engineering to modify crops and livestock.

Genetic modification (GM) involves the alteration of a plant or animal's DNA in order

Table 4.7   Some commercially available genetically modified organisms (GMOs)

GMO	Modification	Source of gene	Purpose of genetic modification
Maize	Insect resistance	*Bacillus thuringiensis*	Reduced insect damage
Soybean	Herbicide tolerance	*Streptomyces* spp.	Greater weed control
Cotton	Insect resistance	*Bacillus thuringiensis*	Reduced insect damage
Escherichia coli K12	Production of chymosin or rennin	Cows	Use in cheese making
Carnations	Alteration of colour	Freesia	Produce different varieties of flowers

Source: After Bruinsma, 2003

to suppress or emphasize certain attributes. Thus genetically modified organisms (GMOs) can be produced that are resistant to viruses, insects or herbicides, that are larger or more productive than in their natural state, or which are designed to appeal to consumers' preferences – by being juicier or brighter coloured (Table 4.7). Supporters of GM argue that it offers the potential of maintaining agricultural productivity without intensive farming. GM crops, they argue, are environmentally friendly because they can be modified to produce their own pest-killing toxins, thus reducing the need for spraying farmland with chemicals. Resistant GMOs of this kind are particularly advocated to be a solution to famine in developing countries by protecting against crop failure due to disease or insects. However, scares over food quality have diminished public confidence in biotechnology and there is considerable scepticism about the safety of GMOs and their likely effect on the environment. Opponents claim that the long-term health consequences of GM are unknown and fear the extinction of traditional crop species – not least because they fear that cross-pollination will transfer GM genes to non–GM plants. As such, the decision whether or not to permit the planting of GM crops has become highly politicized in many countries, forming, for example, a key issue in the 2002 General Election in New Zealand. Moreover, the expansion of GM agriculture would further concentrate power in the commodity chain with large corporations, as the modified seed must be purchased from the patent-holding biotechnology company.

Between 1996 and 2001 the area of GM crop cultivation globally increased 30-fold, from 1.7 million hectares to 52.6 million hectares (Bruinsma, 2003). Yet, 69 per cent of that land is in the United States, and 22 per cent in Argentina, with the remainder spread between just 11 other countries. The result is a highly polarized agricultural geography. GM crops now account for 61 per cent of all upland cotton grown in the United States, and 54 per cent of all soybeans (USDA, 2000), yet in many other countries cultivation is restricted to test sites (as in the UK), or specific, non-food crops (as in France and Spain). GM food is also a major issue in global trade negotiations, with the European Union insisting on the labelling of all produce including GMOs. With restricted trade opportunities and some disappointment at the results of GMO cultivation, there are indications that the rate of growth of GM agriculture is slowing, and that GMO production in North America decreased in the first years of the new century.

Organic farming is often presented as the polar opposite to GM, and as the alternative model for future agriculture. Organic farming prohibits the use of synthetic chemical fertilizers and pesticides, minimizes external inputs and maximizes the use of farm-derived resources and natural products and processes. Advocates claim that organic farming produces better quality and more healthy food, such that whilst there is a lower level of productivity than conventional agriculture, organic produce is able to command premium retail prices. As such, conversion to organic production has become highly attractive to farmers facing economic difficulties in conventional agriculture. The total amount of certified organic farmland in Western Europe and the United States tripled between 1995 and 2000, and by 2000 accounted for 2.4 per cent of agricultural land in the former and 0.22 per cent in the latter (Bruinsma, 2003). Similarly, the value of organic production for the export market in New Zealand increased from US$0.05 million in 1990, to over US$30 million in 2000 (Campbell and Liepins, 2001).

In its early days, organic agriculture was often associated with small-scale, non-commercial farming and self-sufficiency, yet as the significance of organics has grown, so it has become established as another form of capitalist agriculture. Surveys in Denmark and Ontario have shown that more recent converters to organic farming are more likely to be motivated by profit than earlier converters, and less driven by purely environmental concerns (Hall and Mogyorody, 2001; Michelsen, 2001); and there is some limited evidence of 'conventionalization' among organic farmers who have specialized and/or increased their farm size (Hall and Mogyorody, 2001). Moreover, as organic producers move into mainstream markets they have become increasingly dependent on corporate food processors and retailers. One of the most significant boosts to organic farming in the UK was provided by the decision by a supermarket chain, Iceland, in 2000 to switch its entire own brand vegetable range to organics. However, the supermarket's recanting of its policy a year later raised concerns about the sustainability of consumer demand for organic produce and therefore the potential for further expansion of organic farming.

Summary

Agriculture in the developed world has been fundamentally transformed since the beginning of the twentieth century. From a position at the heart of rural life, farming has been pushed to the margins of the rural economy in terms of employment and its contribution to production, but retains a tremendous symbolic power that complicates any efforts to further reform the industry. Much of the change within agriculture has been foisted on to rural areas by external pressures. Indeed, far from being the agents of change, farmers themselves are only one of four groups of key actors that have shaped the evolution of modern agriculture. First, a political-economic analysis of agriculture as a capitalist industry reveals the importance of the owners of capital – including investors, banks and agri-food corporations as well as some landowners – in promoting the 'modernization' of agriculture as a means of maximizing returns. The integration of farmers into 'food chain complexes' dominated by corporations concerned with seed production, food processing and retailing, has left decisions about the future of agriculture increasingly concentrated in corporate hands. Secondly, however, agriculture is not an unfettered free market, but rather is one of the most regulated parts of the global economy. This means that the state is a key actor. Conventionally, state intervention in agriculture has supported capitalist exploitation by absorbing risk through subsidies and price guarantees. Trade policy has also been directed by national agricultural interests and agriculture remains a key concern in trade conflicts (see Chapter 9). More recently, reforms to agricultural policy have directed state support towards non-economic aspects of farming, such as landscape conservation. Arguably this not an anti-capitalist move, but rather recognizes the changing nature of the value of farming in a rural economy driven more by consumption than by production (see Chapter 12). Thirdly, agriculture like all capitalist industries, relies on consumption, and hence consumers are a powerful group of actors. The prices that we are willing to pay for our food, our concern about food quality, our interest or otherwise in where our food comes from, and various preferences for local produce, organic produce, vegetarian diets and so on, all have micro-effects that reverberate back through the commodity chain to influence the profitability of particular farming sectors. Finally, there are the farmers themselves, who whilst influenced by the above pressures must ultimately decide how to respond in the management of their own farm. This is demonstrated, for example, in the reluctance of many farmers to diversify.

The complex web of actors involved in agricultural decision-making therefore means that any account of agricultural change, including that presented in this chapter, necessarily glosses over the detailed dynamics, discrepancies and discontinuities that form the reality of agricultural change as experienced on the ground. Moreover, a focus purely on agriculture artificially separates farming from the wider rural economy and the changes in other sectors. These are examined in the next chapter.

Further Reading

There is a wealth of literature on many diverse aspects of agriculture and agricultural change. As a starting point, the chapter by Brian Ilbery and Ian Bowler, 'From agricultural productivism to post-productivism', in B. Ilbery (ed.), *The Geography of Rural Change* (Addison Wesley Longman, 1998), presents a good overview of the transition from productivism to post-productivism from a predominantly European perspective. To balance, David Goodman, Bernado Sorj and John Wilkinson (1987) *From Farming to Biotechnology* (Blackwell, 1987) presents a largely American narrative of the rise of biotechnology in agriculture. The development of agriculture as a capitalist industry is emphasized in studies of California by George Henderson and Richard Walker, particularly Walker's 'California's golden road to riches: natural resources and regional capitalism, 1848–1940', in the *Annals of the Association of American Geographers*, volume 91, pages 167–199 (2001), and Henderson's *California and the Fictions of Capital* (Oxford University Press, 1998).

The more human side of agricultural change is revealed by Kathryn Marie Dudley in *Debt and Dispossession: Farm Loss in America's Heartland* (University of Chicago Press, 2000), and Andrew O'Hagan in *The End of British Farming* (Profile Books, 2001). For more on the post-productivist debate see Nick Evans, Carol Morris and Michael Winter, 'Conceptualizing agriculture: a critique of post-productivism as the new orthodoxy', in *Progress in Human Geography*, volume 26, pages 313–332 (2002).

Websites

Extensive up-to-date statistics on agriculture are available from a number of websites, including those of the United Nations' Food and Agriculture Organization (FAO) (www.fao.org), the United States Department of Agriculture (USDA) (www.usda.gov/nass), the European Union's Directorate-General for Agriculture (DGVI) (europa.eu.int/comm/agriculture/index_en.htm), the UK Department of the Environment, Food and Rural Affairs (DEFRA) (www.defra.gov.uk/esg/), the Australian Bureau of Agriculture and Resource Economics (www.abareconomics.com) and the New Zealand Ministry of Agriculture and Forestry (www.maf.govt.nz/statistics/).

5

The Changing Rural Economy

Introduction

The transformation of agriculture is only one half of the story of rural economic change over the past century. Other 'traditional' rural economic activities, such as forestry, fishing, mining and quarrying, have experienced a similar evolution in their fortunes and a similar decline in their level of employment. At the same time, employment has risen overall in rural areas in manufacturing, tourism and the service sector. Between 1969 and 1997, the rural counties of the United States lost nearly 750,000 jobs in agriculture, but gained over 827,000 jobs in manufacturing (Isserman, 2000). In rural Canada, six in ten workers are now employed in the service sector (Trant and Brinkman, 1992), as are nearly half the workforce of rural France (INSEE, 1998) and seven in ten workers in rural England (Countryside Agency, 2003).

The shift in the balance of the rural economy from primary industries, based on the exploitation of the natural environment, to the secondary and tertiary sectors is the product of a range of inter-locking processes operating at different scales from the local to global. These include trends within global economic restructuring such as the liberalization of global trade and the increasingly 'foot-loose' nature of economic enterprises as dependence on particular resources in particular places has been diminished by technological advances; as well as more locally contingent factors such as improved infrastructure in rural areas, and higher levels of educational attainment in the rural population. Collectively these factors have altered the relative position of rural areas in the *spatial division of labour* under advanced capitalism through which 'different forms of economic activity incorporate or use the fact of spatial inequality in order to maximize profits' (Massey, 1994). Historically, the opportunities presented by the availability of natural resources, undeveloped land and the structures of rural landownership and employment, were exploited in the development of resource capitalism. More recently, investment has been attracted to rural areas by factors such as lower land prices, taxation and wage levels, greenfield sites for development, and an aesthetically higher quality environment. Equally, however, rural areas must compete on a global scale and major sources of employment like factories and telephone call centres can be suddenly relocated to lower wage economies in the developing world.

This chapter examines the changing rural economy, focusing in turn on the changing circumstances of forestry, fishing and mining, manufacturing industry and the service sector. It discusses the factors that have produced these changes, explores their impact on rural communities, and considers the prospects for the future development of rural economies under advanced globalization.

Forestry, Fishing and Mining: the Fluctuating Fortunes of the Primary Sector

During the first part of the twentieth century, the dominance of agriculture in rural economies was rivalled only by the localized supremacy of other primary exploitative industries including forestry, fishing, mining and quarrying. Often these sectors were interconnected through flows of investment, ownership and employment. Walker (2001), for example, identifies the patterns of cross-investment between mineral exploitation, forestry and agriculture in the development of resource capitalism in California – patterns that were reproduced in other regions. At a different level, workers in many rural communities would divide their employment between mining and farmwork, or fishing and farmwork depending on the season and product demand. In some communities, however, the local mines or quarries, or fishing or forestry, were the only significant sources of employment, particularly where the economic potential of these activities far exceeded that of agriculture. Thus, whilst the decline in employment in these sectors may have had little effect across rural areas as a whole, the localized impact on individual communities has frequently been severe, sometimes creating pockets of extreme deprivation within a relatively prosperous rural region.

In some regions, whole industries have disappeared. The last tin mine in Cornwall, in south-west England, closed in 1998, ending an industry that dated back over 2,000 years and which at its peak in the late nineteenth century employed some 50,000 people in the county. In other regions, significant employment in mining, forestry or fishing has become restricted to fewer and fewer communities, and even in those communities the numbers employed in the dominant industry has decreased. Canada had 80 rural communities in 1976 where over 30 per cent of the labour force were employed in forestry or wood processing (and therefore classified as a 'single industry town'), along with 54 communities dependent on mining and 38 on fishing (Clemenson, 1992). Over the next ten years both forestry and mining experienced economic turbulence. Employment in Canadian forestry fell from over 300,000 in 1980 to 260,000 in 1982, whilst half of the mining sector was shut down temporarily at the height of recession in late 1982. The effect on the communities concerned was dramatic. Two mining towns in Labrador virtually disappeared as their iron ore mines closed – Schefferville, where the population collapsed from 3,500 in 1976 to 320 in 1986, and Gagnon, where only five residents were left by 1986 compared with 3,400 in 1976. In other communities, employment in the main industry slumped (Table 5.1), and for some, like Marathon, Ontario, this meant swapping a sole dependency on pulp-processing for a dual dependency on pulp and mining (Clemenson, 1992). Only fishing prospered in relative terms during this period, with employment in fish processing in Atlantic Canada increasing rapidly in two spells in the late 1970s and mid-1980s.

The fortunes of particular forestry-, mining- or fishing-dependent communities will be

Table 5.1 Percentage of workforce employed in main industry for 172 Canadian communities identified by dependency on fishing, mining or wood in 1976

	Fishing communities			Mining communities			Wood-based communities		
	1976	1981	1986	1976	1981	1986	1976	1981	1986
>30	38	33	34	54	42	24	80	52	37
15–29	0	5	4	0	11	22	0	27	40
<15	0	0	0	0	1	8	0	1	3

Source: After Clemenson, 1992

determined by industry-specific trends and local conditions. However, at a more general scale, there are three key factors that have resulted in job losses in all three sectors. First, the resources being exploited may be exhausted. Minerals in particular are a finite resource and periods of significant mining employment in rural regions are frequently short-lived. Secondly, operations may be abandoned as uneconomic either because of decreased consumer demand or competition. Mining, forestry and fishing are all vulnerable to competition in a globalized economy. Thirdly, environmental challenges are increasingly being mounted to resource exploitation, citing concerns about pollution, landscape degradation and threats to plant and animal habitats. Thus, McManus (2002), in a study of forestry policy in British Columbia and New South Wales, notes that 'the regulation of forestry not only includes urban political power (Victoria and Ottawa in Canada, and Sydney and Canberra in Australia) and commercial power (concentrated in Vancouver, Sydney and Tokyo), but also voter power (largely in Vancouver and Sydney)' (p. 855). The forestry industry (which directly employs some 82,000 people in British Columbia, indirectly supports 300,000 more, and contributes 16 per cent of the province's gross domestic product) must therefore balance commercial imperatives with environmental regulations aimed at reducing production and controlling the nature and location of logging (McManus, 2002).

The potential impact of environmental pressures on forestry-dependent rural communities is illustrated by the case of Catron County in New Mexico. The sparsely populated county of 2,700 residents was reliant on ranching, logging and timber processing until 1990, when the US government severely restricted timber cutting in the area to protect the endangered Mexican spotted owl. One hundred jobs were lost in the closure of the sawmill, with unemployment in the county rising to 10.8 per cent in 1995 – twice the US average – and nearly a quarter of the population falling beneath the poverty line (Walley, 2000).

The fishing communities of Newfoundland and Labrador have come under pressure from all three sources mentioned above. As Kennedy (1997) documents, competition and low prices gradually eroded the fishing industry during the twentieth century, with 16,000 people moving away from coastal communities in the 1960s under a government resettlement programme. Technological modernization, and an extension of the Canadian fisheries jurisdiction to 200 miles, helped a rejuvenation of the industry in the 1970s, but soon the local fishing fleet faced renewed competition from industrial-scale trawlers from Europe and elsewhere in North America. Moreover, intensive fishing was seriously depleting the stock of cod. Under pressure from environmental campaigners, the Canadian government closed the 'northern cod' fishery in 1992 in an attempt to allow stocks to

Table 5.2 Net change in manufacturing jobs in England, Wales and Scotland, 1960–1991

	Number of jobs
London	−979,000
Conurbations	−1,392,000
Free-standing cities	−631,000
Large towns	−388,000
Small towns	−284,000
Rural areas	+238,000
England, Wales and Scotland total	−3,443,000

Source: After North, 1998

replenish. The moratorium led immediately to 20,000 job losses, with a further 10,000 jobs lost the following year. Although compensation payments were made to fishers and processing plant workers, and government schemes introduced to develop alternative sources of employment, including hi-tech industry, aquaculture, tourism and mining, the closures have severely depressed the local economy and intensified problems of poverty and out-migration.

Manufacturing Industry

If agriculture and forestry are commonly associated with rural areas, then manufacturing is perhaps the industry most readily identified with urban areas. In the popular imagination, manufacturing conjures up a picture of a large, smoke-billowing factory towering over endless rows of workers' houses, much in the style exemplified by the British painter L.S. Lowry. Yet, not only does manufacturing have a long history in many small towns and rural communities – particularly in the processing of agricultural, fish and timber produce – but the late twentieth century also saw a net shift in manufacturing employment from urban to rural areas in developed countries. Between 1960 and 1991, rural areas of England, Scotland and Wales made a net gain of nearly a quarter of a million

manufacturing jobs, whilst manufacturing employment decreased in every other type of area (Table 5.2) (North, 1998). Similarly, manufacturing employment in rural counties in the United States increased by 47 per cent between 1960 and 1980 – well above the national average – before fluctuating during the recession of the 1980s (North, 1998; USDA, 2000). As a result, in both the United States and France, manufacturing now employs a greater share of the workforce in rural areas than it does in urban areas (INSEE, 1998; USDA, 2000). However, it should be noted that most manufacturing employment and the majority share of manufacturing output continues to be concentrated in urban areas.

The urban–rural shift in manufacturing has involved two distinctive periods of expansion. First, from the 1940s to the 1960s there was a period of *absolute expansion* as manufacturing employment increased in both urban and rural regions, but more rapidly in rural areas. For example, manufacturing employment in the United States in the 1960s increased by 15 per cent in urban areas, but by 31 per cent in rural areas (North, 1998). Secondly, the 1970s, 1980s and 1990s have been predominantly a period of *comparative expansion* as manufacturing employment has declined more slowly in rural areas than in urban areas – or even, in some cases, increased against the general trend, as in the United States during the 1970s and in the UK during the 1980s (Townsend, 1993).

North (1998) positions these changes in the context of the global restructuring of manufacturing industry. This, he notes, has been characterized by a shift from mass production systems to flexible production, which has enabled firms to become more 'foot-loose' in their location. In an increasingly globalized economy, corporations will seek out locations where the costs of manufacturing can be minimized whilst retaining access to high-profit markets. As such, there has been an overall shift in manufacturing from the established industrial economies of Europe and North America to the Pacific Rim (especially Japan, Taiwan, Malaysia and South Korea) and the developing world. The search for competitive advantage in the manufacturing process, however, has also operated at a domestic level, with rural areas perceived to offer more favourable conditions than urban sites. Variations on this theme form the basis for four explanations of the urban–rural shift discussed by North (1998):

- *The constrained location hypothesis* suggests that firms have become constrained by the quantity and quality of space available in urban areas and thus have relocated to rural sites with space for expansion (Fothergill and Gudgin, 1982).
- *The production cost hypothesis* argues that relocating firms seek to increase profits by taking advantage of spatial variations in production costs, particularly wage costs and land prices that tend to be lower in rural areas (Tyler et al., 1988).
- *The filter-down hypothesis* connects industrial location to the product cycle, suggesting that in the early stages of product development urban locations provide access to skilled labour and specialist knowledge inputs, but that later in the cycle production becomes routinized and may be relocated to rural sites to reduce costs (Markusen, 1985).
- *The capital restructuring hypothesis* takes a broader view, arguing that different phases of capital accumulation produce different requirements for labour and location. It proposes that advances in technology and production processes have reduced the dependency of manufacturing on concentrations of skilled labour and enabled relocation to rural areas where advantage can

be taken of lower wage costs, lower levels of unionization and worker militancy and, often, a captive labour market with few alternative sources of employment (Massey, 1984; Storper and Walker, 1984).

A fifth thesis, the residential preference hypothesis, varies from the above in focusing on new business start-ups as opposed to the relocation of existing firms. The thesis argues that entrepreneurs opt to develop new businesses in rural locations because of the perceived higher quality of life that they afford (Gould and Keeble, 1984). The above hypotheses should not necessarily be regarded as competing models; rather they reflect the complexity of the urban–rural shift in manufacturing as the amalgam of many different processes driven by different imperatives.

Indeed, there are a number of caveats that should be attached to the notion of an urban–rural shift in manufacturing. First, the urban–rural shift has been largely sector-specific. The traditional rural manufacturing industries, such as food processing, timber and paper production, fish canning and textiles, which once dominated single-industry rural towns, have substantially declined, often with devastating consequences for their host communities. Manufacturing growth has been particularly associated with light engineering, high-tech industries and areas of niche goods production, including the production of 'quality' foodstuffs.

Secondly, the urban–rural shift has been spatially selective. Estall (1983) challenged conventional accounts of the manufacturing shift in the United States by demonstrating that growth was greatest in rural counties adjacent to metropolitan areas, and that the regional shift from northern states to both rural and urban locations in southern states was more significant than the urban–rural shift *per se*. Spatial concentration has been particularly marked in the high-technology sector – flagged as one of the rural 'growth' industries. For example, although employment in high-tech industry increased in rural Britain as a whole between 1981 and 1989 by 12 per cent compared with a decrease in urban areas, the growth was focused in three regions. In 1989, three times more people were employed in hi-tech industries in South-East England than in any other region, and much 'rural' growth was concentrated around particular key localities, such as Cambridge (North, 1998).

Thirdly, the urban–rural shift has changed the nature of manufacturing in rural areas. Factories are less integrated with rural communities than previously, they are less likely to use local natural resources, and are less likely to be locally owned. The nature of work performed has also changed as part of a new spatial division of labour. As North (1998) observes, 'it is argued that it tends to be the more routine, less technically advanced, assembly type functions requiring largely semiskilled workers which are drawn to rural and small town locations rather than those functions that require highly technical and skilled workers' (p. 172).

Fourth, many of the factors that have been identified as contributing to the urban–rural shift also make rural manufacturing vulnerable to competition from developing countries and to cut-backs in a recession. Rural areas may, for example, be able to undercut urban locations on wage levels, but not developing countries. Corporate mergers and takeovers have turned rural areas into a branch-plant economy in which the future of a local factory is dependent on a boardroom decision possibly taken on a different continent, where strategy is developed at a global scale. Furthermore, lower levels of unionization mean that corporations may face less resistance to closing rural branch plants than those in more militant towns and cities (Winson, 1997).

It may be easier for corporations to close a factory in a rural community or small town than in an urban area, but the impact on the local community is often more severe as job losses will be proportionately more significant against the size of the local population. This is illustrated by Fitchen (1991) in a case study from rural New York. The factory concerned had started as a knitting mill, but had changed owners and product line a number of times and by the mid-1980s was owned by a St Louis-based subsidiary of a New York City-based company, producing plastic equipment for hospitals. It employed some 500 workers, mainly women, who earned between $7.30 per hour for unskilled assembly line positions to $12 per hour for clerical workers. Of the employees, 155 lived in the immediate local community of around 600 households. In 1989 the factory was closed as production was relocated to Mexico where wages averaged $1.25 per hour. The redundancies affected one in four households in the local community, and of the 365 former employees registered for a job assistance programme, only 20 had found new jobs by the time the plant closed. As Fitchen notes of the sign that appeared to advertise the site for sale, '[the sign] proclaim[ed] a message that captured and epitomized the change in rural manufacturing in this decade: the irony of good facilities and good workers, but no work' (p. 72).

The Service Sector in Rural Areas

The steady growth of the service sector in rural areas appears at first sight to provide a contrast to the fluctuating fortunes of production-based industries. Service sector employment in rural areas increased progressively throughout the twentieth century to become the major source of work in rural regions across the developed world. However, the significance of the service sector tends to get inflated by the large and diverse range of activities included under its

Table 5.3 Service sector employment in England, 2001

	Remote rural (%)	Accessible rural (%)	Urban (%)
Distribution, hotels and restaurants	27.6	25.9	23.7
Banking, finance, insurance, etc.	10.6	17.4	22.0
Public administration, education and health	25.1	22.7	23.7
Other services	4.5	5.1	5.3
Service sector total	67.8	71.1	74.7

Source: Countryside Agency, 2003

Table 5.4 Service sector employment in the United States, 1996

	Rural (%)	Urban (%)
Retail trade	17	17
Government	16	14
Finance, insurance and real estate	5	8
Transport, communications and utilities	4	5
Wholesale trade	3	5
Other services	23	32
Service sector total	68	81

Source: www.rupri.org

umbrella. Service sector employment includes highly paid lawyers, financiers and stockbrokers as well as cleaners, shop assistants and care workers; it includes teachers and truck-drivers, doctors and waiting staff. As Tables 5.3 and 5.4 show for England and the United States, when broken down into industry groupings the impression of dominance is lessened – the largest service sector industry in the rural US, retailing, employs roughly the same proportion of the workforce as manufacturing. Moreover, there are different balances in the employment share of different service sector industries in different rural regions, and, it can be hypothesized, different processes driving their development.

The growth of service sector employment in rural areas can therefore be disaggregated into four components. First, there has been an expansion of the public service sector, including education, health and local government. The development of both the comprehensiveness of coverage of public service provision in rural areas and the extent and quality of the service provided by schools, hospitals and other institutions since the end of the Second World War, has created new employment opportunities in rural areas. Moreover, the significance of a large public sector employer such as a school, a hospital or a prison will be greater in a small rural labour market than in a larger urban labour market. As such, the public sector may account for a quarter or more of all employment in remoter rural areas (see Table 5.3), with, for instance, over 200 rural counties in the United States classified by the USDA as being dependent on government employment.

Secondly, the growth of consumerism has stimulated an expansion in the retailing and leisure services sectors in rural areas as in urban areas. Indeed, rural towns and their hinterlands have frequently been targeted as new markets for expansion by retail and leisure chains, disproportionately increasing service sector employment in such localities. However, investment of this type has also contributed to a spatial restructuring of services in rural areas (see Chapter 7), with the closure of village shops, garages and inns potentially leading to a decrease in employment in retailing and hospitality in some smaller rural communities. Furthermore, the types of jobs created in modern retailing and hospitality are often lower paid, temporary and/or part-time. Over half of all workers employed in distribution, hotels and restaurants in remote rural districts of England, for example, are on part-time contracts (Countryside Agency, 2003) (see Chapter 18).

Thirdly, the increase in employment in distribution and leisure services also reflects the growing significance of tourism in many rural areas. One side-effect of the foot and mouth disease outbreak in Britain in 2001 (see Box 4.3) was to highlight the contribution made by tourism to the rural economy. Some 380,000 jobs were estimated to be dependent on tourism in rural England, with a further 25,000 in rural Wales, and tourists were calculated to spend over £10 billion per year in the rural areas of England and Wales (Cabinet Office, 2000). Similarly, hotels and motels alone employ some 310,000 people in the rural United States (Isserman, 2000). Tourism is often flagged as a means of regenerating rural communities depressed by economic decline in agriculture, primary production or manufacturing, inspired by the example of a few successful initiatives such as that of the former sawmill town of Chemainus on Vancouver Island (see Chapter 12). However, Butler and Clark (1992) warn that, 'the least favourable circumstance in which to promote tourism is when the rural economy is already weak, since tourism will create highly unbalanced income and employment distributions. It is better as a supplement for a thriving and diverse economy than as the mainstay of rural development' (p. 175). Outside of coastal regions and national parks, the potential for tourism to make a significant contribution to rural employment may be limited.

Fourthly, rural areas on the periphery of metropolitan centres have gained in service sector employment through the relocation of financial sector employers and corporate services companies. Murdoch and Marsden (1994), for example, record the relocation of insurance, banking and other financial services companies from London to the country town of Aylesbury, 40 miles from the centre of the city. The dynamics behind this shift are similar to those for manufacturing industry relocation – a lessened need for concentration coupled with the perceived advantages of greater space, lower land, tax and wage costs, and a higher quality environment. Yet, firms in these sectors tend to remain firmly integrated into urban-centred networks and specialist labour markets and thus their reach into more remote rural areas is limited.

The evolution of the service sector in rural areas has therefore been differentiated by region and by industry. Remoter rural areas are more likely to be dependent on tourism or public sector employment, whilst rural areas close to the urban fringe may benefit from the relocation of financial and business services. In addition to service sector workplaces located in rural areas, service sector employment among residents of 'more accessible' rural areas is inflated by workers commuting to nearby towns and cities. This last practice highlights the continuing urban-centric nature of much service sector activity and the infrastructural barriers that still exist to the further expansion of service sector employment in rural areas. For some commentators, however, such barriers could be removed, with the development of information technologies and the advent of 'teleworking'.

Teleworking in the Countryside

People no longer need to commute to the cities. Many jobs can now travel to the workers using today's technologies – revitalising traditional rural communities … And there is a determination of more and more people to achieve a better quality of life, avoiding the stress and pollution of commuting, and playing more active roles day to day in their communities. (Acorn Televillages brochure, quoted by Clark, 2000, p. 19)

The quote above illustrates the aspirations of a new sector within the rural economy that has emerged with the development of information and communications technologies. Advances in computing technology and the development of the Internet, combined with the growth of information-based occupations, are argued to have created an opportunity for individuals to increasingly work from home, using telecommunications to engage with their employers ('teleworking') (Clark, 2000). The geographical flexibility of such work has been furthermore suggested by some authors to herald a de-urbanization of employment (Huws et al., 1990), and the potential has been seized upon by many rural development agencies, who have attempted to promote the development of rural teleworking by providing training and infrastructure, including 'telecottages' or resource centres providing access to information and communications technologies (Clark, 2000).

Clark (2000) identified 152 telecottages operating in the British Isles in 1999, concentrated in peripheral rural regions such as Wales, south-west England and northern Scotland, and mostly in small villages or remote rural locations. Many of the telecottages operated as 'clearing houses' that outsourced work to individual home-based teleworkers, with common areas of work including marketing, secretarial services, translation and publishing. However, as Clark's figures imply, overall levels of employment in teleworking remain low and the growth of the sector in rural areas is restricted by the continuing importance of face-to-face contact in business and by the quality of the rural telecommunications infrastructure.

Summary

There has been a clear quantitative shift in the nature of the rural economy over the past century. Statistics for employment, business type and income generation all demonstrate that the dominance of production-based activities, including agriculture, forestry, fishing, mining and quarrying, in the early twentieth century, has been replaced by a more service-oriented economy. The transition has also been marked by qualitative changes in the nature of the economy, of which three key trends are apparent. First, rural economies at a local scale have become more fragmented, creating a wider range of employment opportunities for rural residents, but also increasing uncertainty. The contemporary rural economy is more fluid than the previous single-industry economies and there are few 'guaranteed' jobs. In order to access better paid employment opportunities, potential employees often need to leave rural areas to acquire the appropriate training or qualifications, whilst lower skilled work is frequently characterized by low pay and temporary contracts. The implications of these changes for people living and working in rural areas are discussed in later chapters (see Chapters 15, 17 and 18).

Secondly, rural economies have become more externally dependent. Traditional industries such as agriculture and mining relied on the export of products to towns and cities, but the farms and mines tended to be locally owned and earned income tended to circulate within the rural economy. The contemporary rural economy is not only dependent on external income (for example in the form of investment, state support, agricultural exports or tourist spending) but much of the profit now flows back to external parent companies and investors. Economic decision-making power has also been concentrated with external actors, such that the degree of control that a rural community has over its economic future has been weakened.

Finally, there has been a discursive shift in the way in which the rural economy is imagined and represented. From being conceived of as a space of production, the rural is now understood as a space of consumption. This includes both consumptive activity *in* the countryside (supporting the service sector), and the consumption *of* the countryside – most notably through tourism but also through residential investment, the marketing of 'rural' crafts and branded speciality foods, and the use of rural locations for film and television (see Chapter 12). The discursive shift is reproduced in government policies and in conflicts, such as between logging and wildlife conservation, where the interests of protecting an idyllized rural environment with consumer appeal are increasingly prioritized over the interests of production (see also Chapter 14).

Further Reading

There are few comprehensive overviews of economic restructuring in rural areas. David North's chapter on rural industrialization in B. Ilbery (ed.), *The Geography of Rural Change* (Addison Wesley Longman, 1998) concentrates largely on manufacturing but contains material of wider relevance, whilst Michael Clark's *Teleworking in the Countryside* (Ashgate, 2000) is a detailed study of teleworking in the UK. For case studies and more information on sector-specific restructuring, see Trevor Barnes and Roger Hayter, 'The little town that did: flexible accumulation and community response in Chemainus, British Columbia', in *Regional Studies,* volume 26, pages 617–663 (1992), for a study of the sawmill closure in Chemainus, Canada, and see Janet Fitchen's description of factory closures in rural New York State in her book *Endangered Spaces, Enduring Places: Change, Identity and Survival in Rural America* (Westview Press, 1991).

Websites

Detailed statistics and commentaries on the rural economies of the UK and the United States respectively can be found in the State of the Countryside reports (www.countryside.gov.uk/stateofthecountryside/default.htm) and on the Rural Policy Research Institute's website (www.rupri.org).

6

Social and Demographic Change

Introduction

In the space of a little under two centuries, the population of rural areas in what is now perceived as the developed world has undergone something akin to a metaphorical rollercoaster ride. From a trend of steady population growth at the start of the nineteenth century, rural areas lost substantial proportions of their populations to towns and cities in the era of rapid urbanization during the late nineteenth and early twentieth centuries, before the flow was reversed in the 1960s and 1970s and the countryside again enjoyed net in-migration. Finally, at the start of the twenty-first century there is a more ambiguous situation with an overall tendency towards rural population gain, but cross-cut by diverse national, regional, local and demographic counter-trends. As the population of rural areas has fluctuated, so the composition of that population has also changed. The rural population today is in general older and more middle class than it was 30 or 40 years ago. Furthermore, these trends have been reproduced by the inflationary effect of participation by middle class in-migrants in rural property markets. This chapter examines these changes in more detail. The first half of the chapter documents the chronology and geography of population change in rural areas and discusses the processes that have driven change. The second half then analyses the recomposition of the rural population, focusing on the rise of the middle classes and the consequence for rural property markets.

From Urbanization to Counterurbanization

Rural depopulation

In 1851 half of the population of England and Wales lived in rural areas. A century later, in 1951, only one-fifth of the population did so. The depopulation of the British countryside that is indicated by these figures was a process of mass migration that was replicated across the globe as industrialization took hold. Between 1851 and 1951, the total population of England and Wales increased by 26 million people (or 144 per cent), yet the population of rural areas *fell* by some half-a-million people (or a 5 per cent decrease) (Saville, 1957). Most of this change occurred during the height of industrialization from the mid-nineteenth century to the 1920s, as migrants were attracted by the prospect of higher wages in urban industry and the greater employment opportunities in towns and cities compared with rural areas, where the early stages of agricultural modernization had reduced the number of farmworkers and where earlier outposts of manufacturing and mining were in decline. The arrival of railways helped to facilitate rural depopulation by increasing the mobility of rural people, and advances in education and communications promoted a social mobility in which migration was also linked to the pursuit of heightened aspirations and the attraction of the potential for independence and freedom in towns and cities compared with the closed and isolated worlds of rural communities (Lewis, 1998; Saville, 1957).

The trend of depopulation was not even. The more peripheral rural counties of England and Wales experienced greater depopulation than those closer to the new metropolitan centres, and this model was repeated at a local scale as smaller and more remote rural communities declined faster than the market towns (Lewis, 1998). The rates and directions of migration also fluctuated over time. The population of Rutland, for example, decreased in every decade between 1851 and 1931, except from 1901 to 1911 when it increased by 3.2 per cent (Saville, 1957). By the 1920s the current of urbanization had begun to slow nationally, as economic depression reduced the employment opportunities in towns and cities, and as the middle classes started to move in the opposite direction to the newly developing suburbs.

The trend of rural depopulation was repeated elsewhere in Europe – albeit often on a later and more rapid timescale. In Ireland, for example, the proportion of the national population living in rural communities of fewer than 1,500 people decreased from 71.7 per cent in 1901 to 63.5 per cent in 1936 to 46.7 per cent in 1971, with out-migration driven by the limited economic and social opportunities available in the countryside (Hannan, 1970).

Beyond Europe, there were in 1900 large parts of rural North America, Australia and New Zealand that were still being settled by Europeans for the first time. Yet, this populating of rural space was outpaced at a national level by the expansion of urban areas, and in the more populous regions a similar process of rural to urban migration could be identified. Urban population growth rates in Canada were consistently more than double those of rural areas from at least the early nineteenth century until the 1930s, with the urban population surpassing the rural population soon after 1921 (Bollman and Biggs, 1992).

The population turnaround

The reversal of the rural to urban migration flow was first observed by population analysts in the United States in the early 1970s. The new phenomenon, labelled 'counter-urbanization' by Berry (1976) (see Box 6.1), was confirmed and documented by a series of studies over the following few years, most notably work by Bourne and Logan (1976) and Vining and Kontuly (1978) that demonstrated the 'population turnaround' was evident not just in the United States (Table 6.1), but also in Canada, Australia and much of Western Europe. In the UK, the fastest population growth in both the 1970s and 1980s was in predominantly rural counties and some 100,000 people were recorded as migrating directly from urban to rural areas in the year prior to the 1981 census (Lewis, 1998; Serow, 1991). Overall, the population of the metropolitan areas of the UK decreased by 6.5 per cent between 1971 and 1981, whilst that of non-metropolitan counties increased by 6 per cent (Serow, 1991). In other European countries the difference was less marked, but present none the less: the net migration rate to rural areas in the Netherlands was around 2 per cent a year in the 1970s; in France it was 1.3 per cent in 1982; and in West Germany, 0.7 per cent during the early 1980s (Serow, 1991). Canada, too, showed a more mixed trend, with urban to rural migration exceeding rural to urban migration from 1971 into the 1980s, but the rate of population growth in rural areas surpassed urban growth rates only for a short period between 1971 and 1976 (Figure 6.1).

Box 6.1   Key term

Counterurbanization: The movement of population from urban to rural areas. It is normally associated with urban to rural migration, but may also be indicated by differential rates of population growth of rural and urban areas. Counterurbanization can involve both decentralization – migration from towns and cities to adjacent rural areas – and deconcentration – inter-regional migration from metropolitan areas to rural districts.

Table 6.1   The population turnaround in the United States, 1960–1973

	Annual population change (%)		Annual net migration (%)	
	1960–70	**1970–3**	**1960–70**	**1970–3**
Metropolitan counties	1.7	1.0	0.5	0.1
Non-metropolitan counties	0.4	1.4	−0.6	0.7
Entirely rural counties	−0.5	1.4	−1.2	1.0
US total	1.3	1.1	0.2	0.3

Source: Champion, 1989

As the Canadian evidence suggests, counterurbanization should be regarded as an amalgam of different processes of population change, rather than as a single, coherent and unidirectional flow. Some authors, for example, have distinguished between *decentralization*, or migration from cities to nearby rural areas, and *deconcentration*, or migration from major cities to rural areas in another region. Decentralization is associated with commuting, whilst deconcentration often involves a more substantial 'lifestyle change'. It is associated in the United States with migration to western states and to the southern 'sunbelt', and in the UK with migration to peripheral rural regions such as the south-west of England, mid–Wales and the northern Pennines. By drawing together analyses conducted over the 20-year period from the mid-1970s, Lewis (1998) proposes that counterurbanization can be identified as involving four common factors. First, there is greater growth occurring at progressively lower levels of the urban hierarchy. Secondly, population increase spreads

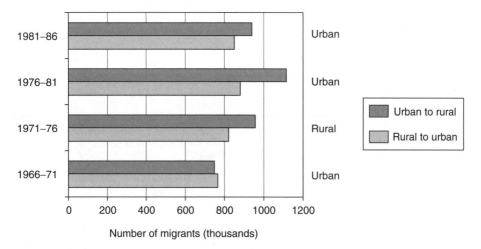

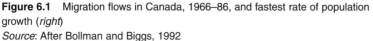

Figure 6.1 Migration flows in Canada, 1966–86, and fastest rate of population growth (*right*)
Source: After Bollman and Biggs, 1992

through extended suburbanization. Thirdly, there are buoyant rates of growth recorded outside metropolitan areas, especially in remote rural areas. Fourthly, there is a shift in population from traditional urban industrial areas to rural locations that are more favoured in environmental terms. These observations suggest that the urban to rural population shift implied by counterurbanization is cross-cut by various regional trends, as will be returned to in a later section.

Vining and Strauss (1977) declared counterurbanization to be 'a clean break with the past', and Berry was similarly bullish about the historical significance of the 'turnaround', stating that:

> A turning point has been reached in the American urban experience. Counterurbanization has replaced urbanization as the dominant force shaping the nation's settlement patterns. (Berry, 1976, p. 17)

The 1980s cast some doubt on the confidence of these early proclamations, with the rate of counterurbanization slowing or even reversing in many countries. A longer-term perspective over the last quarter of the twentieth century, however, suggests that urban to rural migration continues to be the prevailing trend, at least in the United States and England (Tables 6.2 and 6.3), albeit subject to qualifications that will be discussed later in this chapter.

The Drivers of Counterurbanization

Counterurbanization is a product of the economic restructuring of both urban and rural societies, combined with societal and technological changes that mean that people are more mobile physically and socially than in previous generations. As Kontuly (1998) summarizes, studies have proposed a wide range of 'explanations' for counterurbanization, which can be grouped under six key drivers:

- *Economic cyclical factors*, including business cycles, the growth of localized employment in mining, tourism and defence, and

Table 6.2 Net migration rates to rural and urban counties in the United States, 1980–1997

	1980–90 (%)	1990–7 (%)
Rural	−2.8	4.0
Urban	3.8	2.1

Source: www.rupri.org based on USDA ERS statistics

Table 6.3 Population change in rural and urban districts of England, 1981–2001

	1981–91 (%)	1991–2001 (%)	1981–2001 (%)
Rural districts	+7.1	+4.9	+12.4
Urban districts	+1.4	+0.9	+2.4
England total	+3.0	+2.0	+5.0

Source: Countryside Agency, 2003

the cyclic pattern of capital investment in property and business.

- *Economic structural factors*, including the deconcentration of jobs to rural areas in a new spatial division of labour (see also Chapter 5).
- *Spatial and environmental factors*, including social and environmental problems in urban areas, housing availability and costs, and the attraction of rural environmental amenities.
- *Socio-economic and socio-cultural factors*, including changing demographic compositions, the growth of state welfare payments, and changes in residential preferences and social values.
- *Government policies*, including explicit initiatives to promote rural development or to attract in-migrants to rural areas, and the improvement of education, health and other public services in rural areas.
- *Technological innovations*, including improved transport links and telecommunications.

These factors have changed the conditions in which individuals make decisions about where to live. In some cases, new constraints have been introduced, for example the urban to rural shift of many manufacturing and service sector jobs under economic restructuring has meant that employment opportunities may be greater in rural regions than in neighbouring urban regions. In other cases, economic restructuring, social and cultural change and technological innovations have all removed constraints from individuals, such

that residential decisions may increasingly be made on the basis of aspirational factors, including the perceived quality of life in rural areas.

The significance of 'rurality' as a 'pull factor' in migration is suggested by opinion polls that record that a majority of the urban population in countries such as the UK and Canada would prefer to live in the countryside if they were able to do so (Bollman and Biggs, 1992; Halfacree, 1994). Halfacree's studies of in-migrants to villages in Lancashire and Devon in England found that nearly half stated that the rural character of the area was 'extremely important' in their decision to move there, compared with all other factors. Less than one in ten in-migrants said that the rural character was unimportant (Halfacree, 1994). Similar findings were identified by Crump (2003) in Sonoma County, California. Located 50 miles north of San Francisco, Sonoma County is the type of predominantly rural area that has benefited from a combination of population decentralization and deconcentration, with the county's population increasing by 53 per cent between 1970 and 2000. Crump found that for 50 per cent of in-migrants to a more rural district of the county, the 'rural environment' was the most important factor in their migration decision. Even in a suburban district of the county, the 'rural environment' was cited as a 'very important' or 'the most important' factor by over half of in-migrants. Other factors grouped by Crump as relating to the rural surroundings, including the attractive natural environment, nearby open space and 'privacy', were all highly cited by a majority of in-migrants to the rural district.

Crump's grouping of 'rural factors' reflects the constitution of the 'rural' as a social construction, as discussed in Chapter 1, in which different attributes will have more or less importance for different people. The attraction of the rural as a place to live will therefore vary between different in-migrants. Halfacree's (1994) study recorded a wide range of 'key attractions' of rural life cited by in-migrants, including the openness and aesthetic quality of the environment and the 'slower pace of life' and greater 'community feeling', as well as value judgements, for instance of the countryside as a better place to raise children (see Box 6.2). The projection of value judgements on to rural space and rural society introduces a political dimension into migration that can subsequently contribute to the emergence of local conflicts (see Chapter 14). Halfacree, for example, identifies a conservative, racist strand among a small minority of rural in-migrants in England for whom the countryside is attractive as a mono-ethnic and mono-cultural space, whilst, in contrast, Jones et al. (2003) report how the significance of the rural environment as a major attraction for a majority of in-migrants to southern Appalachia, USA, has contributed to a growth of environmentalist activity in the region.

'Aspirational migration' is hence an important component in counterurbanization, but for most in-migrants the attractiveness of the 'rural' will be just one of many factors influencing the multi-stage decision-making process that is followed in the sequence of deciding to move, selecting an area to move to, selecting a community in which to live, and selecting a particular property. Not all rural in-migrants actively *choose* to live in a rural area – Harper (1991) classified over a fifth of migrants to her study area as 'restricted residents' whose residential choices were controlled by the managers of local government or housing association property or whose housing was tied to their job, and more broadly there are many migrants who are effectively compelled to move into rural areas because of employment or family ties. Moreover, many in-migrants are attracted by regional factors that have little to do with the rurality of the district concerned. Walmsley et al. (1995) found in Australia that the climate, lifestyle and environment, and improved employment and housing opportunities, were the key factors driving migration to the rural coastal district of northern New South Wales.

Box 6.2 *The significance of rurality in migration decisions*

Studies by Keith Halfacree of in-migration to rural communities in two English counties, Devon and Lancashire, reveal not just the significance of rural factors in people's decisions to move, but also the wide range of reasons why people consider rural life and rural places to be attractive. As Halfacree illustrates with quotes from his survey respondents, these reasons relate to both the physical quality and the social quality of the environment, as well as to other factors such as privacy, leisure potential and familiarity with the area:

(Continued)

Box 6.2   (Continued)

Physical quality of the environment
   'We wanted to move to ... a more attractive area'
   'Quieter – less traffic. Like the country but not too isolated. Nice to see fields etc.'
   'Wanted more natural surroundings'
   'Space, fewer people, time to breathe and think'

Social quality of the environment
   'To be in a quieter and more pleasant relaxed area'
   'To seek a calmer environment'
   'To get away from it all'
   'Get out of the rat race – better way of life'
   'Pace of life – slower ... More of a community atmosphere'

Other factors
   'Preference for additional land, peace and privacy'
   'Prefer outdoor activities – throughout my life I've liked fell-walking, appreciation of
   country etc.'
   '[I can] walk out of front door into the country, don't have to drive'
   'Wife grew up in a rural area. I've almost always been rural'

*For more see K. Halfacree (1994) The importance of 'the rural' in the constitution of counterurbanization:
evidence from England in the 1980s. Sociologia Ruralis, 34, 164–189.*

Re-appraising Counterurbanization

Counterurbanization has been one of the key concepts employed by rural social scientists over the past couple of decades, yet the evidence accumulated in this period suggests that it needs to be approached critically (Mitchell, 2004). It is clear that the era of persistent rural depopulation has come to an end and that there is now a strong current of urban to rural migration that is contributing to an increase in the population of many rural areas. However, the population dynamics of the contemporary countryside are not as straightforward as the common usage of the term 'counterurbanization' suggests. In particular, there are four key caveats that should be noted.

First, *the emphasis placed on counterurbanization in Anglo-American literature has understated the diversity of national trends.* The counterurbanization thesis has been promoted by researchers in the United States – where the population turnaround was particularly marked – and the UK – where urban to rural migration has been consistently predominant. In many other countries the significance of counterurbanization has been weaker. Kontuly (1998) documents that in many European countries the tide of counterurbanization turned back to a flow of urbanization in the 1980s, or at least a period of indeterminate trend (Figure 6.2). In some countries, including Finland and Portugal, urbanization remained the predominant trend throughout the 1970s and 1980s.

Secondly, *there are regional differences in population dynamics and 'regional' factors may be more important than 'rural' factors in explaining migration.* The regionally uneven nature of counterurbanization is evident in North America, where the 'rural' areas of both Canada and the United States encompass a large and diverse

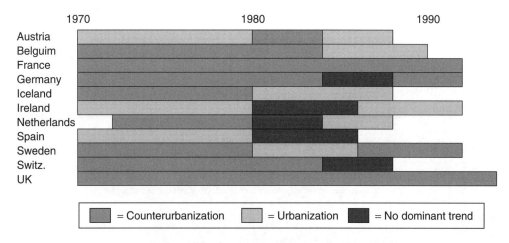

Figure 6.2   Predominance of counterurbanization and urbanization for 11 European countries
Source: Based on Kontuly, 1998

territory. Urban to rural migration in Canada has primarily taken the form of decentralization, with population growth concentrated in those rural districts closest to the metropolitan centres of the St Lawrence valley and southern British Columbia (Bollman and Biggs, 1992). Remoter rural regions of central and northern Canada have conversely suffered significant depopulation, with the population of Newfoundland, for example, falling by 7 per cent between 1996 and 2001, that of the Yukon by 6.8 per cent and that of the North West Territories by 5.8 per cent. In response, the Canadian government has adopted a policy of trying to encourage immigration by foreign nationals directly into rural areas to try to stabilize population decline.

In the United States, population deconcentration is the key factor in counterurbanization, but in a regionally selective manner. Over three-quarters of the population growth in rural areas between 1990 and 1997 occurred in western and southern states, notably Arizona, Nevada, Idaho, Oregon and Washington, stimulated by a combination of

the environment, lifestyle and employment opportunities which also boosted urban populations in the region. Throughout much of the prairie belt, in contrast, the populations of rural counties decreased, sometimes by over 10 per cent, as traditional sources of employment such as farming declined (Figure 6.3).

The polarization of the countryside into zones of population growth and population decline is also evident in France and Australia. Rural population growth in Australia has been focused on the rural coastal strip of New South Wales, Victoria and Queensland, and on sparsely populated regions of Western Australia and Northern Territory where agricultural employment has increased against the trend (Hugo, 1994). At the same time, however, 120 rural municipalities lost more than one per cent of their population in a single year in 1998–9, most of them in the wheat, sheep and dryland grazing agricultural areas of the interior (Kenyon and Black, 2001). Altogether, some 75 rural communities lost more than a fifth of their population between 1976 and 1998, with the population falling by

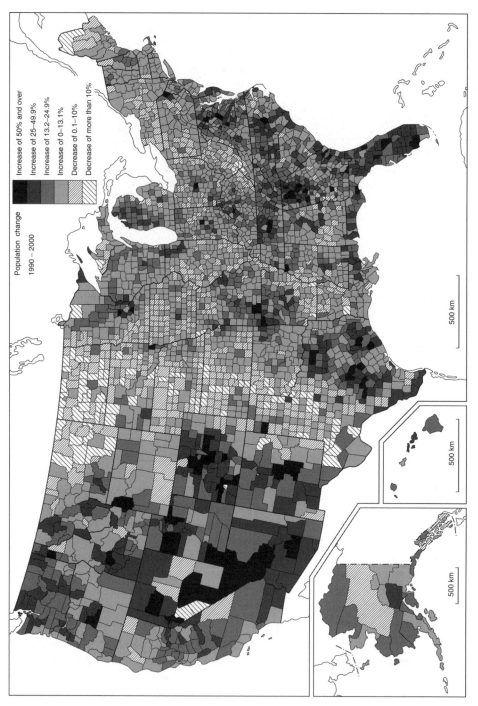

Figure 6.3 Population change by county in the United States, 1990–2000
Source: Based on data published from the US Census

more than a third in the most extreme cases such as Buloke in Victoria and Isisford in Queensland (Kenyon and Black, 2001). Similarly, in France the prevailing trend of counterurbanization masks problems of severe depopulation in many of the smallest rural communities, particularly in the Auvergne, Limousin, Lorraine and parts of Brittany, Normandy and the Pyrenees, where agricultural employment remains comparatively high (INSEE, 1995). Predictions suggest that some 1,500 French villages and hamlets could effectively disappear by 2015 (Lichfield, 1998).

Thirdly, *even in areas of rural population growth there can be pockets of local depopulation.* The UK has the most pronounced and consistent pattern of counterurbanization in the developed world, but even in rapidly populating parts of the British countryside the dynamics of population change can vary starkly from one community to the next. Weekley (1988) showed that nearly half of all rural communities in the English East Midlands with fewer than 1,000 residents in 1981 had decreased in population since 1971; whilst Spencer (1997) found that one in three parishes in South Oxfordshire – one of the fastest growing rural districts – had lost population between 1961 and 1991. The uneven local geography of rural migration is produced by a combination of the residential preferences of individual migrants and the availability of property. Property supply in the UK is regulated by the planning system that controls development, restricting new building and hence the capacity for population growth in valued environments, very small communities and pressured rural spaces (see Box 6.3). As Spencer (1997) argues, planning policy is not formed objectively, but represents the outcome of an asymmetrical power relationship between planning authorities and landowners that has tended to protect less populous communities from growth and can encourage localized depopulation.

Box 6.3 *Planning and counterurbanization in rural Britain*

Property development in the UK is regulated through the town and country planning system. New developments require prior permission from the local planning authority, which is awarded according to the policies outlined in periodically revised local 'plans'. The plans identify land for development and land where development will not normally be permitted, reflecting national and regional guidelines. The operation of this planning system has influenced the geography of counterurbanization in the UK at two levels. First, one of the earliest strategies of the planning system after the Second World War was the designation of 'greenbelts' around major cities in order to control urban sprawl. Development in the 'greenbelts' is heavily restricted, thus encouraging urban out-migrants to 'jump' the greenbelt and move to rural districts further out (for example see Murdoch and Marsden, 1994 on Buckinghamshire). This has helped to promote counterurbanization as opposed to suburbanization as the dominant population trend in the UK. Secondly, at a local level, many councils have adopted planning policies that concentrate new development in 'key settlements'. Population growth hence also tends to be concentrated in these settlements, whilst in other communities new development is restricted, limiting property supply and potentially leading to population stagnation or decline.

(Continued)

Box 6.3   (Continued)

The formulation of planning policies is not an objective process but one that reflects the balance of power in rural localities. As Spencer (1997) argues, planning policy is often biased in favour of landed interests, who may seek to exploit the commercial value of their land through development or alternatively might work to restrict development in order to avoid a dilution of their power base. Middle class residents have also mobilized to oppose development in order to maintain a limit on the supply of property, keeping property values high and hence protecting the exclusivity of certain rural communities (Murdoch and Marsden, 1994) (see the discussion of the 'middle class countryside' later in this chapter). Unsurprisingly, planning policy and development control have become key focal points of political conflict in contemporary rural Britain (see Chapter 14).

For more see Jonathan Murdoch and Terry Marsden (1994) Reconstituting Rurality (UCL Press); David Spencer (1997) Counterurbanization and rural depopulation revisited: landowners, planners and the rural development process. Journal of Rural Studies, 13, 75–92.

Table 6.4   Net migration to (+)/from (–) non-metropolitan areas of the United States (thousands)

Age (yr)	1975–6	1983–4	1985–6	1992–3
18–24	–14.4	–33.6	–39.6	–7.3
25–29	+22.0	–18.2	–26.2	–3.5
30–59	+8.3	–4.5	–1.8	+10.3
60 and over	+7.7	+2.2	+4.8	+6.5

Source: Fulton et al., 1997

Fourthly, *counterurbanization can disguise different migration patterns for different age groups and social groups*. The predominance of counterurbanization in many developed countries for substantial periods of the late twentieth century disguised the persistent net out-migration of young people from rural areas. Even during the heralded 'population turnaround' in 1975–6, over 14,000 more young people aged between 18 and 24 migrated from rural counties of the United States than moved to them, and this net outflow intensified in the 1980s (Table 6.4). Similarly, 44 of the 48 non-metropolitan districts of Australia experienced a net out-migration of 15–24-year-olds between 1986 and 1991 (Gray and Lawrence, 2001).

The out-migration of young people from rural areas is a product of both choice and circumstance. For many young people raised in the countryside, cities still hold an attraction as places of opportunities that are not available in rural communities. Other migration decisions are forced by limited employment opportunities (often reflecting the decline of employment in agriculture and other traditional industries), or, in some areas, an inability to afford inflated property prices. Most significantly, the expansion of higher education means that large numbers of young people leave rural communities to go to college or university and are restricted in their ability to return by a shortage of appropriate graduate-level jobs in many rural areas.

Some will return later in life, as opportunities to do so arise and as their personal circumstances change. Little attention has been paid by researchers to measuring the numerical significance of return migration in counter-

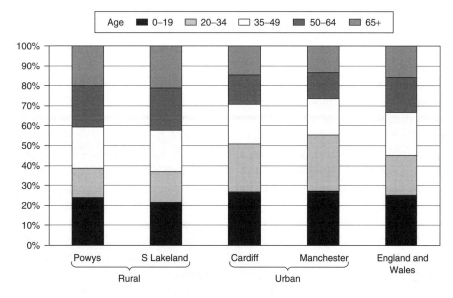

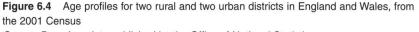

Figure 6.4 Age profiles for two rural and two urban districts in England and Wales, from the 2001 Census
Source: Based on data published by the Office of National Statistics

urbanization, but on the basis of noted observations in several countries it can be concluded that at least a sizeable minority of urban to rural migrants are in fact not 'newcomers' but 'returnees'. Return migrants will normally be more easily integrated into the community, and often have a particular commitment to the place that leads to community service. As Fitchen observes in rural New York State:

These return migrants occupy important roles in the community. They work in planning offices, run employment training offices, and serve as school principals, probation directors, and so forth. In their off-the-job time they are serving as community leaders in various capacities, from scout leaders to coordinators of recycling campaigns. These adults were once youngsters who couldn't wait to leave home to go to college and who vowed that after college they would never return, as there was neither career nor social life for them in their hometown. (Fitchen, 1991, p. 93)

At the other end of the age scale from youth out-migration, the flow of in-migration to rural areas has also been boosted by retirement migration. Fulton et al. (1997), for example, record a net migration of some 6,500 people aged over 60 into non-metropolitan areas of the United States in 1992–3. Much retirement migration is spatially concentrated into rural coastal districts and other 'resort' areas, with particular trends observed in regions such as south-west England and the Australian coastal strip of Queensland and New South Wales. The 190 US counties classified as 'retirement destination counties' (mostly along the west coast and in the Rocky Mountains and Florida), were the fastest-growing localities in the United States in the 1990s, with a net migration gain of over 17 per cent between 1990 and 1997 (Rural Policy Research Institute, 2003).

The effect of these differential demographic dynamics has been to polarize the population profiles of rural and urban areas. Figure 6.4 compares the age profiles of two rural districts in the UK with those of two large cities. In the two rural districts, Powys in Wales and South Lakeland in north-west England, over two-fifths of the population are aged over 50 and around 15 per cent are aged over 70, but there are substantially fewer residents aged between 20 and 35 than the national average. In the two cities, Cardiff and Manchester, in contrast, around one in four residents is aged between 20 and 35, but fewer than a third are aged over 50, and only one in ten is aged over 70. The consequences of these trends for the experiences of both young people and the elderly of living in rural areas are discussed further in Chapter 17.

Migration flows into and out of rural areas are differentiated not just by age, but also by income and social class. The evidence for these trends is more mixed than for age, and suggests that there are a number of complex dynamics that have influenced the migration direction of different social groups at different times. There is, for instance, some evidence of in-migration to rural areas by low income groups. Fitchen (1991) describes a second group of return migrants who left rural communities to find work, but having been unsuccessful have returned home jobless. Hugo and Bell (1998), meanwhile, identify a trend of 'welfare-led migration' in Australia, which reflects the fact that whilst government welfare payments are the same across the country, living costs can be lower in rural areas. However, in much of Europe and many of the fastest-growing parts of rural North America, counterurbanization has been associated with middle class in migration. Fulton et al.'s (1997) analysis, for example, suggests that there was a net in-migration of upper blue-collar and white-collar workers into the rural United States in both the mid-1970s and early 1990s compared with a net out-migration of lower blue-collar and white-collar workers. Where such differential migration occurs, it leads to the recomposition of

social classes in rural communities and can become self-reproducing as property prices soar and middle class in-migrants mobilize to resist development, as will be discussed in the remainder of this chapter.

Class Recomposition in Rural Areas

Traditionally, the class structure of rural society was based on property relations. The ownership of land brought not just status, but also power in a rural economy based on the exploitation of land, through agriculture, forestry, mining and so on (see Chapters 4 and 5). Landowners reaped the profits of land-based economic activity and controlled both the employment opportunities and the housing choices of the rural working class. For its part, the rural working class was also a *tenanted* class, many of whom were dependent on their employers for accommodation. The restructuring of the rural economy in the late twentieth century, however, undermined this class structure. The decline of agricultural employment, together with the expansion of non-land-based economic activities and the growth of public housing provision in rural areas, all diminished the power and status of the landowning class. This did not, though, produce a class-less society, rather it marked a transition to a new class structure based on occupation in which the pivotal position went to the burgeoning 'middle class'.

The growth of the rural middle class was produced both by the restructuring of the rural labour market (see Chapter 5) and by the predominantly middle class character of counterurbanization. Members of the rural middle class are therefore drawn from a wide range of backgrounds and their entry routes into the class will be varied. Moreover, the term 'middle class' now encompasses a large number of different, diverse occupations and employment situations, and a vast spectrum of household income levels, such that it is difficult to attribute the rural middle class with any common set of values or interests. As such, the rural middle class is not a coherent, unified, agent acting to reshape rural communities, but is composed of many different 'fractions' between which tensions and conflicts can arise, becoming dynamics that may drive local-level change in rural areas (Cloke and Thrift, 1987). In particular, rural researchers have focused on the role of a 'fraction' of professional and managerial workers, known as the 'service class' (see Box 6.4).

Box 6.4   Key term

Service class: The 'service class' are a fraction of the middle class employed in professional, managerial and administrative occupations. The term originates in Marxist class analysis and reflects the fact that members of this class are neither the owners of capital nor exploited workers, but rather *service* capital by providing specialist high-order skills and by managing capitalist enterprises. Service class occupations are found in both the private sector (for example, managers, engineers, accountants, lawyers) and the public sector (for example, teachers, doctors, civil servants, planners), but are generally characterized by 'rapid numerical growth, high levels of educational credentials, a considerable degree of autonomy and discretion at work, reasonably high incomes … opportunities for promotion between enterprises and relative residential freedom' (Urry, 1995, p. 209).

The significance of the service class to rural restructuring is derived from five key factors. First, the urban to rural shift in manufacturing employment as a part of a wider restructuring of manufacturing industry (see Chapter 5), and especially the expansion of high technology industry in rural locations, has created service class managerial and technical jobs in rural localities. This has been particularly significant in countries such as the United States and Germany where the emergence of the service class has been associated with the rise of scientific managerialism in private industry (Lash and Urry, 1987). Secondly, the relocation of administrative functions to rural areas by service sector employers (see again Chapter 5) has both promoted job-related in-migration by service class members and created new service class employment opportunities for 'local' residents. Thirdly, the expansion of the public services infrastructure in rural areas has

created more service class jobs, such as teachers, doctors and local government officers, particularly in the UK, where the public sector is an important source of service class employment (Lash and Urry, 1987). Fourthly, as noted in Box 6.4, the service class is characterized by relative residential mobility. Employment opportunities for many service class occupations exist across both urban and rural regions and service class members are usually able to move easily between employers, meaning that they are less constrained in their residential decision-making and more able to follow 'quality of life' motivations (Urry, 1995). The working hours and conditions of service class members also mean that they comprise a disproportionate share of commuters. Fifthly, some analysts have argued that there is a strong identification in service class culture with the countryside and the ideals of the 'rural idyll'. As Thrift contends:

Members of the service class have a strong predilection for the rural ideal/idyll … more than other classes they have the capacity to do something about that predilection. They can exercise choice in two ways. First of all, they can attempt to keep the environments they live in as 'rural' as possible. Such a process can operate at a number of scales. Homes can be covered with Laura Ashley prints and fitted out with stripped pine furniture. Developments that do not gel with service class tastes can be excluded in the name of conservation … Second, they can colonise areas not previously noted for their service class composition … and mould these in their image. (Thrift, 1987, pp. 78–79)

Research in the UK has indicated that some 40 per cent of in-migrants to rural areas between 1970 and 1988 were members of the service class, about twice as great as the proportion of the service class in the pre-existing population (Halfacree, 1992, quoted in Urry, 1995). By the 1990s, Cloke, Phillips et al. (1995) were able to report that nearly two-thirds of residents in three case study areas in the Cotswolds and Berkshire in England and Gower in Wales were members of the service class. Moreover, only among those residents of more than 40 years was the service class in a minority.

The significance of the service class is not just in its numerical strength, but also – as implied above – in the proactive involvement of service class members in local government and community leadership. Members of the service class are well equipped for political activity, with high levels of education, good communication, organizational and other professional skills, strong networks, spare time and money and – crucially – the motivation to defend their investment in the 'rural idyll'. As Cloke and Goodwin (1992) observed, 'Having colonized

[the in-migrant service class] have dominated local politics, and used their power to pursue their own sectional interests which represent very particular ideologies of what rural community and development should be like' (p. 328). In the south-western English county of Somerset, for example, over half of the county council in 1995 was drawn from the service class and reflected this in a programme that included opposition to housing development and hunting (see Woods, 1997, 1998b). Although research on the rural service class has been concentrated in the UK, similar examples can be found elsewhere. Walker (1999), for instance, highlights the leadership role of service class in-migrants in protests against a proposed waste dump in the rural fringe of Toronto.

The service class thesis, however, does have its critics. Urry (1995) acknowledges that there are members of classes other than the service class who also have a strong identification with the countryside, and that there are many service class members who do not participate in rural activities. More forcibly, Murdoch and Marsden (1994) question the supposed dominance of an identifiable 'service class culture' in rural areas, commenting that, 'there is not one "culture" associated with the middle class in the rural areas of Buckinghamshire, although we would agree that these "cultures" are becoming hegemonic' (p. 45). In response, Cloke, Phillips et al. (1995) suggest that this critique is a misreading of the original argument, contending that 'it was never claimed that the service class equates to the middle class, or that all rural areas were becoming dominated by the service class; rather it was claimed that the service class is an increasingly important "fraction" of middle-class residents within selected rural areas' (p. 228).

Either way, the recomposition of the class structure of many rural areas is an indisputable observed fact, and the increasingly middle class nature of many rural communities is reproduced not just by political intervention, but also as a simple result of middle class involvement in the rural property market. A further characteristic of the service class is that members have a relatively high income and are therefore well placed to compete in the increasingly expensive market for rural housing.

Yet, competition of this type forces up property prices still further and excludes lower income potential buyers. In many regions, such as southern England, formerly working class properties such as small, terraced, cottages have increased in value beyond the reach of working class buyers by competition and property improvement as part of a process of rural gentrification (see Box 6.5).

Box 6.5   Key term

Gentrification: The redevelopment of property by and for affluent incomers leading to the displacement of lower income groups who are unable to afford the inflated property prices. Originally coined with respect to the regeneration of urban neighbourhoods such as the Lower East Side of New York and Islington in London, the term has recently come to be applied to rural communities where middle class (or service class) colonization has increased property prices and excluded lower income purchasers.

Gentrification

The gentrification of rural communities involves not only a recomposition of the class structure, such that communities become more middle class, but also the restructuring of the local property market such that lower income households are actively excluded from residence. In urban areas gentrification has been associated with property development where speculators buy run-down or derelict housing, refurbish it and sell it on at a much increased price. This process is to some extent replicated in rural areas, as middle class investors and in-migrants purchase relatively cheap properties, such as farmworkers' cottages, and then enhance their value by redecorating and refurbishing, building extension and modernizing facilities. However, rural gentrification can also occur without property improvement, simply as a product of competition for limited housing stock inflating prices combined with opposition from

middle class residents to further housing development, especially the development of low-cost homes.

The process of rural gentrification and its consequences are best demonstrated by reference to two examples drawn from the UK in the 1990s. The first example concerns four villages in Gower, a rural area close to the city of Swansea on the coast of South Wales (Cloke et al., 1998; Phillips, 1993). The experience of counterurbanization in Gower is typical of that in many parts of the UK. A significant proportion of in-migrants came from nearby towns and cities in South Wales, but many had made longer-distance moves from other parts of England and Wales, including London, the West Midlands and north-west England. They also included a significant number of return migrants, especially from London. Whilst many in-migrants had moved for employment or family reasons, Cloke et al. note that, many of the people we spoke to drew on expectations of rural life which involved some notion that living in the countryside was a way of escaping or minimizing the risks of modern living. In particular notions of community, family, environment and safety (particularly for children) were used frequently as reasons for moving to Gower. (Cloke et al., 1998, p. 179)

Such appeals to the rural idyll were given a more material manifestation in the refurbishment and redecoration of properties that followed the 'ruralist' ideas of lifestyle magazines. Around a third of households in the four villages had carried out substantial improvement and nearly a quarter of householders had purchased their property with the potential resale value in mind. The subsequent inflation of property values produced a 'rent gap' effect as the cost of housing escalated out of the reach of local, low income residents. In a majority of the gentrifying households, the prime earner belonged to the service class; however, the research also identified a significant component of 'marginal gentrifiers', unable to access the mainstream property market but who had purchased and renovated dilapidated dwellings.

The second example relates to the villages of Boxford and Upper Basildon in Berkshire, just over 90 kilometres (50 miles) west of London. The two villages had experienced considerable in-migration with around a third of residents in 1998 having lived in the parishes for less than five years (Phillips, 2002). However, whereas in-migration to Upper Basildon had been facilitated by extensive new housebuilding (with the number of households in the village doubling between 1951 and 1991), the number of households in Boxford had remained more or less constant. Thus in-migrants have had to purchase existing houses, with prices inflated by the limited stock (and protected by – unsuccessful – opposition against proposed new housing development). The limited

property supply in Boxford has also encouraged 'marginal gentrification' through the purchase and modification of formerly public–owned council housing, yet in both communities gentrification has contributed to class recomposition that has seen the strength of the service class more than double in three decades to constitute around half the population in both villages in 1991.

Second homes

One form of gentrification that has a greater impact in rural communities than in urban areas is the purchase of property as second homes or holiday homes by urban-based middle class householders. The extent and status of second home ownership varies between countries and reflects cultural differences. In Scandinavia and North America second home ownership has been commonplace since the 1930s and spans class boundaries. In southern Europe, second home ownership is associated with rural depopulation and out-migrant families retaining property in their native communities. Second home ownership in these countries can be quite extensive and inclusive, with nearly one in four households in Sweden owning second homes in 1970 (Gallent and Tewdwr-Jones, 2000). In the UK and northern Europe outside Scandinavia, however, second home ownership is more restricted and middle class in character and hence is more obviously a form of gentrification. Second homes are bought as an investment, exploiting price differentials between urban and peripheral rural property markets. Yet over time demand for second homes inflates prices and as the type of properties purchased tend to be smaller dwellings that might otherwise go to first-time property owners, the effect can be to exclude local young and low income would-be buyers.

Moreover, as second home purchases tend to be spatially concentrated, often in coastal or winter sports resorts, their seasonal occupancy can contribute to dramatically reducing the permanent resident population of host communities. This has a knock-on effect on community life, including the closure of local shops and services as a result of decreased demand. Such impacts can create tension between local residents and second home owners, particularly if there are also cultural differences between the two groups. For example, the purchase of second homes in Welsh-speaking parts of Wales by non-Welsh-speakers has been accused by some campaigners as a key factor in the declining usage of the language in many communities – although recent research has suggested that in the main second homes in Welsh-speaking counties constitute only around 4–5 per cent of the total housing stock (Gallent et al., 2003).

The potential for conflict also arises in rural France, where there are some two million second homes. Many of these are owned by French urban-dwellers, but a significant minority have been purchased by Britons; over 200,000 are estimated to own homes in France, although this figure includes full-time residents as well as holiday home owners (Hoggart and Buller, 1995). The British are attracted by lower property prices in France and by the romanticized appeal of sparsely populated rural France compared with the more urbanized British countryside. As such, British buyers often purchase properties in need of renovation in areas of population decline and therefore operate outside the mainstream property market, thus avoiding conflict with local communities despite holding to a very different cultural conception of rurality (Gallent and Tewdwr-Jones, 2000; Hoggart and Buller, 1995). Tensions are far more likely to arise between rural populations and French second home owners, for whom a 'place in the country' is an escape from the city, and who are less likely than the British to integrate with the permanent local community.

Summary

The social restructuring of rural areas has progressed in tandem with economic restructuring throughout the past century. The shifting spatial division of labour, including the decline of traditional industries such as agriculture and the new employment opportunities in the expanding service sector, has variously exerted push and pull influences on migration patterns between towns and countryside at different times. Wider societal trends have also been significant, including, amongst others, increased private vehicle ownership, technological advances, the expansion of higher education, and longer life expectancy. Combined together these various factors produced a dominant flow of out-migration from rural areas during the first part of the twentieth century, which has in many regions been reversed to a trend of counterurbanization over the past three decades. However, there are considerable regional and local differences in migration patterns that are contributing to the increasingly diverse rural population geography. Moreover, differences in the migration patterns between different age groups and social classes are reshaping the demographic structure of the rural population. The population of many rural communities is getting more elderly as young people leave the countryside for education and employment and older people move in on retirement. Many communities are also becoming more middle class, a trend that can be self-reproducing as middle class competition for housing inflates property prices beyond the reach of local low income households.

As the rural population has been recomposed, so the nature of community life has changed. The solidarity of rural communities where residents shared common values and reference points and could often trace their family's presence in the village back over centuries, has been exploded by the dynamics of population change. The impact of this on the structure and coherence of communities, and particularly on the demand for services and facilities that were traditionally the focal points for community life, are explored in the next chapter.

Further Reading

Paul Boyle and Keith Halfacree's edited volume *Migration Into Rural Areas* (Wiley, 1998) provides a good overview of relatively recent research on rural population change, including chapters on counterurbanization, welfare-led migration, class recomposition and gentrification, and examples from the UK, the United States, Australia and Europe. A good critical review of the literature on counterurbanization can also be found in Clare Mitchell, 'Making sense of counterurbanization', *Journal of Rural Studies*, volume 20, pages 15–34 (2004). For more on the service class and rural change see John Urry's chapter 'A middle-class countryside?', in T. Butler and M. Savage (eds), *Social Change and the Middle Classes* (UCL Press, 1995), and for more on rural gentrification see Martin Phillips, 'Rural gentrification and the process of class colonisation', in *Journal of Rural Studies*, volume 9, pages 123–140 (1993), and Phillips, 'The production, symbolization and socialization of gentrification: impressions from two Berkshire villages' in *Transactions of the Institute of British Geographers*, volume 27, pages 282–308 (2002).

Websites

Detailed population statistics are available on national statistical and census office websites, including those for the United States (www.census.gov), the UK (www.statistics.gov.uk/census2001/default.asp) and Australia (www.abs.gov.au). The New Zealand census website has a specific section on rural New Zealand (www.stats.govt.nz/census.htm) whilst the Canadian census website includes detailed maps of internal migration patterns (www12.statcan.ca/english/census01/release/index.cfm).

7

Changing Communities:
Restructuring Rural Services

Introduction

'Community' is one of the most powerful words to be associated with rurality. For many early sociologists, the idea of 'community' encapsulated the essence of the difference between rural life and urban life (see Chapter 1). Ferdinand Tönnies, for example, contrasted the pre-eminence of *gemeinschaft*, or 'community', in rural areas based on 'close human relationships developed through kinship ... common habitat and ... co-operation and co-ordinated action for social good' (Harper, 1989, p. 162), with that of *gesellschaft*, or society, in urban space, where relationships were based on formal exchange and contract. Although later writers have critiqued the overly simplistic nature of this dualism, 'community' remains a strong element in lay discourses of rurality and is a commonly used term in rural policy documents. However, it is far from clear what 'community' means in each of these contexts. In lay discourses, 'community' is often used to imply frequent, high-quality social interaction between individuals, strong social networks and a shared sense of identity (Bell, 1994; Jones, 1997), but such qualities exist more as ambiguous abstractions than anything concrete and measurable. In policy discourse, 'community' may variously be a shorthand term to refer to an administrative territory, or to the public, or a normative concept of a self-organizing group of people. Even in academic discourse the meaning of the term 'community' can be elusive.

As such, communities are best envisaged as multi-dimensional entities. The first part of this chapter discusses one such approach, which conceptualizes a 'community' as comprising the four elements of people, meanings, practices and spaces/structures (Liepins, 2000a). An advantage of adopting this perspective is that it highlights the ways in which the different dimensions of a community are inter-dependent and co-constitutive, thus the impact of social and economic restructuring on any element of the community will have wider implications. The closure of shops and facilities that had acted as meeting places for community members, for example, may change the patterns of everyday practice in the community, the structure of social interaction in the community and the meanings that members attribute to a community. The second part of the chapter develops this line of thought further by focusing on the changing patterns of service provision in rural communities, with examples from the United Kingdom, United States and France. The chapter then proceeds to examine issues of accessibility in rural areas that have continue to reinforce the importance of the geographical community for many rural residents, before finally considering some of the strategies that have been adopted to overcome problems of peripherality and isolation in rural service provision.

Conceptualizing Community

The meaning of the term 'community' can be elusive even within academic discourse. Liepins (2000a) argues that four main approaches to 'community' have been used in rural studies, all of which are imperfect. The first two, the *structural-functionalist* approach – which identified communities as discrete and stable entities with observable characteristics – and the *ethnographic/essence* approach – which sought to discover and document the lived 'essence' of communities – can both be critiqued because they take the existence of communities to be a given and therefore can say little about how communities are produced. The third approach, the *minimalist* approach simply involves reference to 'community' as a way of denoting a scale of enquiry or a loosely specified social collectivity. Fourthly, researchers have focused on the *socially constructed meanings and symbolism* attached to the term 'community'. However, this last approach is criticized for downplaying the significance of material practices and physical elements in constituting communities, and for detaching symbolic representations of community from the social relations that produce them.

To move beyond these conventional approaches, Liepins (2000a) proposes a fifth perspective, which recognizes communities to be 'social collectives of great diversity'. She argues that, 'at least in a temporary sense, "community" can be conceived as a social phenomena [*sic*] that unifies people in their ability to speak together even while being located in many positions and holding a variety of contrasting identities' (p. 27). Moreover, Liepins suggests that the spaces in which a community is enacted can be conceived of to include 'both the material sites filled by communal activities, and the symbolic and metaphoric spaces in which people connect "in community" even while existing in different physical or social locations' (p. 28). This latter point means that communities need not necessarily be geographical entities (one might think, for example, of 'the agricultural community', or 'the business community' or 'the gay community'), but Liepins's definition also allows for community to be approached in geographical terms even when the population of the territory concerned has undergone substantial restructuring.

The model developed by Liepins represents a community as comprising four elements: people, meanings, practices and spaces/structures. People are positioned at the centre of the community because 'community' is created through social collectivity and connection, and participate in 'community' through their engagement with the three components of meanings, practices and spaces/structures (Figure 7.1). First, people

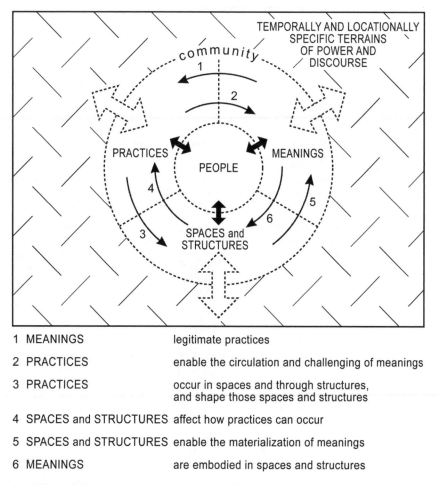

1 MEANINGS   legitimate practices

2 PRACTICES   enable the circulation and challenging of meanings

3 PRACTICES   occur in spaces and through structures, and shape those spaces and structures

4 SPACES and STRUCTURES affect how practices can occur

5 SPACES and STRUCTURES enable the materialization of meanings

6 MEANINGS   are embodied in spaces and structures

Figure 7.1 The constitutive components and dynamics of 'community'
Source: Liepins, 2000a

create the symbolic representation of a community by formulating meanings about their connections and identities. Significantly, Liepins argues that such ideas need not be universally held by all members of a community, and that communities are constituted as much by contested meanings as by shared meanings. Secondly, communities are given a more material manifestation through the practices and activities in which members participate. These include routine everyday interactions with neighbours as well as formal events, such that,

the circulation of meanings and memories through newsletters and meetings, the exchange of goods and services at a local store or health clinic; the creation and maintenance of social groups and rituals; and the operation of local government boards are all examples of ways in which we may trace practices of community. (Liepins, 2000a, pp. 31–32)

Thirdly, cultural and economic dimensions of community life occur in particular spaces and through particular structures, which may be read as the metaphorical and material

Table 7.1 Key characteristics of the three case study communities

	Duaringa (Queensland)	Newstead (Victoria)	Kurow (South Island)
Population	< 500	< 800	< 1000
Type of farming	Beef, grain and cotton	Mixed sheep/cropping	Predominantly sheep
People (listed alphabetically)	Aboriginals Farming families Service employees Local government employees	Commuters Farming families Lifestylers Service employees	Beneficiaries Farming families Service employees
Locational aspect	Main highway, 1 hour from a regional city (Rockhampton)	Minor highway, 1.5 hours from a State capital (Melbourne)	Minor highway, 1.5 hours from a regional city (Dunedin)
Services	Petrol station/shop Post office Hotel Primary school Shire offices	Bakery Post office Hotels Primary school Butcher Farm supplies	Petrol station Post shop Hotels Area school Supermarkets Farm supplies Transportation companies
Key trends and concerns	Declining employment Depopulation	Loss of local council Increasingly diverse population	Contraction of local economy Depopulation

Source: After Liepins, 2000b

embodiment of 'community'. These include the physical sites of schools, halls, street corners and parks which serve as the 'meeting places' of a community, alongside other 'structures' such as newspapers and websites that facilitate social collectivity.

Liepins contends that these four elements of 'community' are mutually constitutive. As Figure 7.1 indicates, meanings legitimate practices that in turn enable the circulation and challenging of meanings. Practices also occur in spaces and through structures, and shape those spaces and structures, whilst spaces and structures affect how practices can occur. Finally, spaces and structures enable the materialization of meanings and meanings are embodied in spaces and structures (Liepins, 2000a).

Communities in practice: three case studies

Liepins (2000b) demonstrates the application of the model through case studies of three rural communities in Australia and New Zealand. The three communities – Duaringa in central Queensland, Newstead in central Victoria, and Kurow in South Island, New Zealand – share a broadly common macro-economic and political context, including the decline of agriculture, neo-liberal government policies, a historical dependency on farming, and locations that are a similar distance from a larger city. However, they also represent different local social, economic and cultural contexts, different priority issues, and different local responses to wider processes of change (Table 7.1).

In all three case studies, residents made reference to 'community' in describing the place in two ways. First, meanings of community were expressed that positioned the locality within a wider context. Thus reference was made to topographical features – 'the small community nestled in under the hill' (Kurow) – or, more commonly, to the historic function of servicing agriculture such that the community identity was associated with a farming identity. Secondly, residents acknowledged the heterogeneity of their communities. In Kurow and Duaringa this was given a negative meaning, suggesting that fragmentation posed a threat to 'community'. In contrast, residents in Newstead, which had experienced the most significant social recomposition through the in-migration of commuters and 'alternative lifestylers', felt that the diversity and tolerance of the community was positively promoted as part of its identity.

The meanings of community are reproduced through the community practices in which its members interact with each other. The post office, garage, school, shop and hotel or bar all formed key places of community practice in the case studies:

> To them, [the community] is their local centre. [Going to] the butcher, the baker, and the milk bar and that kind of thing, that nucleus is there for the people to go down into the town and get what they want.
> (Newstead resident, quoted in Liepins, 2000b, p. 333)

> You find out about things going on in Duaringa through the Post Office and the school. Going to the Post Office, and, there's the school newsletter I would have to say. And word of mouth. I see a fair few people at the school. (Duaringa resident, quoted in Liepins, 2000b, p. 333)

Additionally there are regular communal events that help to promote practices of community, including a market, summer festival, flower show and ball in Kurow; a 'bullarama', charity golf day, flying doctor race day and the Dawson river mud trials in Duaringa; and a market, school fete and an Australia Day concert in Newstead. Events such as these, and the services and facilities mentioned above, also form some of the spaces in which community occurs (see also Figure 7.2). As Liepins comments,

> these sites are not just material spaces but also form a locus of interaction, whether it be through sport, 'community' days or leisure activities. In each instance the space itself is a resource through which people engage in different forms of 'community' interaction. It is a site at which 'community' becomes socially embedded and visible (however temporarily) within the social and cultural life of the 'community'. (Liepins, 2000b, p. 336)

Thinking about a community in terms of its meanings, its practices and its spaces and structures provides a useful route in for the analysis of the processes and consequences of rural change in two ways. First, it highlights how processes of social and economic restructuring impact on communities in both material and immaterial ways. The decreasing importance of agriculture, for instance, will impact on the spaces and structures of a rural community as sites such as livestock marts become less significant, will alter community practices as interaction through young farmers' clubs or agricultural shows diminish, and will change the meanings of community as the identification with farming weakens. Secondly, the approach reveals how change may occur through the different components of a community and the shifting dynamics between meanings, practices and spaces and structures. In particular, the stress placed on the importance

Figure 7.2 Events such as this community dance advertised in Lompoc, California – which brands itself as the seed-growing capital of the world – are part of the practices through which communities are constituted
Source: Woods, private collection

of community services and facilities in Liepins's case studies suggests that their disappearance, or any alteration in their nature, would significantly change the meanings, practices and spaces and structures of the community concerned.

The Disappearance of Rural Services

There was a time when the idyllized image of a rural community almost existed. Every small town would have its bank, its post office and its store. Every village would have its church, its shop and its pub. No more. The rationalization and closure of both private and public services in rural communities has been one of the most visible manifestations of contemporary countryside change. Like many of the trends discussed in this book, the disappearance of rural services is a product of both overarching global social and economic

processes, and national and regional factors. First, there are economic forces within capitalism that mean that independent traders have become less common as local companies are bought up by larger corporations, which then seek to rationalize their networks of outlets, closing those that are judged to be unprofitable. Any smaller, independent, enterprises that remain find it difficult to compete effectively against the larger corporations and many are driven out of business. Second, there are social forces that have altered consumption habits. A more mobile population is less dependent on shops and facilities within their place of residence and those commuting to work may even find it more convenient to shop away from the community in a mall or larger town. Technological advances such as refrigeration mean that consumers bulk buy on more infrequent shopping trips, and that supermarkets

Table 7.2 Percentage of rural parishes in England having key public and commercial services, 2000

	All parishes	Population		
		100–199	**500–999**	**3,000–9,999**
Post office	54	22	67	93
Bank or building society	9	n/a	n/a	n/a
General store	29	7	26	78
Small village shop	29	10	35	52
Petrol filling station	19	4	16	64
Public house	75	63	58	92
Primary school	52	13	71	94
Village or community hall	85	72	93	96
Youth club	51	23	58	91
Doctor's surgery	14	1	7	64

Source: Countryside Agency, 2001

are able to offer extensive choice across a range of goods imported from around the world. Third, however, there are national and regional factors that reflect cultural differences and rural settlement geography in shaping the actual trend of service provision. To illustrate these, this section presents three national perspectives on the changing fortunes of rural services, from England, France and the United States, respectively.

England

Most rural parishes in England have neither a general store nor a village shop. Only just over half have a post office, and only a similar proportion have a primary (elementary) school. Fewer than one in five parishes have a doctor's surgery (Table 7.2). These figures, recorded in 2000, are the product of a process of concentration by which many public and commercial services have effectively disappeared from smaller rural communities to be centrally provided in larger villages and small towns. As Table 7.2 indicates, parishes with fewer than two hundred residents will typically have only a public house and a village hall as community facilities. Those with a population between 500 and 1,000 will typically be served by a post office, primary school and

youth club as well as a public house and a village hall, whilst nearly all larger parishes have a post office, primary school, public house, community hall and youth club. This clear relationship between community size and services not only suggests that there are rough population thresholds necessary for certain facilities to be viable, but also that facilities in larger settlements serve both the immediate community and that of neighbouring smaller villages.

The evolution of this pattern of service provision is illustrated by the case study of Crewkerne, a town of around 6,000 people in south-west England that acts as a service centre for 18 neighbouring villages with populations ranging between 50 and 2,000. In keeping with many similar towns, Crewkerne has experienced considerable growth under counterurbanization, with its population increasing by 23 per cent between 1971 and 1986. Four of the neighbouring villages also experienced population growth of 25 per cent or more over this period, but in three the population fell and, overall, six of the villages had a smaller population at the end of the twentieth century than at the beginning. Yet both expanding and depopulating villages lost facilities and services. In 1902, there were

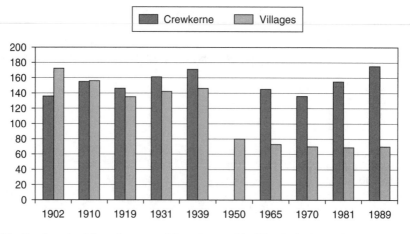

Figure 7.3 Number of public and commercial services and facilities in the town of Crewkerne and surrounding villages in south-west England, 1902–1989
Source: Woods, original research

more services in total in the villages than in Crewkerne. Even the smallest settlement had both a village store and a blacksmith, whilst the largest village, with a population of 1,300, boasted two schools, seven public houses, a police station, a post office, a laundry and 20 shops or retailers including grocers, bakers, butchers, coal dealers and a tobacconist.

The first wave of concentration came in the 1910s, 1920s and 1930s as a combination of the introduction of the first bus services – enabling village residents to travel more easily into the local town – and significant depopulation of some of the smaller settlements shifted the balance of facilities from the villages to the town (Figure 7.3). The second wave occurred after the Second World War, driven by a combination of increased car ownership, shifting employment patterns, the restructuring of public services following the establishment of the welfare state, and the effects of rationing in reducing the economic viability of smaller enterprises. The impact on individual communities was severe. In one village, for example, the range of services that had in 1939 included two public houses, two bakers, a post office, a

school, a tobacconist, an insurance agent and a cycle repairer, was reduced by 1953 to only the post office and one public house. This pattern of concentration continued, at a slower pace, throughout the latter half of the century (Figure 7.4), such that by 2000 the range of facilities even in the largest village had been reduced to two public houses, a post office, a school, two general stores, a petrol filling station, a village hall and a fish-and-chip takeaway.

At the start of the twenty-first century a third wave of concentration is under way as the development of supermarkets and edge-of-town shopping complexes in larger nearby towns has drawn trade from Crewkerne itself, leading to the closure of food shops and other mainstream retailers. The only functions to show any expansion in numbers, either in Crewkerne or the surrounding villages, are those linked to leisure consumption – notably bars, restaurants, specialist retailers such as pottery shops and community halls.

France

Two key factors have produced a different pattern of rural service provision in France

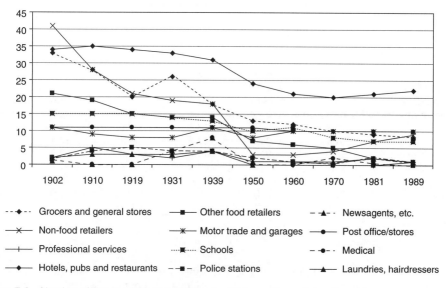

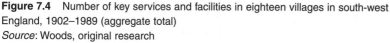

Figure 7.4 Number of key services and facilities in eighteen villages in south-west England, 1902–1989 (aggregate total)
Source: Woods, original research

from that in England. First, French consumers continue to exhibit a greater preference for purchasing fresh produce, and particularly local produce, than their English counterparts. This has helped to maintain both local markets, which are still an important community space for most French small towns, and specialist food retailers such as butchers and bakers. Secondly, local government at the community level in France – the *commune* – has extensive powers, including responsibilities for the police, social services and primary education. In consequence, many rural communities in France have a level of service provision that is significantly higher than that in equivalently sized villages in England (Table 7.3). Yet, as Table 7.3 also shows, the coverage of key services in rural *communes* decreased during the 1980s, and there is evidence that the downward trend intensified in the 1990s. In the Poitou-Charentes region, for example, 131 *communes* lost their last grocery store between 1988 and 1995

(Soumagne, 1995); whilst across France, 12 per cent of banks in isolated rural *communes* closed between 1988 and 1994 (INSEE, 1998).

The closure of rural services in France is particularly associated with depopulation. Lichfield (1998) describes the village of Vallières in Creuse, a settlement of 500 residents and falling. Despite its small population, Vallières in 1998 boasted five cafés, two restaurants, two grocers, two butchers, a bakery, a hardware store, two electrical goods shops, two general shops, a bank, a garage, a chemist and a post office. Yet, as Lichfield continues, 'another six shops or bars around the main square are closed. The patisserie down the street has a large sign saying "à vendre". The menu outside the hotel two doors away reads "hotel fermé définitivement"' (p. 12).

United States

The connection between depopulation and the disappearance of rural services is also

Table 7.3 Percentage of rural *communes* in France having key public and commercial services, 1988, and change between 1980 and 1988

	Rural centres		**Villages around rural centres**		**Isolated rural communities**	
Bakery	79.6	(–1.0)	26.8	(–2.3)	35.7	(–2.0)
General store	81.5	(–8.0)	34.1	(–11.1)	45.9	(–9.6)
Supermarket	64.1	(+13.6)	1.0	(+0.5)	5.4	(+2.6)
Clothes shops	63.3	(+2.2)	3.1	(–0.3)	12.4	(–0.2)
Post office	71.2	(–0.1)	18.3	(n/a)	31.6	(n/a)
Bank	61.4	(+0.2)	2.7	(=)	12.4	(+0.2)
Doctor's surgery	74.8	(n/a)	10.2	(n/a)	20.3	(n/a)
Primary school	96.9	(n/a)	67.4	(n/a)	61.7	(n/a)
Sports ground	83.5	(+4.4)	35.4	(+6.7)	36.9	(+3.0)
Library	89.1	(+5.4)	58.8	(+7.3)	60.5	(+6.4)
Cinema	41.6	(–2.4)	0.3	(–0.3)	4.1	(–0.9)

Source: INSEE, 1998

well established in the United States. McPherson County in Nebraska is typical of many parts of the American rural heartland – since 1920 the county has lost two-thirds of its population as well as 19 post offices and 58 school districts (Gorelick, 2000). More generally, the rationalization of service provision in rural communities has been driven by economic strategies aimed at maximizing profits in the private sector and minimizing costs in the public sector. In the private sector the traditional small town 'main street' of independent retailers has been eroded by the development of corporate chains, super-markets, malls and out-of-town commerical centres (Vias, 2004). In rural areas these are frequently located at highway intersections and can draw custom from a wide area, pulling trade away from shops and services in rural communities. Wal-Mart, the largest supermarket in the US, built its market position by retailing primarily in rural areas, establishing a presence in over 1,100 rural towns (Farley, 2003). In the public sector the provision of health facilities, social services, libraries and schools (see also Box 7.1) has been squeezed by a prevailing emphasis on cost-effectiveness. As Fitchen (1991) comments,

because the cost-effectiveness model judges a program's worth not in terms of what it does for people or communities but how much it costs per person served, it aggravates the effect of the higher cost of rural services ... it contributes to the increasing centralization of services and leaves some rural populations unserved. (p. 155)

For example, between 1980 and 1988, 161 of the United States's 2,700 rural hospitals closed as a result of marketplace pressures and cost-containment measures (Fitchen, 1991). The professionalization of public services, with greater expectations of qualifications and level of training of employees such as librarians and ambulance personnel, has also created recruitment problems in rural areas. Over 22 million people in rural America live in areas that have been officially designated as 'Health Professions Shortage Areas' or 'Medically Unserved Areas' (Rural Policy Research Institute, 2003).

The loss of rural services has not, however, been universal. Resort areas in the rural US, such as parts of the Rocky Mountains, have often experienced an expansion of local services and facilities along with a growing population. Yet many of the new ventures are aimed at leisure-based consumption rather than serving the everyday needs of the local community. One such community is Ridgway, Colorado, whose population doubled in the two decades from 1975 to just over a thousand people and which additionally attracts tourists and seasonal residents. New services include a library and a chiropractor, as well as a bakery, a hardware store, fast-food outlets and a grocery store. As Decker describes:

> Old-timers no longer recognize the town. In the old rail yard there stands a building housing a washeteria, a real estate office, and an office supply shop. At the town's main intersection, two convenience stores with gas pumps have sprung up, as has the town's second liquor store. New shops and boutiques sit on the site of the old roundhouse. Local cappuccino cowboys can now find comfort at a coffeehouse while they ride herd on their stocks in the Wall Street Journal ... New shops up the street offer Guatemalan clothing and furnishings, fresh flowers, motorcycles, antique furniture and prints, lingerie, saddles, quilts, western wear, and 'collectables'. Four restaurants serve everything from pasta, enchiladas, and lobster tails to alfalfa sprouts and zucchini bread. (Decker, 1998, p. 93)

New developments of this kind will impact on the sites and structures of the community just as much as the closure of shops and services.

Box 7.1   Rural schools

For many rural communities the village school is more than just an educational establishment, it is also a focal point of community life. Fundraising events for schools and school-gate conversations between parents both serve as sites and structures through which community is practised. Friendships formed between children at school can shape the social networks of a rural community for decades. School halls are used as venues for community gatherings. Moreover, the presence or absence of a school can influence the attractiveness of a village to in-migrants, with families with school-aged children being less likely to move to villages without a school, thus contributing to the disproportionate ageing of that community. It is therefore unsurprising that proposals to close rural schools are highly contentious and are usually met with resistance (Figure 7.5). As Mormont (1987) observes, 'the village school constitutes a symbol ... of local autonomy. Their closure was to become the focus for a fairly substantial opposition, insofar as inhabitants not only felt deprived of a service to which they considered they were entitled, but also of a local institution with which they could identify' (p. 564).

Yet, the rationalization of rural schools has been a feature of recent education policy in a number of countries including the United States, the UK, Canada, New Zealand, Ireland, Germany, Sweden and Finland (Ribchester and Edwards, 1999; Robinson, 1990). In France, over 1,400 rural *communes* lost their school between 1988 and 1994 (INSEE, 1998); whilst 415 small rural schools closed in the United States between 1986–7 and 1993–4 (NCES, 1997).

Rural schools are particularly vulnerable to a cost-effectiveness analysis because population and demographic trends have meant that many have very low (and often decreasing) enrolments. Some 2,700 primary schools in England (15 per cent) have

(Continued)

Box 7.1   (Continued)

fewer than 100 pupils, as do over 9,000 schools in the United States (10 per cent of all schools) – predominantly schools in rural areas. Enrolments decreased by a tenth or more between 1996 and 2000 in 38 per cent of rural schools in the United States (Beeson and Strange, 2003). As the number of students in a school falls, the cost per student of running the school increases because of the fixed costs of buildings and staff. Thus, the New Zealand Education Minister was reported justifying the closure of rural schools in 2003 as a switch of resources from buildings to 'things which directly influence education' (*Manawatu Evening Standard*, 17 June 2003). Rural school closures may also result from difficulties in implementing national education strategies (such as tests or common curricula), teacher recruitment shortages, and changes in the organization of local education authorities, or, controversially, on the basis of pedagogic arguments.

More recently, pedagogic arguments in favour of small schools have been advanced to slow or halt rural school closures. However, schools in rural areas continue to confront challenges from higher unit costs, shortage of resources, and falling enrolments. Rural schools are often heavily dependent on volunteer help and in many cases costs have been cut by sharing resources and amalgamating the administration of schools or school districts.

For more on rural education provision in Britain, see Chris Ribchester and Bill Edwards (1999) The centre and the local: policy and practice in rural education provision. Journal of Rural Studies, 15, 49–63.

Figure 7.5   Protests against the proposed closure of Llangurig village school, Mid-Wales, Summer 2003
Source: Woods, private collection

Table 7.4 Percentage of rural households in England within 2 km, 4 km and 8 km of key public and commercial services, 2000

	Rural households within distance of nearest facility		
	2 km	**4 km**	**8 km**
Post office	93.5	99.5	–
Bank	58.1	78.4	96.7
Cashpoint	61.1	79.3	96.2
Supermarket	60.9	79.0	96.0
Primary school (elementary school)	91.6	99.0	–
Secondary school	57.2	78.2	96.3
Doctor's surgery	66.1	85.8	98.5
Hospital	–	44.7	74.1
Jobcentre (employment exchange)	–	42.5	72.4
Benefits Agency office	–	15.7	36.4

Source: Countryside Agency, 2001

Accessibility to Services and Rural Public Transport

Changing patterns of rural services provision impact on communities by changing the sites and structures through which community takes place, but they also change communities by creating new divisions between those residents who are able to easily access services located outside the village or town, and those who are more constrained in their mobility. At one level, the rationalization and concentration of services in rural areas reflects increased levels of mobility. People are willing to travel further to shop or to access key services, and for the majority who now work outside their place of residence, shopping or using services in the town in which they work may be more convenient. As such, it could be argued that the spatial restructuring of rural services is simply part of an upscaling of people's everyday lived community. However, the restructuring of service provision in this way excludes a significant element of people in rural communities

who do not have control over their own mobility.

This impact is best measured not by a count of those communities with or without key services, but rather by the distance that people have to travel to access key services and facilities. Evidence from Britain and France suggests that these distances can be quite considerable. More than nine in ten rural households in England are within 2 kilometres (1.25 miles) of a post office and a primary school, but fewer than two in three rural households are within the same distance of a bank, cashpoint, supermarket, secondary school or doctor's surgery (Table 7.4). Residents of isolated rural communities without key services in France had, on average, to travel 6 km to the nearest bakery, 7 km to the nearest post office, 10 km to the nearest supermarket, 10 km to the nearest bank and 18 km to the nearest clothes shop in 1988 (INSEE, 1998). In every case the distance had increased significantly since 1980.

The relatively high population density of England means that only a small percentage of rural households have to travel significant distances to access key services, but these households tend to be geographically concentrated in the most rural parts of the country. There are, for example, some 29,000 rural households over 4 km from the nearest post office, mostly in the northern uplands and parts of south-west England and East Anglia. Similarly, large parts of the northern uplands, the Welsh Marches, Dorset, Devon and Cornwall are more than 8 km from the nearest supermarket. In less densely populated countries, including the United States, Canada and Australia, distances from remote communities to services of this kind may be measured in hundreds of kilometres. A basic rule of thumb applies in both contexts, however – remote and isolated rural communities are more likely to have key services and facilities than equivalently sized settlements nearer to urban areas, but are likely to be a greater distance from those services that they do not possess (see also Box 7.2).

Distances of a few kilometres are easily travelled by car, but for those rural residents who are unable to drive or who do not own a car, any journey that involves leaving their own settlement can be difficult. One in 14 rural households in the United States do not own a vehicle, yet 80 per cent of rural counties do not have a public bus service and 40 per cent of the rural population live in areas with no form of public transport (Rural Policy Research Institute, 2003). Only half of rural settlements in England, and a third of rural communities in France, have a daily bus service. Public transport provision has been eroded by the same calculations of cost-effectiveness as other public services, with routes withdrawn when

passenger numbers fall below a sustainable threshold. In some countries, governments have sought to underwrite unprofitable routes, but the rise of neo-liberal policies in the 1980s and 1990s has challenged such strategies, with public transport in the UK, for example, undergoing 'deregulation' in the late 1980s with a consequential restructuring of rural bus services. A number of communities have responded by experimenting with 'alternative' transport initiatives to replace commercial services. In 2001, 20 per cent of rural parishes in England had on-demand 'dial-a-ride' transport schemes, whilst 17 per cent were served by community minibus or taxi schemes. Yet, in 16 per cent of English rural communities an important form of public transport is provided by the 'supermarket bus' that takes residents free-of-charge to a supermarket in a neighbouring town, thus helping to redirect trade from smaller independent retailers to supermarkets.

Notably, many of the services that are most distant from the majority of rural communities – including hospitals, jobcentres and benefit offices – are those whose users are least likely to have access to their own private transport. This produces a double disadvantage that is a distinctive feature of social exclusion in rural areas (see Chapter 19). Similar issues arise with access to financial services, as the concentration of bank branches has produced new geographies of financial exclusion. Efforts to rechannel banking services through other outlets such as post offices can themselves be undermined by the closure of sub-post offices which are fully compensated for by increases in car ownership in the communities affected.

Box 7.2   *Isolated rural communities: the islands of western Ireland*

For many remote rural communities, isolation is part of their narrative meaning. The community defines and describes itself by reference to its insularity and poor accessibility, and employs such meanings to construct community practices that promote self-sufficiency. The small island communities off the west coast of Ireland are a prime example of this, including nine islands studied by Cross and Nutley (1999) – Arranmore, Beare, Cape Clear, Clare, Inishbofin, Inisheer, Inishturk, Sherkin and Tory. The islands have populations ranging from 78 to 596, and all experienced considerable depopulation during the twentieth century – although four recorded population growth between 1981 and 1991. Service provision on the islands is inevitably finely balanced between the level of trade that can be sustained by the small populations and the needs that result from the difficulties of access to the mainland. All nine islands in 1991 had a grocery store, a pub or club, a nurse and a primary school. All but the smallest had a resident priest and five had a hotel. However, only two islands had a secondary school and only the largest had a resident doctor. Even basic provisions could be difficult to obtain. Daily newspapers were available on only four islands, and fresh milk was not available on two islands. Islanders were hence heavily dependent on travel to the mainland in order to access most services, yet transport links to many of the islands were poor. Daily ferry services operated all year round to just three of the islands, with five more served daily only in the summer. The smallest island, Inishturk, had no ferry service, 'with residents having to depend on the weekly mail boat or one of the island's fishing boats' (p. 322). Despite government support for transport links, including mail boats and helicopter services, the communities have little option than to accept a lower level of service provision than that expected elsewhere.

For more details see Michael Cross and Stephen Nutley (1999) Insularity and accessibility: the small island communities of Western Ireland. Journal of Rural Studies, 15, 317–330.

Overcoming Isolation: from Mail Order to the Internet

There are numerous isolated farms and settlements in remote rural areas that have always been beyond the reach of mainstream commercial and public services. For such households the vital link to healthcare and education, as well as to shopping, has been provided by the postal service, telecommunications, mobile services and, more recently, the Internet. As early as 1872, the Patrons of Husbandry, a social and educational organization for American farmers also known as the Grange, had launched a mail order service for its members. By 1900, the Grange's business had been eclipsed by that of the Sears-Roebuck Catalog, selling everything from shoes to cars to rural households across the United States. In Europe, mobile and peripatetic services were extensively adopted after the Second World War as a means of delivering a wide range of services to rural communities, including libraries and healthcare, groceries and even cinema

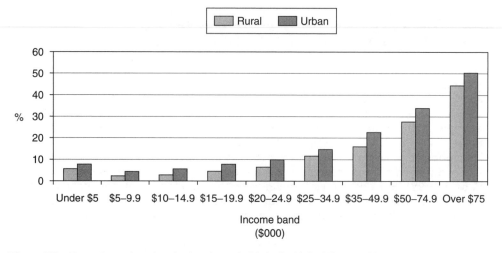

Figure 7.6 Percentage of rural and urban households in the United States with access to on-line services, 1999, by income band
Source: Based on Fox and Porca, 2000

performances. Similarly, air transport was enrolled to help provide healthcare cover to large rural areas of Australia. However, such services are subject to assessments of cost-effectiveness just as static facilities are, and as the rural population has become more mobile itself the demand for peripatetic services has fallen and many have been axed.

In the late twentieth century, the possibilities offered by new telecommunications technologies were eagerly exploited as a tool for overcoming rural isolation. New Zealand had launched a correspondence school for rural children in 1922, but made increasing use of radio broadcasts and tape cassettes to expand the service in the 1970s. Australia similarly developed 12 'schools of the air' after 1951 using two-way radio communications. One such school, based at Port Augusta, South Australia, was in 1978 teaching some 80–90 pupils living up to 700 km away (Nash, 1980).

More recently, attention has turned to the potential of the Internet to provide rural residents with access to health advice, education

and training, banking services, entertainment, information sources, and, of course, on-line shopping. Yet, as with the development of teleworking (see Chapter 5), the usefulness of the Internet in rural service provision is restricted by the IT infrastructure in rural areas. A pilot scheme to connect schools in rural North Carolina to a fibre-optic system in 1995, for example, found that it cost a typical high school $110,000 to $150,000 to buy the equipment and up to $50,000 in annual telephone bills – significant demands for smaller rural schools (Marshall, 2000). At a household level, computer ownership by rural households in the United States is, at 40 per cent in 1998, lower than in urban areas and the percentage of rural households with access to online services is lower than for urban households in every income band (Figure 7.6) (Fox and Porca, 2000). Studies have identified a similar 'digital divide' in Britain, with fewer than 6 per cent of the population in peripheral rural areas participating in on-line shopping – less than in any other part of the country (Figure 7.7).

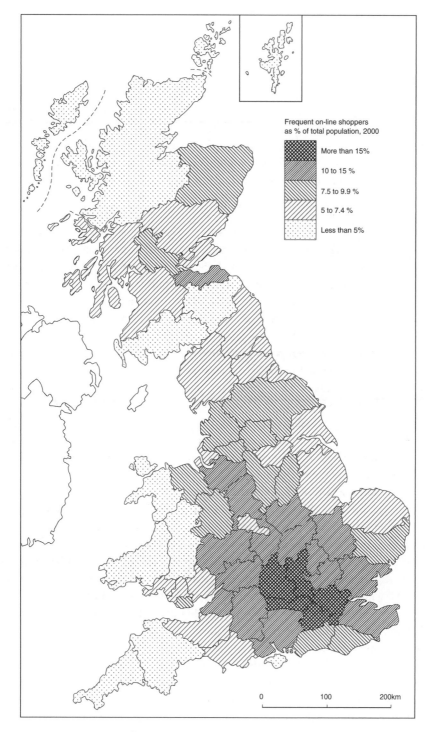

Figure 7.7 Frequent on-line shoppers as percentage of adult population, by county in
Great Britain, 2000
Source: Based on CACI, 2000

Summary

Rural communities are changing. Social and economic trends, including the decline of agriculture, the shifting spatial division of labour, depopulation and counterurbanization, and increased levels of mobility, have all impacted on the structure and coherence of rural communities. Change has also been driven from within communities, as the restructuring of public and commercial services and facilities has altered the dynamics of community life. Communities are about more than just services, but changes in the provision of rural services have had particular consequences for communities that can be analysed by returning to Liepins's model of the community as constituted by meanings, practices and sites and structures.

First, the rationalization of service provision can impact on the meanings of community that circulate in the local population. Small towns that have defined themselves as service centres for a surrounding rural area have this meaning challenged as shops, banks, hospitals and other key services are closed. Similarly the sense of independence of smaller villages may be undermined if key functions are lost, and further integration into the service field of a larger urban settlement may lead to the rural identity of the community being contested.

Secondly, as Liepins describes in her case studies, everyday interactions in shops and post offices and events associated with schools and community halls are central to the practice of community. The closure of shops, schools and other community facilities hence removes a whole stratum of community practices and reduces the capacity for regular interaction between community members. Thus, thirdly, the closure of rural services can be linked to a relocation of the sites and structures of community from the public spaces of shops, schools, post offices, main streets and village squares to the private spaces of residential gardens, backyards and porches. This relocation assists the fragmentation of communities as residents retract to interact only with their immediate neighbours and social networks rather than as previously through more open and inclusive sites of communal interaction.

Therefore, whilst the notion of a historically homogeneous rural community may be contested, it is indisputable that contemporary rural change has promoted a flowering of multiple communities in rural space. A single village or small town may have many different communities existing alongside each other (sometimes overlapping, sometimes not), each with their own meanings, practices and sites and structures that construct boundaries of exclusivity. Rural residents may also participate in different communities at different scales – in their own hamlet, in the wider parish, in the catchment area of the local town and so on – but the ability of individuals to participate at progressively higher scales will be determined by their access to transport, such that structures of social exclusion are created. Finally, individuals may additionally identify with particular communities of interest which they may consider to be more important than any affiliation with a geographical community.

These shifting patterns of community are significant for measures to respond to rural change, as any attempt to engage 'communities' in the delivery of rural development or other policy initiatives can no longer expect to connect with discrete geographical communities, but must be sensitive to and inclusive of the multiple rural communities that now exist.

Further Reading

The approach of thinking about rural communities in terms of their meanings, practices and sites and structures is developed further by Ruth Liepins in 'New energies for an old idea: reworking approaches to "community" in contemporary rural studies', in the *Journal of Rural Studies*, volume 16, pages 23–35 (2000); and 'Exploring rurality through "community": discourses, practices and spaces shaping Australian and New Zealand rural "communities"', in the *Journal of Rural Studies*, volume 16, pages 325–341 (2000). The first paper positions the model within a discussion of approaches to 'community' in rural studies, whilst the second paper focuses on the case studies in Australia and New Zealand, examining the impact of rural change on communities. Detailed accounts of service provision in rural areas tend to be fairly specific to a particular type of service and a particular country or region, and many are now quite dated. However, Sean White, Cliff Guy and Gary Higgs, 'Changes in service provision in rural areas. Part 2: Changes in post office provision in Mid Wales: a GIS-based evaluation', in the *Journal of Rural Studies*, volume 13, pages 451–465 (1997), provides an empirical study of the provision of rural post offices, whilst Alexander Vias, 'Bigger stores, more stores, or no stores: paths of retail restructuring in rural America', in the *Journal of Rural Studies*, volume 20, pages 303–318 (2004), discusses trends in retail provision in rural areas of the United States.

Websites

Detailed statistics from the 2001 survey of rural services in England are available on the Countryside Agency's website (www.countryside.gov.uk/ruralservices/). Resources on the site include GIS maps of distances to key services and the full Excel datasheets of the survey returns.

8

Environmental Change and Rural Areas

Introduction

If 'community' is one keyword associated with rurality (Chapter 7), then 'nature' is another. It may be argued that there are no truly 'natural' places left, that all rural areas have been shaped by human intervention to a greater or lesser extent, but the predominance of 'natural' features and materials in the rural landscape continues to be its most visually distinctive characteristic. The popular association of the countryside with 'nature' and the 'natural environment' explains in part why rural landscapes and places are valued in modern society, and why the 'rural idyll' has such appeal, yet it also emphasizes the vulnerability of the rural environment. We may value the countryside as a 'place of nature', but we often do not treat the natural environment of the countryside very well. Indeed, many of the key processes of social and economic change in rural areas over the past century have had significant, negative, environmental impacts.

This chapter examines environmental change in rural areas by focusing on three particularly prominent trends: the degradation of the environment by modern agriculture, including pollution, poisoning and the destruction of habitats; urban encroachment and the expansion of the built environment within rural areas, again producing pollution and the destruction of habitats; and the rural dimensions of global climate change, including the probable impact on the geography of agriculture and tourism. The level of concern that is attached to each of these trends, and therefore the responses that are considered to be appropriate, are influenced by the philosophy of nature that one adopts. From some perspectives, nature is regarded as resilient and able to adapt to change, from others, nature is seen as fragile and in need of protection. Thus, these different perspectives are discussed further in the first part of the chapter, which examines in more detail the association of rurality and nature.

Rurality and Nature

The identification of the countryside with nature is an offspring of the fundamental dualisms in western culture between nature and society and nature and civilization that have historically informed the separation of town and country in literature, art and government policy. The alignment of rurality with nature has also produced moral geographies in which the countryside is held to be a purer, nobler and more treasured space than the city (see Bunce, 1994; Macnaghten and Urry, 1998; Short, 1991). Furthermore, these various elements have been drawn into the lay discourses by which individuals define their own 'rural identity' and understand places as rural (see Chapter 1). Bell, for example, highlights the importance placed on nature in the lay discourses of the residents of his anonymized case study village of 'Childerley':

> Although the villagers are by no means sure that the village of Childerley is a place of nature, they have no doubt that such places exist. Moreover, they do not doubt that there are country ways of living and people who follow those ways. A close association with nature, they find, is the surest way to identify what those ways and whose those people are. The moral foundation of country life ... rests upon this rock. (Bell, 1994, p. 120)

This rather romanticized association of rurality and nature is built on three core components. First, the *rural landscape is perceived as a natural landscape*. It is distinguished from the urban landscape by the pre-eminence of ecological features, including flora, fauna and a relatively unmodified physical geomorphology. Although the concept of 'landscape' itself implies a fusion of the ecological and the human, the presence of human artefacts is tolerated in this discourse of the rural landscape only if they are essentially biological (for example, crops, forest,

pasture, orchards), or employ local natural resources in small-scale constructions that conform to the prevailing aesthetic of the landscape (for example, drystone walls, stone cottages, isolated farmbuildings) (Woods, 2003b).

Secondly, *rural activities are defined as those that use and work with nature*. Thus, farming, forestry, fishing, hunting and crafts such as basket-making are all held to be intrinsically 'rural' in a way that, for example, manufacturing industry, accountancy and skateboarding are not. Thirdly, there are perceived to be *rural people, who can be identified by their knowledge of and sensitivity towards nature*. True rural people, it is conjectured, are in tune with the changing of the seasons, understand the weather and have an innate knowledge of local plants and wildlife (Bell, 1994; Short, 1991).

Like many elements in the social construction of rurality (see Chapter 1), the above associations are idealized notions that are difficult to demonstrate empirically. Yet they are powerful ideas because they inform a popular conflation of the protection of nature with the protection of the countryside that has shaped the ways in which environmental change in rural areas is perceived and responded to.

On the one hand, a discourse of nature as pure, idyllic and vulnerable has been drawn on to position the rural environment as needing protection from damaging human intervention. Human activity in rural space is considered acceptable only insofar as it works with nature and constructs artefacts in the landscape that conform to the natural aesthetic (as described above). Developments that introduce large quantities of alien material (such as tarmac or metal) or modern technology into the landscape, or which appear disproportionate in scale to the morphology of the landscape, are considered to be unnatural and out of place (Woods, 2003b). Similarly, technological innovations in agriculture that employ synthetic chemicals, or that involve the manipulation of nature (GM crops, for example), are positioned as harmful to the environment. From this 'natura-ruralist perspective', the disconnection of the human realm from the natural world that is a central characteristic of modernity (see Chapter 3) has eroded sustainable forms of rural living and produced environmental problems that are now perceived as threatening the character of the countryside.

On the other hand, a utilitarian perspective on the rural environment conceives of nature as being both wild and resilient. From this perspective, the rural in its 'natural' state is a wilderness that requires taming through road-building, bridge-building, electrification and so on in order to make it hospitable for human activity. At the same time, rural space is also represented as offering the opportunity for the harnessing of 'natural' resources for human service – through mining and quarrying, forestry and

agriculture, the creation of reservoirs and the generation of hydro and wind power. Resilient nature is considered to be able to withstand the impact of such developments, and to adapt to scientific innovations in agriculture (Woods, 2003b).

The two perspectives offer contrasting approaches as to how environmental change in rural areas might be evaluated. They provide different guidance as to which changes should be represented as 'problems' and on the appropriate remedial action. However, both perspectives would recognize that the rural environment is changing and that these changes have resulted from a range of factors including the practices of agriculture, forestry and primary production; the impact of urbanization and building development; and the consequences of tourism and leisure activities; as well as environmental processes originating outside rural space (Box 8.1).

Box 8.1   Factors in environmental change in rural areas

Agricultural practices
- Use of pesticides
- Use of chemical fertilizers
- Increasing yields
- Removal of hedgerows
- Destruction of habitats
- Specialization – reduction of plant species

Urbanization and building development
- Loss of open space to housing, etc.
- Construction of roads, etc.
- Increased pollution
- Demand for drainage, water, sewerage
- Noise and light pollution

External processes
- Acid rain
- Removal of water for drinking, etc.
- Global warming
- Downstream pollution

Forestry and primary production
- Deforestation
- Afforestation of open moorland
- Planting of non-native species
- Spoils of mining and quarrying
- Flooding of land for reservoirs

Tourism and leisure activities
- Demand for facilities, accommodation, car parks, etc.
- Erosion of footpaths, etc.
- Damage to trees, plants, walls, etc.
- Litter
- Disturbance of wildlife

Agriculture and the Rural Environment

Modern capitalist agriculture turned the tables on nature. Traditional farming had been dependent on nature, restricted by soil type, climate and topography and at the mercy of the weather, pests and disease. For the pioneers of modern agriculture, however, these constraints and risks represented wasted capital and they began to harness new technologies to control, manipulate and modify environmental conditions. From long-established techniques such as irrigation and selective breeding, through 'improvements' to slopes and soils, to advanced biotechnology and the application of agrichemicals, agricultural practices were developed that changed the environment in order to enhance productivity (see also Chapter 4).

The first major warning that agricultural modernization of this type could lead to serious environment problems came in the 1930s when over-grazing, the conversion of grassland to arable land, and drought conspired in the American prairie to produce the catastrophe of the 'dust bowl' (Box 8.2). The experience of the dust bowl resulted in the replanting of grasslands in the prairie states and the introduction of government programmes for soil conservation, but fundamentally the agricultural practices that had contributed to the problem – changes in land use, the removal of vegetation, overstocking and the over-exploitation of water tables – not only continued but intensified under productivism.

Box 8.2   *The dust bowl*

The great plains of the central United States are natural grassland. However, in the early part of the twentieth century they were transformed by industrial agriculture. First came large-scale cattle ranching, with over-grazing thinning the vegetation cover. Then farmers moved into the more lucrative arable sector, ploughing up the grassland. Across the southern plains of Kansas, Colorado, Nebraska, Oklahoma and Texas, some 11 million acres (4.4 million ha) of grassland were ploughed for arable crops between 1914 and 1919. Between 1925 and 1930 another 5.3 million acres (2.1 million ha) were converted (Manning, 1997). The motive was economic. As Worster (1979) comments, 'by that time the Western wheat farmer was no longer interested in merely raising food for himself and his family. More than any other part of the nation's agriculture, he was a cog in an international wheel. As long as it kept turning, he would roll along with it. But if it suddenly stopped he would be crushed' (p. 89).

   The change in land use removed vegetation and loosened soil. This could be tolerated in the unusually wet years of the late 1920s, encouraging expansion into the most environmentally marginal regions, particularly as farmers were pressurized by a severe economic depression. In 1931, however, the rains failed. Average yearly precipitation across the region from 1931 to 1936 was only 69 per cent of normal levels. In the drought conditions the soil dried to dust, and with little vegetation to hold it together, the soil was rapidly eroded by strong winds that whipped up fierce dust storms. The worst affected area was the region where the Oklahoma panhandle intersects with the states of Kansas, Colorado, New Mexico and Texas, but between 1935 and 1940 areas of severe wind erosion periodically extended to cover the entire western half of Kansas, large parts of south-east Colorado and the cotton-growing region of northern Texas (Worster, 1979).

(Continued)

Box 8.2 (Continued)

At the height of the storms, in spring 1935, the University of Wichita in Kansas measured a cloud of some five million tons of dust suspended over 30 square miles of the city (Manning, 1997). The worst single storm, on 14 April 1935 – Black Sunday – travelled from northern Kansas to Texas, blacking out daylight for more than four hours as it passed. The next day a report in the *Washington Evening Star* coined the term, 'the dust bowl of the continent' (Worster, 1979). Confronted by the combination of drought and dust storms, crops failed or were destroyed and cattle starved. Buildings and farm structures were damaged by drifts of dust and the incidence of respiratory diseases increased significantly. The effects of the dust bowl compounded the earlier agricultural depression to create acute levels of poverty, particularly in the Oklahoma panhandle, northern Texas and south-west Kansas. Over 3 million people left the region during the 1930s – many migrating to California. Some counties in the worst affected zone lost between a third and a half of their population (Worster, 1979).

By 1940 the dust storms had become more infrequent. The return of 9 million acres of abandoned farmland to nature helped to stabilize environmental conditions and government-led soil conservation programmes worked to restore grassland and plant shelterbelts of trees. Despite these efforts, soil erosion has continued to be a serious problem in the region.

For more on the dust bowl, its causes and its consequences, see Richard Manning (1997) Grassland (Penguin); Donald Worster (1979) Dust Bowl: The Southern Plains in the 1930s (Oxford University Press).

The second major warning came in 1962 with the publication of a ground-breaking book, *Silent Spring*, by an American scientist, Rachel Carson. Carson argued that the increasing use of inorganic chemicals in agriculture – as pesticides, herbicides, insecticides and so on – risked making the Earth an unfit place to live. She demonstrated how toxic chemicals passed through the food chain, devastating wildlife populations, and explored the potential threats to human health. In particular she highlighted the extreme toxicity of the chemical DDT, introduced in 1943 and used in pesticides, which Carson proved was responsible for significant numbers of deaths of birds, fish and mammals that were not its intended targets. Above all, Carson attacked the culture of biotechnology and the belief that nature could be controlled:

The 'control of nature' is a phrase conceived in arrogance, born of the Neanderthal Age of biology and philosophy, when it was supposed that nature exists for the convenience of man ... It is our alarming misfortune that so primitive a science has armed itself with the most modern and terrible weapons, and that in turning them against the insects it has also turned them against the earth. (Carson, 1963, p. 243)

Silent Spring had a dramatic impact on agricultural policy. The use of DDT was banned and measures taken to control the worst excesses of pesticides. Yet, again, the agricultural practices that contributed to the problem remained fundamentally unchanged. Farmers continue to use pesticides and other chemicals and biotechnology companies continue to attempt to control nature.

The rural environment has been significantly changed by the practices of industrial and productivist agriculture, and still is being changed. These impacts can be grouped into

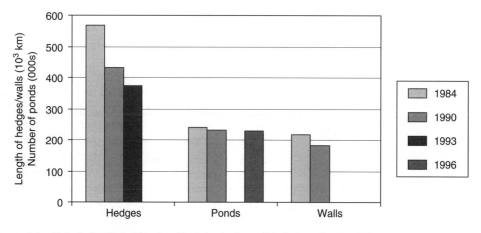

Figure 8.1 Extent of selected farmland features in Great Britain (pond and wall figures are for England, Wales and Scotland; hedgerow figures are for England and Wales only)
Source: After Cabinet Office, 2000

three dimensions – the destruction of habitats and loss of flora and fauna; the pollution of watercourses; and soil erosion, flooding and the lowering of aquifers.

Loss of habitats, flora and fauna

The extent to which the loss of wild plants and animals is considered to be an environmental problem created by agricultural practice is a matter of perspective. Some destruction of flora and fauna is deliberate on the part of farmers as they seek to eradicate pests and weeds and has always, in one form or another, been part of farming. The difference introduced by industrial agriculture is that the chemicals employed as pesticides and herbicides are more indiscriminate than biological or manual methods and can have unanticipated effects elsewhere in the food chain. Similarly, the destruction of habitats is for many farmers an acknowledged and accepted side-effect of efforts to improve productivity. Drawing on the idea of resilient nature, they will argue that nature can withstand the loss of the occasional hedgerow, pond or meadow. Environmentalists, however, contend that the aggregate loss of such

habitats seriously depletes the populations of native species.

Modern agriculture primarily impacts on wild plant and animal populations through three processes of agricultural 'modernization', each aimed at increasing farm productivity or income. First, habitats are lost through the modification of farmland. The pursuit of higher productivity leads farmers to minimize the amount of unused land on farms, whilst the effective use of machinery such as combine harvesters favours large, uninterrupted, fields. Together these factors have provided a rationale for the removal of hedgerows that previously formed field boundaries. Between 1945 and 1985, 22 per cent of hedgerows in England and Wales were removed or otherwise lost, with some 8,000 kilometres of hedgerow lost each year during the 1970s (Green, 1996). A further third of remaining hedgerows disappeared between 1984 and 1993 (Figure 8.1). Around a third of native British plant species have been recorded in hedgerows, but as Green (1996) notes, only about 250 species occur regularly in hedges and none of these is threatened with extinction as a result of hedgerow removal. More serious, Green suggests, is the impact of the loss of breeding grounds for animals leading to smaller populations. Three in four species of British lowland mammals breed in hedgerows, as do seven in ten native species of bird and four in ten species of butterfly (Green, 1996).

Secondly, habitats are also lost through changes in land use for economic reasons. The higher rate of return from arable farming than from pastoral farming has encouraged the conversion of large areas of pasture into cropland. In Europe, conversion to cropland was supported by grants under the Common Agricultural Policy, and even after subsidies were withdrawn market forces have continued to dictate the trend. Some 122,227 hectares of permanent grassland (or 4.1 per cent of the total) were lost in England and Wales between 1992 and 1997 – the equivalent to the area of a hundred soccer pitches disappearing every day (Wilson, 1999). Consumer fashions can also have an influence. The area of orchards in England and Wales fell from 62,000 hectares in 1970 to 26,000 hectares in 2002 as supermarket purchases have switched from native apples and pears to cheaper imported fruit (DEFRA, 2003).

Thirdly, plants and animals have been affected by the use of chemical pesticides and herbicides. As Carson noted, the introduction of new chemicals, including DDT and other chlorinated hydrocarbons, into agricultural use passed lethal toxins into the food chain. The impact on birds and predator mammals is summarized by Green:

> In Britain there were mass deaths of seed eating and other farmland birds including pigeons, pheasants and rooks, and of their predators particularly raptors and foxes, especially in the corn growing areas of East Anglia. The population of golden eagles collapsed and the peregrine falcon became a rare species all

over the country: by 1963 its UK population was only 44% of the 700 pairs breeding in 1939. In other parts of the world the decline was even greater: in the USA its population fell by 85%. Research by the Nature Conservancy in [the UK] was instrumental in substantiating that the cause was the new pesticides. Dieldrin (used as a seed dressing to give protection against the wheat bulb fly) and Aldrin (used in sheep dips) were being passed along the food chain to predators. (Green, 1996, p. 208)

In addition to poisoning, DDT and similar pesticides harmed bird populations by thinning the eggshells of some species, reducing rates of successful reproduction. Shell thinning in the eggs of the South Carolina brown pelican, for example, contributed to a decrease in the breeding population from more than 5,000 pairs in 1960 to 1,250 pairs in 1969 (Hall, 1987). Other species have suffered from the effect of pesticides and herbicides in reducing their food supply (Green, 1996).

The above processes have also worked collectively to damage habitats. For example, the disappearance of 97 per cent of wildflower meadows in the UK since the 1960s is a result not just of conversion to arable land but also of the application of herbicides to remaining grassland and of poor land management. Similarly, hedgerows that have been left *in situ* have been depleted by chemicals, either directly applied or drifting from adjacent fields, such that

> where hedges do survive on farmland, their wildlife now is usually very limited. A few coarse herbicide-resistant weeds such as cleavers and others such as cow parsley, hogweed, false oat and sterile brome, which are favoured by fertiliser at the expense of less competitive species, are often all that remain of once-rich floras. (Green, 1996, p. 206)

Table 8.1 Change in population of selected British farmland birds

	% change	
	1968–99	**1994–9**
Grey partridge	–85	–33
Corn bunting (farmland habitats)	–88	–38
Lapwing (farmland habitats)	–40	–2
Skylark (farmland habitats)	–52	–10
Linnet (farmland habitats)	–47	+2
Kestrel	–4	+2

Source: British Trust for Ornithology (Common Birds Census), www.bto.org/birdtrends

Furthermore, the impact on wildlife is often intensified by the combination of these processes. The populations of many bird species, for instance, have been hit not just by direct chemical poisoning, but also as a result of the loss of nesting sites in hedgerows and the depletion of food supplies by the use of insecticides and herbicides. As Table 8.1 shows, the numbers of many farmland birds have decreased dramatically (see also Harvey, 1998). Overall, the population of 12 common farmland bird species in England fell by 58 per cent between 1978 and 1998.

There is some evidence that more recent changes in agricultural policy and practice, including the introduction of agri-environmental schemes (see Chapter 13) and the growth of organic farming (Chapter 4), have begun to reverse the decline in wildlife populations. Studies in the UK have indicated that 30 species of bird, spider, earthworm and wildflower out of 92 monitored were present in greater numbers on organic farms than on conventional farms, and that populations of butterflies are increasing on farms with agri-environmental projects. However, such recoveries remain comparatively small compared with the scale of population loss over the past 50 years.

Pollution of watercourses

The intensive use of chemicals in agriculture has also increased the pollution of watercourses draining farmland. Some of this again results from pesticides, which enter watercourses either by surface run-off or by leaching through the soil. Once in rivers and lakes, pesticides can act to reduce reproduction levels in fish and other aquatic organisms, as well as lowering water quality to below fit standards for human consumption. In 1993 concentrations of the herbicide atrazine were found to exceed EU drinking water standards in 11 per cent of samples taken from rivers in England and Wales (Harvey, 1998).

The most serious form of agriculture-related pollution, however, is by nitrates and phosphates from inorganic fertilizers. The annual use of nitrogen-based inorganic fertilizers in the UK increased from 200,000 tonnes in 1950 to 1,600,000 tonnes in 1985 (Winter, 1996) and there are similar levels of usage elsewhere in the European Union. The application of nitrogenous fertilizers to crops significantly increases productivity, but it also increases the productivity of the most competitive weeds, such as nettles, which consequently monopolize hedgerows and verges, dominating less competitive species. When washed into watercourses on eroded soil particles, nitrates produce a similar result:

The accidental eutrophication of these waters has exactly the same ecological effect as does its deliberate use on grassland to increase crop yield. The more vigorous waterweeds are favoured and they rapidly reduce the diversity of the ecosystem by outcompeting other plants which are lost, with their associated animals. In water this effect is magnified by the kills of fish and other aquatic animals that result from deoxygenation brought about by the aerobic microbial breakdown of greatly increased plant production. Thousands of coarse fish, sea trout and swan mussels died from these causes on the River Rother in Sussex in 1973 and piled up in a stinking mass following a draining scheme in the valley. (Green, 1996, pp. 211–212)

Some 300,000 tonnes of nitrates are leached into British rivers each year, with particularly high concentrations in watercourses draining intensively farmed arable areas (Harvey, 1998). In the United States, over 10 kg per hectare of nitrates may find their way from the croplands of Iowa, Illinois, Indiana and Ohio into the Mississippi river system, with the cumulative build-up of nitrogen eventually resulting in a 15,000 km^2 'hypoxic zone' in the Gulf of Mexico in which there is insufficient oxygen in the water during summer to support normal populations of fish and shellfish (USDA, 1997). Whilst nitrate pollution is associated with arable farming, similar effects can result from the pollution of watercourses with livestock slurry and silage effluent. Cattle slurry is 80 times more polluting than raw domestic sewage and silage effluent up to 170 times more polluting (Lowe et al., 1997).

Agricultural pollution can pose a threat to human health when drinking water supplies are contaminated. There is a strong correlation between poor quality water and areas of intensive farming. Parts of England with drinking water supplies below the acceptable European Union standard in the late 1980s included the major arable farming areas of East Anglia, the Vale of York and Salisbury Plain (Ward and Seymour, 1992); whilst the water at only two out of 50 sampling sites on rivers in Brittany, France, was deemed to be of passable quality in 1999 (Diry, 2000).

Soil erosion and aquifer depletion
In spite of the experience of the dust bowl and the soil conservation programmes launched by the US government in response, soil

erosion remains a major problem in rural areas. A degree of soil erosion is natural, but modern farming practices can intensify the process beyond tolerable levels. In particular, soil erosion is aggravated by the removal of vegetation – including the conversion of pasture to arable fields, the destruction of hedgerows and deforestation – the creation of larger fields, the abandonment of rotation farming for specialization, and the use of large machines that need to be worked up and down slopes rather than along contours (Green, 1996; Harvey, 1998; USDA, 1997). Around 2.8 billion tonnes of soil were eroded from cropland in the United States in 1982, and whilst conservation programmes succeeded in reducing this total to 1.9 tonnes in 1992, erosion rates were still more than twice the tolerable level in around 9 per cent of arable land, including large parts of Texas, eastern Colorado, Montana and the central plains of North Carolina (USDA, 1997).

Agricultural practices that provoke soil erosion are counterproductive in that one of the major results is reduced soil productivity. Soil erosion also contributes to the destruction of habitats, as native plants can no longer survive on denuded top soils, to the pollution of watercourses by pesticides and nitrates, and to localized flooding. In southern Europe, soil erosion associated with the conversion of traditional forms of cultivation to intensive arable production has contributed to creeping desertification, particularly in southern Italy, south-central Spain and upland Greece. Attempts to maintain productivity levels in such conditions are supported by irrigation, which can in turn produce environmental problems of aquifer depletion if the rate of extraction exceeds the rate of replenishment by precipitation. Severe groundwater depletion has been recorded in a number of parts of the United States, including the massive Ogallala or High Plains aquifer that provides water for some 8 million hectares (or 5 per cent of the total farmed area in the United States) from Texas north to South Dakota and Wyoming, and where over-extraction has resulted in water table declines of over 30 metres (100 feet) in the worst affected areas (USDA, 1997).

Urbanization and the Physical Development of the Countryside

Rural environmental change also occurs through the physical development of the countryside. The construction of buildings, roads, car parks, airports and power stations, along with other permanent structures, is perceived to introduce an unnatural, urban, presence into rural space. In addition to this discursive impact, such developments have a measurable environmental impact through the removal of vegetation, disruption of hydrological systems and destruction of habitats. The physical development of the countryside can, depending on the circumstances, be either driven by the consequences of rural social and economic change, or imposed by external actors. Generally, however, developments are linked to one of four processes.

First, there is continuing urban encroachment on rural space. The area of 'urbanized space' in the United States more than doubled from 10.3 million hectares in 1960 to 22.6 million hectares in 1990, and was predicted to exceed 25 million hectares by 2000 (Heimlich and Anderson, 2001). This rate of expansion is far greater than that of urban population growth and reflects social trends towards smaller households and residential preferences for low density housing being met through contiguous suburban development. One major effect is to squeeze the capacity for agriculture in the urban–rural fringe (which currently accounts for around a third of total US agricultural production). Between 1982 and 1992 nearly 1.7 million hectares of

cropland in the United States were converted to developed land, 68 per cent of it for residential use, and low density urban sprawl is estimated to reduce the value of agricultural production in the Central Valley of California by $2 billion each year (USDA, 1997). Other environmental impacts include the destruction of habitats, the loss of aesthetically valued recreational land, and local problems of waste disposal, water supply and disruption of drainage systems, the latter of which can lead to flooding and landslides (Rome, 2001).

National and local governments have adopted a number of initiatives to restrict urban sprawl, including planning controls (see Chapter 13), and the purchase of 'greenbelt' land for protection under public ownership (Rome, 2001). The consequence, however, can be for development simply to 'leapfrog' over protected areas into the surrounding rural areas. Thus, secondly, population growth in rural areas has generated demands for development within the countryside itself. Around 80 per cent of new housing development in the United States between 1994 and 1997 was located outside urban areas (Heimlich and Anderson, 2001). Similarly, in rural parts of the UK, such as Dorset, significant new housing development has taken place in rural communities, particularly in small towns (Table 8.2 and Figure 8.2). This trend is anticipated to continue. Land use planning policy in the UK has projected that 2.2 million new houses will need to be built in rural areas by 2016, in turn provoking a fierce political debate (see Chapter 14).

Thirdly, the changing social and economic character of the countryside has created demands for the development of new infrastructure, including new roads, car parks, sewerage systems and shopping facilities. Pressure for such developments is produced not just by population growth and new house growth, but also by the rise in commuting, the relocation

Table 8.2 New houses built in Dorset, England, 1994–2002, by population size of parish

Population of parish in 1994	No. of parishes	No. of new houses built 1994–2002	% of all new houses built in county	% of total population of county	Mean no. of new houses per parish
Under 250	121	202	1.3	3.5	1.7
250–499	52	484	3.2	5.0	9.3
500–999	38	959	6.4	7.9	25.2
1000–2499	27	1555	10.3	9.3	57.6
2500–4999	10	1392	9.3	10.0	139.2
5000–9999	13	4267	28.4	26.7	328.2
10000–19999	4	3063	20.4	15.2	765.8
Over 20000	2	3122	20.8	22.3	1561.0

Source: Dorset County Council

Figure 8.2 New housing in the village of Burton Bradstock, Dorset, built in the local vernacular using reconstituted stone
Source: Woods, private collection

of industrial plants and offices, and the expansion of tourism (Robinson, 1992). Major infrastructure such as highways and railways are also routed through rural space to connect urban centres. The visual disruption to the rural landscape and the physical destruction of habitats have emerged as keys sites around which environmental protests against new roads have been mobilized in locations as diverse as Newbury and Twyford Down in England, Wyoming County in New York State, the Interstate 69 route in Indiana and

Thüringen in Germany. More subtle environmental effects of development are the increases in light pollution and noise pollution in rural areas. For example, a British pressure group, the Campaign to Protect Rural England (CPRE), has claimed that the extent of 'tranquil areas' in England – defined by distance from major sources of noise pollution such as significant roads, airports and power stations – decreased by 21 per cent between the 1960s and 1990s (Figure 8.3).

Fourthly, rural locations have continued to be favoured as sites for large-scale, noxious and otherwise sensitive land uses, whose development is either easier, or faces less resistance, in less populated regions. These include airports, reservoirs, power stations, prisons and military camps. As well as the environmental impact of the development itself, in some cases the nature of the land use concerned may also introduce new environmental risks. For example, rural Tooele County in Utah contains a magnesium factory, a private low-level nuclear waste burial site, three toxic chemical stores and a military depot that stores half of the United States's chemical weapons. In 1999, a conflict developed between the State government and local tribal authorities over the construction of a facility in the Skull Valley Goshute Reservation for the interim storage of high-level nuclear waste, intended for eventual disposal at a proposed (and equally controversial) dump at Yucca Mountain, Nevada. Whilst tribal leaders argued that the facility was needed to create jobs on the reservation, the State expressed wider public concerns about pollution by radioactive material (Wald, 1999).

Climate Change

Rural environmental change is not just the result of human activities within rural space but is also influenced by global scale environmental processes, such as global climate change. There is now a significant scientific consensus that human activity has increased atmospheric concentrations of 'greenhouse gases' – carbon dioxide, methane, nitrous oxide, chlorofluorocarbons and ozone – and that as a result the global climate is changing and is likely to change dramatically over the course of the next century. The key impacts identified by the Intergovernmental Panel on Climate Change (IPCC) include an increase in the overall global mean temperature of between 1.4 to 5.8 degrees celsius by 2100, higher maximum temperatures and increasing minimum temperatures over most land areas, more intense precipitation events and an increase in the sea level globally of 10–50 cm by 2050 (IPCC, 2001).

Rural areas contribute to climate change through the production of 'greenhouse gases' (particularly methane), and can also help to moderate climate change through carbon sequestration by agricultural crops and forests (Bruinsma, 2003; Rosenzweig and Hillel, 1998). Moreover, the economies and societies of rural areas are vulnerable to the environmental consequences of climate change. Although the modelling of climate change impacts is an imprecise science and different models vary in their predictions, a number of likely consequences can be identified with respect to agriculture, tourism and human communities.

Agriculture

Increased concentrations of carbon dioxide in the atmosphere should in theory increase photosynthesis and stimulate greater productivity for agricultural crops, yet the IPCC and other commentators have argued that this benefit is likely to be offset by negative impacts including crop damage from higher temperatures and extreme events, drought, soil degradation and changing ranges of pests and diseases (IPCC, 2001; Rosenzweig and

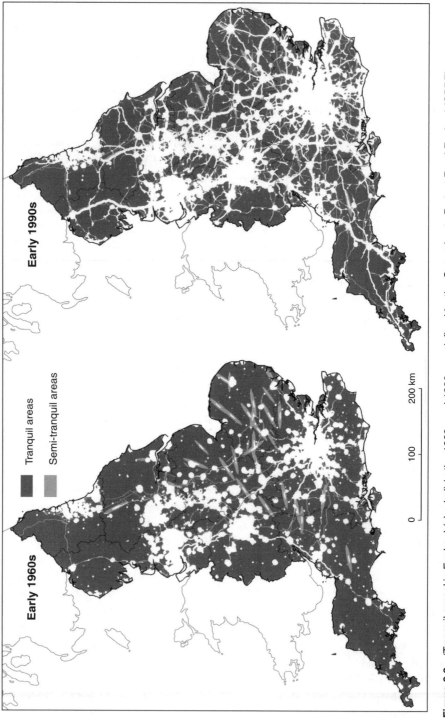

Figure 8.3 'Tranquil areas' in England (shaded) in the 1960s and 1990s, as defined by the Campaign to Protect Rural England (CPRE)

Source: CPRE

Hillel, 1998). The agricultural impact of climate change is therefore likely to be spatially differentiated. Crop productivity is most probable to increase in high-latitude regions such as Canada, Scandinavia and Russia, whilst productivity is predicted to decrease most substantially in tropical regions (Rosenzweig and Hillel, 1998). In effect this will mean that poorer developing countries will suffer most from climate change, whilst some developed nations may be in a position to benefit from new trade opportunities. However, even within developed states there are likely to be significant changes in the geographies of agricultural production.

In New South Wales, Australia, increased temperatures, reduced soil moisture, more frequent heavy rainfall and decreased river flow are all projected to have a negative impact on agricultural production. The area of arable land is expected to be reduced as a result of drought and soil degradation and increased carbon dioxide concentrations are predicted to reduce grain quality. Similarly, increased incidences of heat stress in cattle are projected to reduce dairy production in New South Wales by around 4 per cent by 2030 (Harrison, 2001). In Europe, the cultivation of olives and citrus fruit is projected to move northwards as Mediterranean zones become more prone to drought (Rosenzweig and Hillel, 1998). New crops such as navy beans could be introduced in the UK, offering farmers an opportunity for diversification (Holloway and Ilbery, 1997). The most economically significant impact, however, is likely to be in North America where more frequent droughts and heat waves could substantially reduce crop production in the prairie regions, especially the southern plains (Rosenzweig and Hillel, 1998). The extended drought of 1988 is regarded by some commentators as a foretaste of future problems and saw crop yields in the US grain belt drop by 40 per cent. Other regions, however,

including parts of Canada, the Great Lakes area and the Pacific states, could see increases in arable production as they become regarded as more favourable environments.

Tourism

Climate change poses challenges for both winter and summer tourism in rural regions. Temperature increases are already reducing snow cover in mountainous areas, threatening the winter sports industries in New Zealand, the Alps and the Rocky Mountains. Summer tourism, meanwhile, is likely to be affected by problems of water supply and heat stress in areas such as southern Europe, and by sea level rise and exposure to typhoons in rural coastal zones of Australia. Rural economies that have diversified from agriculture into tourism may therefore find that further economic restructuring becomes necessary. At the same time, however, patterns of more consistently dry and warm summers could help to boost countryside tourism in more temperate parts of northern Europe and North America, thus providing new opportunities for economic diversification (IPCC, 2001).

Human communities

Aside from the economic challenges to agriculture and tourism, climate change also can have a direct impact on the everyday lives of people in rural areas. The low population densities of some rural regions reflect the already harsh environmental conditions and many are particularly exposed to extreme weather events such as storms, tornados, flooding and drought, all of which are predicted to increase with global warming. Additionally, the culture of some indigenous communities in remote rural regions is threatened by the impact of climate change on wildlife. Both of these processes are starkly evident in Alaska, where temperatures are increasing at ten times the global average.

Since 1960, average winter temperatures in Alaska have risen by 4.5 degrees celsius, with the consequence that snowfall has decreased, glaciers are retreating and the tundra is melting. The thawing of the permafrost has caused problems of subsidence and landslips damaging buildings and roads at a cost of over $30 million per year. The environmental changes have also dried up streams and rivers – starved of seasonal meltwater – and disrupted the feeding patterns of wildlife such as caribou and polar bears, reducing their numbers. These changes in turn threaten the traditional hunting- and fishing-based culture of the Gwich'in people living above the Arctic Circle (Campbell, 2001).

Significantly, however, many of the strategies promoted by campaigners in order to alleviate the human contribution to climate change are also challenging to aspects of rural life. For example, punitive taxes on petrol and diesel aimed at reducing consumption of fossil fuels have a disproportionate impact in rural areas where many residents are dependent on the use of private vehicles to access employment, schools and key services – as demonstrated by farmer-led protests against high fuel taxes in Europe in September 2000. Furthermore, any substantial transition to renewable energy sources depends on the construction of a large number of renewable power generation plants, notably hydroelectric stations and 'windfarms', in rural locations that can meet their resource demands. Such developments inevitably have an impact on the immediate local environment as well as conflicting with aesthetic appreciations of the rural landscape (see Box 8.3).

Box 8.3   The environmentalist's dilemma: wind power generation in rural locations

The harnessing of wind power is a crucial element in the transition to renewable energy. Commercial wind power generation was pioneered in Denmark in the early 1980s, closely followed by California, where from the installation of the first 'windfarm' in 1981 nearly 16,000 wind turbines were in operation by 1991 (Gipe, 1995). In the UK, wind energy is targeted to produce 10 per cent of the national electricity supply by 2010 (Woods, 2003b).

Although in some places single wind turbines have been constructed to supply individual communities, most commercial wind power is generated by large-scale installations predominantly located in rural settings. However, such developments have increasingly been contested by local protest movements in the UK, Germany and the United States. As Brittan (2001) notes, objections to wind turbines are frequently aesthetic, but in many cases they also highlight ecological damage to the immediate local environment.

One such case concerned proposals to construct a 39-turbine wind power station at Cefn Croes in the Cambrian Mountains of Wales in 2000. The proposed windfarm was at the time the largest to be built in the UK and was promoted by supporters, including Friends of the Earth, as a significant contribution to renewable energy generation and to the alleviation of global warming. However, a vociferous protest campaign, supported by the local Green party and the Campaign for the Protection of Rural Wales, emphasized not only the visual impact on the landscape, but also the effect on local wildlife (Woods, 2003).

For more information see Michael Woods (2003b) Conflicting environmental visions of the rural: windfarm development in Mid Wales. Sociologia Ruralis, 43 (3), 271–288.

Summary

Nature is at the heart of popular understandings of rurality, yet the natural environment of rural areas has been degraded by the human exploitation of rural space. Modern agriculture has become distanced from nature to the extent that practices such as removing hedges and the use of chemical pesticides and inorganic fertilizers have been blamed for falling populations of plant and animal species. Tourists, attracted by 'natural' rural landscapes, have contributed to environmental problems of erosion, pollution and the loss of land to building developments. Similarly, counterurbanization – motivated in part by lay discourses of the rural as a 'natural' space – has created demands for housing developments, and new roads and facilities, and has contributed to light pollution and the loss of 'tranquil areas'.

At the same time, rural areas have also suffered the consequences of global environmental change, including global warming. These have the potential to significantly alter patterns of agricultural production and tourism as well as causing damage to property and infrastructure and threatening the cultural practices of indigenous peoples. As such, the processes of rural environmental change have a cyclical character. They are produced or intensified by human activities and in turn they have an impact on human activity. The question of how human societies should respond to rural environmental change, however, generates different answers depending on one's perception of nature. From a utilitarian perspective, a certain amount of environmental change is not concerning, as nature is perceived to be resilient enough to adapt. In contrast, from a natura-ruralist perspective, environmental change has already resulted in irreparable damage to nature and urgent action is required to halt or reduce further change. Finding appropriate courses of action, though, inevitably involves compromises. Measures to protect wildlife habitats, for example, may involve an unprecedented degree of regulation of farming, whilst initiatives aimed at alleviating climate change, such as constructing wind power stations, can have a significant impact on the immediate local environment. Thus, although numerous conservation programmes and measures have been introduced (see Chapter 13), the appropriate response to rural environmental change remains a key source of conflict in the countryside (see Chapter 14).

Further Reading

Bryn Green's *Countryside Conservation* (Spon, 1996) and Graham Harvey's *The Killing of the Countryside* (Vintage, 1998) discuss in detail many of the changes to the rural environment, particularly those related to agriculture, albeit from a strongly British perspective. Adam Rome, in *The Bulldozer in the Countryside* (Cambridge University Press, 2001), meanwhile provides a historical overview of urban expansion into the American countryside and the rise of the environmental movement in response. For an overview of the potential impact of global climate change on agriculture see Cynthia Rosenzweig and Darrell Hillel, *Climate Change and the Global Harvest* (Oxford University Press, 1998).

Websites

A number of reports on climate change are available on the Internet, including those by the National Assessment Synthesis Team in the United States (www.gcrio.org/National Assessment) and the Department for Environment, Food and Rural Affairs in the United Kingdom (www.defra.gov.uk/environ/climate/climatechange).

Reports and (subjective) accounts of other impacts on rural environments are available on the websites of a number of pressure groups, including the Campaign to Protect Rural England (www.cpre.org.uk) and Scenic America (www.scenic.org).

Part 3

RESPONSES TO RURAL RESTRUCTURING

9

Rural Policy and the Response to Restructuring

Introduction

As the previous chapters have described, the rural areas of the developed world have experienced considerable social, economic and environmental change over the course of the past few decades. The next part of this book focuses on the responses to these changes that have been adopted by policy-makers and rural communities. Subsequent chapters examine strategies for rural development, reforms to the way in which rural areas are governed, the repacking of the countryside for the new consumption-based economy, initiatives to protect the rural environment, and the emergence of rural political conflicts. In each of these cases the responses to rural restructuring have been shaped by a number of actors both within and outside rural space, including local residents, employers, tourists, corporations and – most significantly – governments. The practices that will be described in the next few chapters with respect to rural development, conservation, governance and the commodification of the rural all either involve or have been influenced by the adoption of particular policies by government. Similarly, rural political conflicts are commonly targeted at particular government policies. As such, this chapter aims to provide an introduction to Part 3 of the book by examining the processes through which 'rural policy' is made. It starts by exploring what is meant by 'rural policy' and discussing how the way in which governments approach rural policy has been affected by restructuring. The chapter then proceeds to describe how policy is made, before concluding by concentrating on one of the key policy challenges – the reform of agricultural trade – as an illustration of how different governments have adopted different policy responses to a similar problem.

The Enigma of Rural Policy

One of the first things to note about 'rural policy' is that it is a very elusive and enigmatic entity. Government documents and websites tend to make relatively few explicit references to 'rural policy' as such, and, indeed, relatively few countries have a government department for 'rural affairs' (or a similar title) – with the UK and the Republic of Ireland being the notable exceptions. To some extent the elusiveness of 'rural policy' reflects the fact that there are a whole range of policies – in health, education, transport, law and order, and so on – that apply to rural areas as much as they apply to urban areas, but which do not get branded as 'rural policy'. Yet, there are other policy areas that are primarily or largely concerned with rural space and rural activities: agriculture, forestry, rural economic development and, arguably, substantial elements of land use planning and conservation policy. The fact that until recently little connection had been made between the development of policy in each of these areas reveals a great deal about the power of vested interests in the countryside and the way in which policy-makers have perceived of the character and needs of rural societies and economies.

Bonnen (1992), for example, has argued that the United States failed to develop a coherent rural policy because of the emphasis placed on agriculture and the strength of the agricultural lobby. He contends that the government policies adopted in the late nineteenth and early twentieth centuries were not focused on rural communities or people, but were 'industrial policies' aimed at supporting agriculture, forestry and mining. It was thus around these interests that the institutional structures for governing rural America were developed, including notably the US Department of Agriculture (USDA). As the

role of the state in regulating and subsidising agriculture increased, so rural policy in the United States effectively became farm policy, and the farmers' unions became increasingly powerful as the gatekeepers to the policy-making process (see also Browne, 2001a, 2001b). Although lobby groups did emerge to represent, for example, rural electricity cooperatives and rural health providers, their attentions tended to be directed at representing rural interests within their own policy field, not in seeking to promote a more inclusive rural policy or in working together to form a 'non-agricultural' rural lobby. Furthermore, once agricultural interests had control over the bulk of government spending on rural areas, they were not inclined to support policy shifts that would see funds redirected to tackling rural poverty or regenerating declining communities. As Browne (2001a) observes, 'within this institutional structure, an alternative national rural policy was regarded as far from a good thing. Passing it might bring cuts in farm program funding and even skepticism about the whole policy base' (p. 49).

A similar privileging of agriculture shaped British rural policy as it was developed after the Second World War. Although a number of wide-ranging reports on rural Britain had been produced during the wartime period, their recommendations were implemented through several different Acts of Parliament that helped to create a segmented policy structure. In this structure, agriculture, conservation, land use planning and economic development were all treated as parallel but separate policy fields that were the responsibilities of different government departments and different government agencies, engaging with different pressure groups (Winter, 1996). Thus, agriculture was the responsibility of the

Ministry for Agriculture, Fisheries and Food; conservation was the responsibility of the Department for the Environment and of the Countryside Commission; whilst economic development was the responsibility of the Department of Trade and Industry and the Rural Development Commission.

The structuring of the 'rural' policy process in this way had four key consequences. First, agriculture was elevated above other rural interests by being awarded a dedicated government department and ministerial position of its own. Secondly, as the most visible 'rural' presence in government, agriculture departments came to be regarded as representing 'the countryside' whilst in fact they only represented one sector of rural society. Thirdly, agricultural policy-making operated within a vacuum and was protected from having to take account of non-agricultural interests. Fourthly, the non-agricultural countryside became subordinate to the agricultural and was marginalized in policy terms. Problems such as rural poverty were effectively constructed as 'non-issues' because there was no one speaking up for them within the policy-making process.

Towards integrated rural policy

The segmentation of rural policy has become more and more unsustainable as social and economic restructuring has progressed. The decline in the economic significance of agriculture has brought its privileged position into question, not least as it has become apparent that agricultural policy has been part of the problem – encouraging a strategy of agricultural modernization that has degraded the rural environment, drastically reduced the farm labour force and resulted in over-production (see Chapter 4). Yet, such is the entrenched nature of the agricultural policy

process that it has often taken crises such as the BSE and foot and mouth epidemics in the UK and the BSE scare in Germany to force significant changes in the policy structure. At the same time, rural policy-makers, practitioners and commentators have all become increasingly aware that many of the problems resulting from rural restructuring can be addressed only by integrating the various diverse strands of government policy relating to the countryside.

One of the most notable initiatives to develop an integrated rural policy was the production in 1995 and 1996 by the UK government of 'Rural White Papers' for England, Scotland and Wales. These documents were integrated statements of the government's policies on a wide range of issues relating to rural areas, from agriculture to telecommunications, housing to village halls, and forestry to sport. Moreover, they explicitly recognized the diverse character of the contemporary countryside and the need to develop a more integrated rural policy:

> We constantly have to find the balance between competing interests and conflicting concerns. Farmers and foresters who look after 80% of the land; the enthusiasts for our flora and fauna, for ancient buildings and traditional crafts; those who are building the new businesses to replace the old; the incomers who seek to realise their rural idyll; the ramblers and lovers of country sports; those who delight in birds, train horses, or ride the bridleways; landowners and people whose roots run deep into rural England – these and many more have interests that need to be accommodated in a living and working countryside. (DoE/MAFF, 1995, p. 6)

However, critics have argued that the White Papers continued to be influenced by the thinking of the old policy structure, that they did not fundamentally challenge existing agricultural policy, and that they failed to develop a truly integrated approach. As Hodge (1996) contends, 'we may feel that there is a chapter missing at the end of the Rural White Paper. One that draws out the inter-relationships between the areas of rural policy and between rural conditions and wider social, economic and environmental change' (p. 336). None the less, the 1995–6 Rural White Papers marked the beginning of a tran-sition towards an integrated rural policy in the UK, that has been subsequently developed by the publication of a second Rural White Paper for England in 2000, and the merger of the Ministry for Agriculture, Fisheries and Food with the Department of the Environ-ment in 2001 to form the Department for the Environment, Food and Rural Affairs (DEFRA).

An alternative approach was taken in Australia, where the government convened a 'Regional Australia Summit' in 2000 to bring together representatives of various rural interests. The summit and its 12 work-ing groups considered a wide range of issues, including infrastructure, health, community well-being, facilitating entrepreneurship, adding value to farming communities, new industries, community leadership, education and sustainable resource management. The outcomes from discussion of these themes were integrated into a final report structured around the three 'strategic areas' of commu-nity empowerment, economic and business development in regional communities, and equity of services in regional communities. As such, the Australian approach could be argued to be more inclusive and delibera-tive than the British approach, and to have involved the type of strategic analysis that

Hodge (1996) bemoans the absence of in the British White Papers. Yet, although the Australian government was committed to taking forward the summit's recommendations, its ephemeral nature and its semi-independent status means that the summit did not in itself change the policy-making structure.

The British and Australian cases both high-light some of the challenges that continue to confront the development of integrated rural policy. Both initiatives were arguably more symbolic than substantive and in both cases many of the actors, institutions and attitudes associated with the old segmented policy structure remain involved in the new struc-ture and may continue to press for their own particular sectoral interests. Therefore, in order to understand how an evolving rural policy might produce responses to rural restructur-ing, it is first necessary to step back and con-sider how policy is made.

The Policy-making Process

The precise way in which rural policy is made will vary between different countries, and dif-ferent scales of government, depending on the constitutional structure of the state, the pre-vailing political ideology and the relative strengths of the various institutional actors involved. Essentially, however, the policy-making process will in all cases involve negotiation between the state institution responsible for formulating the policy – whether national, supranational, regional or local scale – the agencies responsible for implementing the policy, and the pressure groups campaigning for or against particular policy outcomes (see Box 9.1). The nature of the relationship between these various actors may be close-knit or loose, stable or unstable, consensual or conflictual, as is described by a number of different models of policy-making.

Box 9.1   *Institutional actors in rural policy-making*

Global

World Trade Organization (WTO) – Supranational body of 146 member states, responsible for negotiating international trade agreements with the aim of lowering tariffs and eliminating trade barriers. Influences rural policy through agreements on trade in agricultural products, forest products, etc. WTO policy-making tends to be dominated by the major industrialized nations, particularly the United States.

Supranational

European Union (EU) – Organization of 25 member states in Europe. Responsible for Common Agricultural Policy (CAP) and for funding rural development through European Structural Funds. Policy is made by the Council of Ministers – comprising ministers from the member states – and managed by the Commission – comprising commissioners appointed by member states, each of whom heads a Directorate-General (DG). The Directorates-General for Agriculture, Environment and Regional Policy have most involvement with rural policy.

North American Free Trade Agreement (NAFTA) – Agreement facilitating free trade between the United States, Canada and Mexico, including many agricultural and forestry products. No independent policy-making structure.

Cairns Group – An alliance of 18 agriculture exporting states including Australia, Canada, New Zealand, South Africa and Argentina. Formed in 1986 to lobby for global free trade in agriculture.

National

Government departments – Responsible for proposing, introducing and enforcing policy. A range of different departments may have an interest in rural policy (including departments of health, education, transport, etc.), but the most important for rural policy are departments of agriculture and/or rural affairs. Notable examples include:

- *United States Department of Agriculture (USDA)* – Founded 1862, now has over 100,000 employees. Responsible for agriculture and agricultural trade, rural development, food safety, natural resources and the environment in the United States.

- *Department of the Environment, Food and Rural Affairs (DEFRA)* – UK government department, formed 2001 by merger of former agriculture ministry and environment department. Responsible for agriculture, forestry, fisheries, food safety, rural development, animal welfare and environmental protection.

Government agencies – Responsible for implementing rural policies. Agencies typically have a specific remit for conservation, national parks, forestry, rural development, etc. Notable examples include:

(Continued)

Box 9.1 (Continued)

- *US Forest Service* – Manages the 155 national forests and 20 national grasslands in the United States.
- *Natural Resources Conservation Service (NRCS)* – Works with farmers, ranchers and landowners in the United States to develop voluntary conservation schemes.
- *Countryside Agency* – UK government agency responsible for supporting rural community development, countryside protection and recreational access to rural land in England.

Sub-national

Sub-national regional governments may have some responsibility for areas of agriculture, planning, conservation, rural development, health and education. Examples include the United States and the Australian states, Canadian provinces, German Länder, and the devolved governments in Scotland, Wales and Northern Ireland in the UK.

Pressure groups – Agricultural

Agricultural pressure groups have traditionally been the most influential non-governmental actors in shaping rural policy. Notable examples include:

- *Farm Bureau* – US farmers, union, with 5 million members, formed in 1919 with a federal structure of county- and state-level associations.
- *National Farmers' Union (NFU) (US)* – US farmers' union, with 300,000 members, formed in 1902.
- *National Farmers' Union (NFU) (UK)* – Largest British farmers' union, with 90,000 members, formed in 1908.
- *National Farmers' Federation* – Formed in 1979 as a coalition of state farmers' organizations and commodity councils to provide a single voice for Australian agriculture.
- *Federated Farmers of New Zealand* – Farmers' union, with 18,000 family and individual members, formed in 1945.
- *Fédération Nationale des Syndicats d'Exploitants Agricoles (FNSEA)* – Largest French farmers' union.
- *Comité des Organizations Professionnelles Agricoles (COPA)* – Coalition of farmers' unions representing agricultural interests at the European Union level.

Pressure groups – Other

Rural policy increasingly engages a wide range of other pressure groups, including conservation bodies, pro-hunting lobbies, rural poverty campaigns and industry organizations. Notable examples include:

- *Countryside Alliance* – British pro-hunting group, with 100,000 members, increasingly campaigning on a range of rural issues, including farming and service provision.

(Continued)

Box 9.1   (Continued)

- *Campaign to Protect Rural England (CPRE)* – British conservationist group established in 1926 to resist urbanization. Formerly known as the Council for the Preservation of Rural England.
- *Friends of the Earth* – International environmental group involved in rural policy through interests in the rural environment and GM crops.
- *Sierra Club* – American conservation group founded in 1892. Now has over 700,000 members in the United States.
- *National Rural Health Association* – Industry organization representing healthcare providers and associated groups in the United States.
- *Rural Coalition* – Federation of progressive rural pressure groups in the United States and Mexico, including farmworkers' groups, local environmental organizations and organic farming groups among others.

Conventionally, political analysts distinguished between *pluralist* and *corporatist* models of policy-making. In the *pluralist* model, policy is open to influence by a large number of groups responsive to grassroots members. Government is seen to be passive, simply allocating resources and making policy according to the relative strengths of competing pressure groups at any one time. The *corporatist* model, in contrast, involves a close relationship between the state and a limited number of interest groups representing major economic interests. The state plays an active role in driving policy, but the interest groups are fully involved in policy-making and implementation, delivering benefits to their members. Policy-making hence takes the form of bargaining, resolving conflicts between different economic interests (Marsh and Rhodes, 1992). Commentators such as Grant (1983) and Winter (1996) have suggested that the closed structure of agricultural policy-making in the mid-twentieth century was a form of corporatism; however, this has been critiqued by Smith (1992), who argues that farmers were rarely involved in the implementation of policy, that the agricultural policy structure was not intended to resolve conflict and that bargaining was limited.

Instead, Smith and others have promoted the alternative model of *policy networks*. In contrast to the polarity of the pluralist/corporatist dualism, the policy networks model recognizes that there are different degrees of interaction between government and interest groups, ranging from loose 'issue networks' at one extreme to closed, tight-knit, 'policy communities' at the other (see Box 9.2). Whereas issue networks have a large, unstable membership, fluctuating access to policy-makers and variable influence, policy communities are characterized by a limited, stable membership, frequent high-quality access and consistent influence (Table 9.1). In between these two extremes are stable and restricted 'professional networks' representing specific professions such as medical doctors; 'inter-governmental networks' of sub-national government organizations; and 'produce networks' representing the interests of producers in policy-making, whose membership and influence varies with economic trends (Marsh and Rhodes, 1992).

Box 9.2   Key term

Policy network: A cluster of organizations, including state institutions and interest groups, connected by relationships of interdependence for resources and therefore involved in policy-making with respect to a specified policy field.

Policy community: The most close-knit form of policy network, involving a limited number of participating actors in a stable interdependent relationship, exercising tight control over a policy field.

Issue network: The loosest form of policy network, involving a wide range of groups with limited interdependence in fluctuating participation in the policy-making process with respect to a particular policy field.

Table 9.1   Characteristics of policy communities and issue networks

	Policy community	**Issue network**
Number of participants	Very limited. Some groups consciously excluded	Large, unstable membership
Interests of participants	Economic and/or professional interests predominate	Encompasses range of affected interests
Frequency of interaction	Frequent, high-quality interaction on all matters related to policy issues	Contacts fluctuate in frequency and intensity
Continuity	Membership, values and outcomes are consistent over time	Access fluctuates significantly
Consensus	All participants share basic values and accept legitimacy of the outcome	A measure of agreement exists, but conflict is ever-present
Distribution of resources within network	All participants have resources; basic relationship is an exchange relationship	Some participants may have resources, but they are limited and basic relationship is consultative
Distribution of resources within participating organizations	Hierarchical; leaders can deliver to members	Varied and variable distribution and capacity to regulate members
Power	A balance of power between members. One group may dominate but all groups must gain if the community is to persist	Unequal powers, reflecting unequal resources and unequal access. Some participants benefit at expense of others

Source: After Marsh and Rhodes, 1992; Winter, 1996

The policy network approach is based on two key assumptions. First, that policy-making is undertaken through a large number of different segments, each of which is accessible to only a limited range of groups; and secondly, that policy-making involves an inter-relationship between state institutions and interest groups who are dependent on each other for resources. Government, for example, can provide finance, but may be dependent on interest groups to supply the cooperation of workers on the ground. Policy networks provide a structure for these relationships by defining the roles of the actors involved, deciding which issues will be included on the policy agenda and which are excluded, and setting the 'rules of the game' which shape the behaviour of the participating groups.

Furthermore, the differentiation between types of policy network recognizes that interest groups have differing degrees of influence and access to government policy-makers. As such, a distinction can be made between 'insider groups', who operate within policy communities, are regarded as legitimate by government and are consulted on a regular basis; and 'outsider groups', who are excluded from policy communities and whose contact with government is less frequent and less influential (Grant, 2000; Winter, 1996). In the context of rural policy, the large farmers' unions and established conservation groups such as the CPRE and the Sierra Club may be regarded as 'insider groups', whilst 'outsider groups' include smaller farm unions, farmworkers' unions, militant rural protest groups and campaigners on rural poverty issues. Environmental groups and consumer groups have arguably moved from being 'outsiders' to at least a partial insider status.

When applied to rural policy, the policy networks framework reveals that a broad transition has occurred in the past two decades from a policy-making structure organized around a number of specialized and autonomous policy communities, to a more open, less stable series of issue networks with broader remits. This is most clearly evident in the case of the agricultural policy community in the UK and the United States.

The agricultural policy community

Agricultural policy-making in post-war Britain and the United States has been described by Smith (1993) as 'the paradigm case of a closed policy community' (p. 101). In the United States the policy community consisted of the USDA, the congressional agriculture committees and the three main farmers' associations – the Farm Bureau, the NFU and the Grange. This close-knit group effectively controlled agricultural policy from the 1930s to the 1970s, overseeing the rise of productivism (see Chapter 4). At the same time they excluded from the policy-making process smaller farm unions, environmental and consumer groups, non-agricultural members of Congress, and even the Bureau of the Budget and the White House. The participants in the policy community shared a common ideology that shaped the direction of policy, and were dependent on each other for resources. The USDA had the capacity to deliver the outcomes desired by the farm unions, but relied on the consultative structures of the unions. The unions in turn provided electoral support to the members of Congress who supported their agenda (Smith, 1993).

In the UK, a similar 'primary' policy community comprising the Ministry of Agriculture, Fisheries and Food (MAFF) and the National Farmers' Union (NFU) was intimately involved in day-to-day policy-making, but also drew in members of a 'secondary community', including the Country Landowners' Association, food processors and the farm workers' union, for consultation on specific issues. Excluded completely, however, were environmental, consumer and animal welfare groups. As in the United States, the British agricultural policy community shared a common ideological commitment to productivism and represented an interdependent relationship in which MAFF provided a single decision-making centre and the NFU provided the support of farmers in delivering policy outcomes (Smith, 1992, 1993). After the UK joined the European Community (now the EU) in 1973, a further dimension was added through adherence to the Common Agricultural Policy. The established policy community continued to control agricultural policy within the UK but also had influence through their representatives in the policy community that operated at a European level, involving the Directorate-General for Agriculture (known as DG VI) and COPA, the European's farmers' organization (Smith, 1993).

The products of the agricultural policy communities included state intervention in agriculture, the unquestioned expansion of farm production and the prioritizing of agriculture over consumer and environmental interests. Yet, they also arguably included overproduction, environmental degradation and food quality scares, all of which were allowed to happen because the policy community rationalized that there was no need to listen to dissident voices. As Smith observed,

> By excluding groups which disagree with agricultural policy, the community is able to say that there is a consensus on agricultural policy. Consequently, the consensus demonstrates that there is only one possible agricultural policy, and hence that there is no need for consumer representation, as the community ensures that agricultural policy is in their interests. (Smith, 1992, p. 32)

From policy community to issue network

The exclusivity of the policy communities became their fatal weakness as the absence of non-agricultural rural interests limited their ability to adapt to the changing character of the countryside. As the economic significance of agriculture declined, and that of tourism and the service sector increased, so the privileging of farming over other economic interests became less justifiable. The 1980s farm crisis and later problems of falling farm incomes created tensions within the agricultural community, bringing the legitimacy of the farmers' unions involved in the policy

community into question. Meanwhile, the rise of neo-liberal governments in the 1980s created a political climate that was unsympathetic to the emphasis on state intervention that was a core tenet of the policy community ideology.

The US agricultural policy community began to disintegrate into an issue network in the 1970s, as tensions emerged between the participating farm unions, and as the White House – an outsider in this context – began to exert more strongly its own agricultural agenda. In the UK, Smith (1989) argues that the 1980s saw a turn back towards pluralist policy-making as the assumptions of the policy community were questioned by both the governing neo-liberal Conservative party and the increasingly vocal environmental lobby. Certainly, the production of the Rural White Papers in 1995–6 marked a clear break with the closed policy community. A consultation exercise elicited some 380 responses from interest groups and individuals (Table 9.2), whilst a further 1,300 people contributed to an associated debate on rural policy run by a television programme. However, it would be erroneous to conclude that rural policy in the UK is now pluralist. The government's handling of the 2001 foot and mouth outbreak, for example, demonstrated the continuing influence of the National Farmers' Union. Thus, in the UK, as in the US, rural policy-making might be regarded more as an issue network, in which a large range of different groups may participate, but where some groups continue to have greater access to policy-makers than others do.

The Challenges for Rural Policy

The trends towards a more integrated rural policy and towards the involvement of more open networks in policy-making are both signs in themselves of the challenges confronting governments in adopting appropriate

Table 9.2 Responses received in the consultation exercise for the 1995
Rural White Paper for England, by category of respondent

Category	Number	Examples
Conservation bodies	51	CPRE, National Trust, Royal Society for the Protection of Birds
Businesses	50	Booker Countryside Ltd
Individuals	49	
Local government	47	
Academic institutions	26	Centre for Rural Economy
Professional bodies	26	NFU, Country Landowners' Association, Small Farmers' Association, Institute of Chartered Foresters
Voluntary organizations	26	Women's Institute, Rural Voice
Government agencies	20	Rural Development Commission, English Nature, National Rivers Authority
Others	57	

Source: DoE/MAFF, 1995

responses to rural restructuring. By dismantling the previous segmentation of rural policy and bringing a wider range of actors into the policy-making process, governments have acknowledged the need for new thinking, and even for fundamental changes in the way in which rural areas are governed and regulated. The challenges that confront contemporary rural policy-makers are diverse and numerous, and include issues concerning the regeneration of rural economies, the conservation of rural environments, support for rural communities and community services, and the alleviation of rural poverty and deprivation. The policy responses adopted to many of these challenges are explored in subsequent chapters (see Chapters 10, 11, 13, 14). The remainder of this chapter, however, focuses on the key issue of reforming agricultural policy, and in doing so considers the ways in which national political contexts can lead to the adoption of different strategies in response to the same broader problems.

The background to agricultural policy reform has been discussed at length in Chapter 4. As detailed in that chapter, the productivist policies that had guided agriculture through the mid-part of the twentieth century had resulted in a situation where more agricultural goods were being produced than could be sold at appropriate market prices in the developed world. Furthermore, the squeeze on agricultural prices, combined with other factors such as heavy farm debts from investment in modernization, has led to periods of severe economic depression in agriculture, including the 'farm crises' in the United States in the mid-1980s and late 1990s, in New Zealand in the 1980s, and in the UK from the late 1990s. The key challenges facing government with respect to agriculture are therefore:

1 To find new markets for agricultural products, including export markets, whilst protecting the domestic market for home-produced goods.
2 To reduce government expenditure on agriculture whilst protecting the viability of small and marginal farms that have been dependent on subsidies.
3 To balance the economic interests of agriculture with environmental and consumer concerns.

The regulation of global trade in agricultural products is a particularly tricky issue in the resolution of these questions. For major exporters such as Australia and New Zealand, the expansion of free trade would provide additional markets for their produce and support the economic health of their farms. In Europe and the United States, however, the picture is more complex. Some large farmers and agri-food companies would benefit from more free trade, but it could also pose a threat to smaller farmers who are more dependent on domestic markets and who can often not compete effectively with exports without government assistance. With this concern in mind, the political strength of the agricultural lobbies in Europe and the United States means that these governments have tended to adopt a much more cautious attitude to the deregulation of agricultural trade than those of Australia and New Zealand. As such, the adoption of three distinctive strategies for agricultural reform can be identified in New Zealand, the European Union and the United States respectively.

New Zealand: deregulation

Agriculture is a major contributor to the New Zealand economy, accounting for some 57 per cent of the nation's exports in the mid-1980s (Cloke, 1989b). In the post-war period, the agricultural export industry was supported by substantial state intervention in farming, including subsidies to encourage increased production and state-owned marketing boards. Yet, in the early 1980s New Zealand farming faced similar problems to those experienced by agriculture elsewhere in the developed world and therefore faced similar pressures for reform to agricultural policy. The response in New Zealand, however, was dictated by a sharp shift in the prevailing national political ideology that followed from the election of David Lange's Labour

government in 1984. Lange and his finance minister, Roger Douglas, introduced a series of neo-liberal policies aimed at restructuring the role of the state, similar to those followed by the 'New Right' governments of Reagan in the United States and Thatcher in the UK. Differently, however, Lange's reforms were also applied to agriculture. Price supports and subsidies for fertilizers, pesticides, water supply and irrigation were withdrawn or reduced; tax credits and concessions were ended; and subsidised farm interest rates increased to commercial levels (Cloke, 1989b; Cloke and Le Heron, 1994; Le Heron, 1993). In part, these reforms were driven by a broader concern to reduce the national debt, as the NZ$2.5 billion spent on farm support between 1980 and 1985 had been largely financed by overseas borrowing (Cloke and Le Heron, 1994).

Notably, the reform package received broad support from the agricultural lobby, although the Federated Farmers Union was divided internally, with smaller farmers concerned at the effect of the loss of subsidies. To some extent these concerns were realized, as the immediate results of deregulation included a fall in farm incomes, increased farm debt, changes in farm practice and the folding of some farm businesses. However, the rate of closure of less economic farms was slower than that of up to 8,000–10,000 per year predicted at the time (Cloke, 1989b) and supporters have argued that the reforms strengthened the ability of New Zealand's agricultural exporters to compete in global markets. Yet, New Zealand agriculture has not escaped the economic pressures faced by farming internationally in the 1990s, and as Le Heron and Roche (1999) note the proclaimed 'deregulation' of the 1980s has in fact turned out to be more of a 'reregulation' to meet the demands of the export market, global agri-food corporations and purchasing supermarkets.

European Union: diversification

Agricultural policy reform in the EU has been driven by the triple concerns of over-production, environmental degradation and the financial cost of the Common Agricultural Policy (see Chapter 4). Although periodic reforms have been attempted since the early 1980s, the pressure for fundamental change became more acute as the enlargement of the EU into Eastern Europe neared. The ten new members who joined in 2004 have substantial agricultural populations and to include them within the CAP on the existing terms would significantly increase EU spending on farm support. The EU's policy-making structure requires the unanimous agreement of member states on major reforms; thus, whilst some reformers have called for New Zealand-style deregulation, this is blocked as being politically unacceptable in countries such as France and Ireland, with many small farmers who benefit significantly from EU subsidies. The compromise, therefore, has been to use EU funds to encourage the diversification both of individual farm businesses and of the wider rural economy. This has involved two key elements.

First, *modulation* progressively caps direct payments to larger farmers, in theory making available funds that can be directed towards rural development initiatives (Lowe et al., 2002). Following the 2003 agreement on CAP reform, payments to larger farmers will be reduced by 5 per cent by 2007. Secondly, *decoupling* ends the connection between farm subsidies and production, such that under the 2003 reforms farms will receive a single payment that is based on previous receipts and is not linked to production. However, France negotiated concessions in the 2003 agreement and CAP reform continues to be a highly contentious political issue, with many critics arguing that the reforms have not yet gone far enough.

United States: protectionism

Agricultural policy in the United States is regularly reviewed on a five-yearly basis with the passing of new Farm Bills (Dixon and Hapke, 2003). In theory, this should present a greater opportunity for reform in the US than exists in many other countries and the advent of the 2002 Farm Bill (formally known as the Farm Security and Rural Investment Act) was greeted with speculation that significant reforms could be introduced to a policy that was still essentially productivist with a heavy emphasis on subsidies and price supports. The optimism stemmed in part from the increasing pressure on the United States (and the EU) to reduce subsidies and remove tariffs as part of WTO negotiations on an agreement on free trade in agriculture. However, international pressures were in the event trumped by domestic political pressure from two directions. First, agri-business interests exercised influence through their close connections with the Republican administration. Secondly, the mainstream farm groups lobbied members of Congress through their rural constituents. With the 2002 congressional elections replicating the knife-edge result of the 2000 presidential election, rural areas had assumed a greater electoral significance and legislators were mindful of the potential consequences of reforming subsidies or tariffs for their agricultural constituents (see Box 9.3).

Box 9.3   *The local impact of policy reform – Louisiana sugar*

Sugar cane has been farmed in Louisiana since the eighteenth century and the industry currently employs some 27,000 people across 25 parishes in the south of the state. In 2002, sugar growers earned around 46 cents per kilogram (21c per lb) for their produce – more than double the world market rate, and kept artificially high by a price-support policy. Under this policy, the US government controls demand by a quota system for domestic production (Louisiana's allocation was 1.4 million tons in 2004), and limits on imports, which are also subject to tariffs. The government also intervenes to buy sugar from farmers at 40 cents per kilogram (18 per lb), when the market price falls below this rate. However, under NAFTA, tariffs on Mexican sugar are due to be phased out by 2008. When this happens, surplus stocks of Mexican sugar that are currently traded on the world market are likely to be exported instead to the US, where they could earn more whilst still undercutting domestically produced sugar. As such, imports from Mexico are predicted to account for 16 per cent of US sugar consumption by 2011, and the price of US sugar is forecast to halve by 2012. The impact on areas such as southern Louisiana, where sugar growing is the main economic activity, could be severe, leading to the closure of farms and processing plants, unemployment and out-migration. As such, sugar was a major issue in the 2002 US Senate election in Louisiana.

Source: John M. Biers (2003) Bittersweet future. The Times-Picayune, 9 March, pages F1-2.

The outcome of the 2002 Farm Bill was hence to further entrench subsidies, with support for agricultural production actually increasing. Taken together with the continued imposition of tariffs on selected imported foods (including the 100 per cent increase in the tariff on Roquefort cheese in 1999 that provoked Bové's protest against McDonald's in France – see Chapter 3), the Farm Bill suggests that the United States has followed a policy of protectionism in response to the problems of agriculture. Yet, the agricultural interests protected are those of agri-business. Sixty per cent of the farm payments will go to just 10 per cent of farms, and family farmer groups had lobbied for more substantial reform. At an international level, Sumner (2003) argues that American protectionism will make an agreement in the

Table 9.3 Subsidies as a percentage of agricultural turnover, late 1990s

Switzerland	76
Japan	69
European Union	42
United States	16
Australia	9
New Zealand	3

Source: The Guardian, 26 November 1999

WTO more difficult as US negotiators will have less room for manoeuvre, developing nations may follow the US's example and adopt protectionism, and the attentions of the pro-free trade Cairns Group will continue to be focused on the US rather than on states with higher subsidy levels such as the members of the EU, Switzerland and Japan (Table 9.3).

Summary

The social and economic restructuring of the contemporary countryside has required changes in the rural policies of governments. These include not just reforms to the way in which agriculture is supported and regulated but also, as subsequent chapters will illustrate, new strategies for rural economic development and new initiatives to conserve the rural environment, among others. As part of this policy review, the compartmentalizing of policies for agriculture, conservation, land use planning, rural development and so on, that had been a feature of rural government in many countries during the twentieth century, has begun to be dismantled in favour of a new emphasis on integrated rural policy. At the same time, the closed policy communities that controlled these segmented policy fields have disintegrated, with a far wider range of interest groups now being involved in rural policy-making as part of more open, but also less stable and less coherent, issue networks. This has in turn opened the door for radical, fundamental policy reforms to be considered, and governments and interest groups alike are learning from each other as part of an international policy transfer of ideas. Yet, as the case of agricultural policy reform demonstrates, even if governments in different countries face similar problems and share similar ideological positions, the responses they adopt may be very different as domestic political considerations intervene.

Further Reading

Accounts of the development of rural policy in the United States and the UK respectively are provided by William P. Browne, *The Failure of National Rural Policy: Institutions and Interests* (Georgetown University Press, 2001), and Michael Winter, *Rural Politics* (Routledge, 1996). Winter's book also discusses the different models of policy-making. The agricultural policy communities in the UK and the United States are examined in detail by Martin J. Smith, *Pressure, Power and Policy* (Harvester Wheatsheaf, 1993). A background to the contemporary debates on agricultural policy reform is provided by Richard Le Heron, *Globalized Agriculture* (Pergamon, 1993), who examines the deregulation of agriculture in New Zealand and issues in agricultural policy in the United States and the European Union, as they stood in the early 1990s.

Websites

Information about the 2002 US Farm Bill can be found on the USDA website, including the full text of the Act (www.usda.gov/farmbill/index.html). A detailed but critical summary and commentary is provided by the Rural Coalition, a pressure group that had campaigned for more radical reform (www.ruralco.org/html2/farmbillreport.html). Details of the reforms to the European Union's Common Agricultural Policy agreed in 2003 are available on the website of the EU's Directorate-General for Agriculture (europa.eu.int/comm/agriculture/mtr/index_en.html). A brief summary is also available from the UK Department of the Environment, Food and Rural Affairs (www.defra.gov.uk/farm/capreform/agreement-summary.htm).

10

Rural Development and Regeneration

Introduction

Governments take an interest in the economic development of rural areas for a number of possible reasons. First, there is the welfarist rationale that the state has a duty to support basic levels of social well-being and to promote equity between its citizens. The state therefore intervenes to improve the living conditions of people in rural areas and invests in infrastructure to provide public services. It also acts to stimulate economic development when the decline or withdrawal of established economic activities produces significant unemployment or poverty. Secondly, there is an economic rationale that the capitalist state operates to support business in the accumulation of capital. This might involve providing infrastructure that allows businesses to develop in rural regions and to exploit rural resources, and absorbing risk through, for example, providing low interest loans to businesses and involvement in training the workforce. Thirdly, there is a 'stewardship' rationale that the state acts in the interest of society as a whole to ensure that rural land and resources are properly maintained and wisely used. Finally, there is a rationale of the spatial control of the population. The depopulation of rural areas in the early twentieth century demonstrated that the natural reaction to economic depression is for people to move to the places where the jobs are. However, large-scale population movement of this kind creates instability and demands on the state for a reconfiguration in the provision of public services. From a managerial perspective, it is better for the state to invest in economic development in depressed regions, hence reducing the 'push-factors' for out-migration.

This last rationale conflates *rural development* and *regional development*. Government support for the economy in rural areas can involve both, but they are different with different objectives and cast at different scales. For example, the European Union's Structural Funds include both support for *rural development* through schemes such as LEADER (see Box 10.2 later in this chapter), aimed at helping the readjustment of rural economies following the decline of agriculture, and support for *regional development* through schemes such as the Objective 1 programme, aimed at increasing the GDP of the poorest regions of the EU, for which a number of rural regions qualify (again see section below).

A further terminological distinction can be made between 'development' and 'regeneration'. Although the terms tend to be used interchangeably, they in fact imply distinct processes. 'Development' suggests a process of progressive change or modernization. Thus, the provision of electricity to rural parts of the United States, for example, was clearly a rural development project. 'Regeneration', on the other hand, suggests a more cyclical process – that there has been a buoyant economy that has fallen into decline and requires remedial action to return it to its previous condition. Initiatives aimed at reversing the decline of rural small towns, or at replacing lost jobs in agriculture or manufacturing, might therefore be more properly described as strategies for rural regeneration. Moreover, it can also be argued that the distinction between 'development' and 'regeneration' corresponds with a 'paradigm shift' in policy that has replaced an emphasis on 'top-down' rural development, characterized by large, state-led, infrastructure projects, with an emphasis on 'bottom-up' rural regeneration, characterized by small, community-led initiatives drawing on indigenous resources. This chapter explores both sides of this transition, first briefly discussing state-led top-down development before more extensively examining the bottom-up regeneration of rural economies.

State Intervention and Top-down Development

The state has a long history of intervention in rural development. In North America, Australia and New Zealand, state involvement in the creation of communications links and other infrastructure that supported the European settlement of rural regions were *de facto* exercises in rural development, often associated with the facilitation of resource capitalism. Similarly, the establishment of the Land Grant Colleges in the United States in the 1860s was part of an early rural development strategy based on agriculture. Meanwhile, in Europe, the UK government set up a Rural Development Commission as early as 1910, with an initial remit of supporting the development of small rural industries. It is therefore unsurprising that when governments were confronted with rural economic change in the wake of decreasing farm employment, they chose to rationalize the problem as being a need for the modernization of rural areas and selected state investment in infrastructure as the primary vehicle for delivery.

One of the earliest and largest projects of this kind was the Tennessee Valley Authority (TVA) in the south-western United States. Launched in 1933 as part of President Roosevelt's 'New Deal' to counter the economic depression, the TVA involved the construction of a series of nine dams along 1045 km (650 miles) of the Tennessee River between Knoxville, Tennessee and Paducah, Kentucky, along with eight power plants, two chemical plants and eleven dams on tributary rivers (Martin, 1956). As well as providing a system for flood control, the TVA project aimed to stimulate economic development in three main ways. First, the power generated by the TVA plants provided energy to support the industrialization of the region. Second, the chemical plants produced nitrate fertilizers to assist agricultural modernization. Third, the construction projects and the

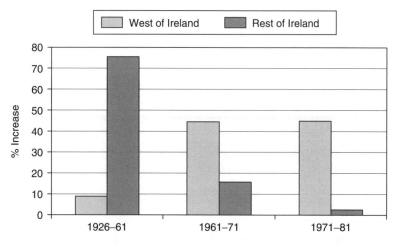

Figure 10.1   Increase in manufacturing employment in the Republic of Ireland, 1926–1981
Source: After Robinson, 1990

management of the scheme created a significant number of new jobs in themselves.

These outcomes had a significant impact in a region where in many counties over 50 per cent of families were receiving welfare payments in 1933. Some 170,000 people in the region who were unemployed in 1940 had been re-employed by 1950, with over 20,000 new jobs created in the chemical industry between 1939 and 1947, and 12,000 in the primary metals industry (Martin, 1956). Overall, the TVA project resulted in a shift in the employment structure of the region from agriculture to industry, trade and services. Yet, as Lapping et al. note, the success of the TVA as a strategy for rural development was mixed:

> The cities grew, and the rural areas provided the necessary labor and natural resources for further urban development. TVA, then, actualized 'growth pole' development theories, wherein development was concentrated in the larger communities and the rural areas became organized around these development poles or centers. Opportunities and wealth supposedly 'trickled down' from the cities into the hinterland, and in this manner rural incomes should rise and the quality of life should be enhanced. Although this did take place (at least to some degree), the region as a whole remained poor and subject to further environmental degradation and economic dislocation. (Lapping et al., 1989, pp. 32–33)

Manufacturing industry also formed a cornerstone of top-down rural development in Europe. In the Republic of Ireland, for instance, the Industrial Development Authority (IDA), set up in 1949, operated a strategy of site acquisition and advance factory building to stimulate the development of manufacturing industry in rural western Ireland. Between 1972 and 1981, around half of the advance factory floor space built in Ireland was located in 11 western rural counties, particularly in the 'growth poles' of Galway and Shannon Airport (Robinson, 1990). In consequence, during the 1960s and 1970s, manufacturing employment in the west of Ireland increased at a rate – around 45 per cent – that was substantially higher than the rest of Ireland (Figure 10.1).

A similar strategy was followed in the UK by the Development Board for Rural Wales (DBRW) (formed in 1976, replacing a predecessor established in 1957), and the Highland and Islands Development Board (HIDB) (formed in 1965), two state-controlled agencies responsible for promoting economic development in the peripheral rural areas of Wales and Scotland respectively. Both bodies purchased land for industrial estates and constructed advance factory units to attract manufacturing investment and, as in Ireland and the Tennessee Valley, much of the resulting economic growth was concentrated in 'growth poles', such as Newtown in Wales and Inverness and Fort William in Scotland, whilst other parts of the region continued to experience problems of economic decline and out-migration. To some extent, DBRW and HIDB attempted to compensate for this effect by providing grant support for smaller-scale rural enterprises in tourism, fishing and craft industries, but these activities essentially remained within a paradigm of top-down intervention rather than stimulating bottom-up initiatives (Robinson, 1990).

The strategy of top-down state intervention in rural development did have numerous successes, including the creation of millions of new jobs to replace those lost as part of agricultural modernization; the slowing down of, and in some regions, reversal of out-migration from rural areas; the improvement of communications and utilities infrastructure; and the relative prosperity brought to those towns selected as 'growth poles'. However, top-down rural development can also be critiqued on two key grounds.

First, top-down rural development tends to be dependent on external investment. It rarely seeks to nurture growth from within the indigenous rural economy and because external investors want a return on their investment, profits generated by the new factories and other new employers are often exported out of the locality rather than kept within the local economy. Rural areas may also be left more vulnerable to corporate decision-making informed by wider economic trends, as the companies that had been courted by the development agencies may subsequently close their branch plants if more favourable conditions can be found elsewhere.

Secondly, the top-down nature of the strategy can create a democratic deficit. Although some programmes, such as the TVA, have involved grassroots participation, more generally the input of local people is perceived to be limited. This can mean that the developments introduced and the jobs created are not those wanted by local people. It can also create a risk of corruption, or of rural development funds being diverted to 'vanity projects' (see also Box 10.1). The European Union's Structural Funds for economic development were restructured in the late 1980s for precisely this reason, following concerns that the funds 'were actually financing "the wrong actors"' (Smith, 1998, p. 227).

Box 10.1   Rural development in Japan

One of the key concerns of rural development policy in Japan has been that the increasingly urbanized population is losing touch with its rural roots. To counter this, the Japanese government in the late 1980s gave every town and village council a lump sum of 100 million yen (approximately £0.5 million or US $0.8 million) which they were instructed to use 'creatively' on projects to revitalize the 'hometown' spirit of rural Japan. This was in essence a top-down rural development strategy, and resulted in

---

Box 10.1   (Continued)

numerous expensive vanity projects. In Tsuna on the island of Awaji, the mayor spent the money on a 63 kg (139 lb) ingot of gold – at the time the largest block of solid gold in the world – to be displayed in the town as a tourist attraction. Another town cashed the money and created a pyramid of banknotes; others built theme parks and some paid for holidays for residents. The strategy was modified in 2001 with a proposal for a 'village restoration' programme in which 600 billion yen would be used to support projects to promote exchanges between rural and urban residents.

Source: Jonathan Watts (2001) Rural Japan braced for new riches. Guardian, 27 September, p. 19.

Bottom-up Rural Regeneration

The shift to a bottom-up approach in rural development has involved both a change in the way in which rural development is managed, and a change in the type of activities that are promoted through development initiatives. In contrast to the state-led management of top-down strategies, bottom-up rural development is led by the local communities themselves. Communities are encouraged to assess the problems that they face, to identify appropriate solutions, and to design and implement regeneration projects. They usually have to apply to draw down public funds for projects, often as part of a competition, and are frequently required to piece together resources from different sources through partnership working (see Chapter 11 for further discussion). As such, the role of the state changes from being the provider of rural development to being the facilitator of rural regeneration (Edwards, 1998; Moseley, 2003).

Similarly, the form and focus of rural development also changes. In most cases, the emphasis is no longer on attracting external investment, but rather on enhancing and exploiting local endogenous resources – also known as *endogenous development* (Ray, 1997). In many cases, the immediate focus of a project may not be economic development, but community development – aiming to build the capacity of the community to regenerate its own economy. Community development is seen as a necessary component of rural development, not just so that communities may be able to take responsibility for regeneration without relying on the state, but also in order that economic development does not lead to social polarization within rural localities (Edwards, 1998; Lapping et al., 1989; Moseley, 2003).

Significantly, the bottom-up approach has received support from both rural development professionals and neo-liberal politicians seeking to restructure the state. For the former, the bottom-up approach means the empowerment of local communities and the development of regeneration strategies that are in tune with local needs and the local environment. For the latter, the bottom-up approach means that responsibility for rural development is shifted from the state to its citizens, in line with the broader-scale 'rolling back of the state' from areas of activity, and that the state can reduce its expenditure on rural development.

EU rural development and endogenous development

The most extensive programme of rural development is that operated by the European Union through its Structural Funds. As Box 10.2 shows, EU support for rural development is delivered through two different mechanisms. First, a number of rural regions qualify for assistance under the objectives of the EU's regional policy (Figure 10.2). These include 'least favoured' regions funded under Objective 1, such as southern Italy, western Ireland, Cornwall and west Wales in the UK, and large parts of rural Scandinavia; and 'converting' regions funded under Objective 2, including large parts of rural France, Italy and England (additionally, some regions that no longer qualify for full level assistance receive 'transitional support', including parts of rural Scotland and rural Ireland). Funding under both objectives is also available to urban regions and rural areas account for around 29 per cent of population covered by Objective 2.

Box 10.2   EU programmes for rural and regional development

Regional development

Support for regional development is provided under the three objectives of the Structural Funds:

- *Objective 1: to develop the least favoured regions of the European Union.* To qualify, regions must have a GDP of 75 per cent or less of the EU average, or a population density of fewer than eight people per square kilometre (predominantly in Scandinavia). Regions eligible for Objective 1 funding in the 2000–06 round include 22 per cent of the EU's population and will receive assistance from a total fund of over 135 billion euro.
- *Objective 2: to revitalize areas facing structural difficulties, including industrial, rural, urban and fisheries-dependent regions.* To qualify, rural areas must have a population density of fewer than 100 inhabitants per square kilometre *or* a proportion of the workforce employed in agriculture that is at least double the EU average, *and* an unemployment rate higher than the EU average *or* a declining population. Regions eligible for Objective 2 funding in the 2000–06 round include 18 per cent of the EU's population and are supported by a total fund of 20 billion euro.
- *Objective 3: supports education, training and employment.* Objective 3 is not territorially limited, except that Objective 1 regions are not included.

Prior to 2000, a specific objective of the structural funds (Objective 5b), supported restructuring rural regions. This was absorbed into the new Objective 2 for the 2000–06 round of funding.

Rural development

LEADER + (Liaison entre actions de développement de l'économie rurale) – one of four 'Community Initiatives' (the others being INTERREG, EQUAL and URBAN). LEADER is specifically aimed at supporting rural development. Its three 'actions' include support for bottom-up territorial development, support for inter-territorial and transnational cooperation, and the networking of all rural areas.

Originally established in 1991, LEADER is now in its third incarnation (LEADER+) and is delivered through local action groups which may be established in any rural area. A total of 2,105 million euro are available from the European Agricultural Guarantee and Guidance Fund (EAGGF) for 2000–06, which is expected to be match-funded by 2,941 million euro from other public and private sources.

For more information see europa.eu.int/comm/regional_policy/index_en.htm (Objective 1 and Objective 2) and europa.eu.int/comm/agriculture/rur/leaderplus/index_en.htm (LEADER+).

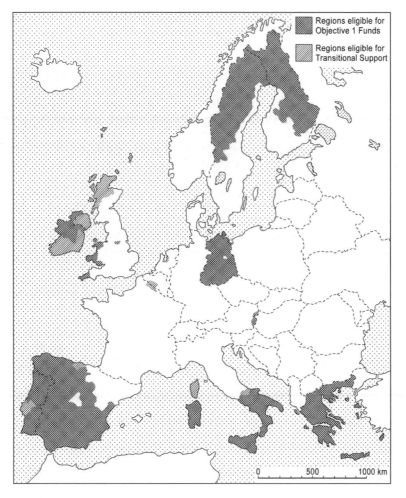

Figure 10.2 Regions receiving EU funding under Objective 1 of the Structural
Funds, 2000–06
Source: Based on information from the European Commission

Second, rural development is also supported by the LEADER community initiative, which falls within the remit of the Directorate-General for Agriculture. Now in its third incarnation, the LEADER programme is delivered through 938 local action groups in the EU's 25 member states. Whereas the regional policy programmes still retain an element of top-down development (for example by financing transport infrastructure), LEADER is firmly rooted in the bottom-up approach. Over 86 per cent of expenditure in the programme is directed at 'integrated territorial development strategies of a pilot nature based on a bottom-up approach' (European Union, 2003). The principles that guide the LEADER programme (and EU rural development policy more broadly) were articulated in the Cork Declaration issued by the participants in the European Conference on Rural Development held in the Irish Republic in 1996 (Box 10.3), which provides a clear statement of the philosophy of bottom-up rural regeneration.

151

Box 10.3   The Cork Declaration

Support for diversification of economic and social activity must focus on providing the framework for self-sustaining private and community-based initiatives: investment, technical assistance, business services, adequate infrastructure, education, training, integrating advances in information technology, strengthening the role of small towns as integral parts of rural areas and key development factors, and promoting the development of viable rural communities and renewal of villages ... Policies should promote rural development which sustains the quality and amenity of Europe's rural landscape (natural resources, biodiversity and cultural identity), so that their use by today's generation does not prejudice the options for future generations ... Given the diversity of the Union's rural areas, rural development policy must follow the principle of subsidiarity. It must be as decentralized as possible and based on partnership and cooperation between all levels concerned (local, regional, national and European). The emphasis must be on participation and a 'bottom-up' approach, which harnesses the creativity and solidarity of rural communities. Rural development must be local and community-driven within a coherent European framework.

Extract from The Cork Declaration: A Living Countryside, issued by the European Conference on Rural Development, November 1996.

Table 10.1   Main concerns of LEADER I groups

	No. of groups
Promoting rural tourism	71
Training and human development	40
Adding value to agricultural production	38
Supporting small firms and craft industries	34
Developing a more balanced portfolio	34

Source: After Moseley, 1995

Ray (2000) describes the LEADER programme as a 'laboratory' for endogenous rural development, as each LEADER group was intended 'to search for innovative ideas that not only would assist socio-economic viability in the locality but also serve a demonstrative function for other participating territories' (p. 166). By searching for innovation in rural development through grassroots experimentation, the LEADER programme, argues Ray, encapsulated the core principles of endogenous development – community-driven, territorially focused development that maximizes the 'retention of benefits within the local territory by valorizing and exploiting local resources – physical and human' (p. 166). The consistency with which this vision was delivered in practice by local LEADER groups has been questioned by empirical studies. In Germany, for example, LEADER has been described as a conservative force, moderating more radical rural development ideas (Bruckmeier, 2000); whilst Storey (1999) raises concerns about the extent of local participation in LEADER schemes in Ireland. None the less, in general the type of projects supported by LEADER do suggest a qualitative shift in the nature of rural development, with many focused on capacity building and product valorization (Table 10.1) (the same applies to projects funded by EU regional development programmes, see Ward and McNicholas, 1998). As Box 10.4 illustrates, many projects have a strong environmental component and as such contribute not just to endogenous development but also to sustainable development (see also Moseley, 1995).

Box 10.4   Examples of projects supported by LEADER

Garfagnana, Italy: Introduction of green forestry engineering techniques based on the use of endogenous resources and natural materials to help revitalize local forestry cooperatives. Some 120 new jobs were created in forestry cooperatives in the region between 1995 and 1999.

Waterford, Republic of Ireland: Use of constructed wetlands as a means of cleansing dirty water from farmyards. The lagoons were planted with vegetation and stocked with fish and are intended to form a tourist attraction as well as helping to reduce pollution.

Les Combrailles, France: Development of a housing scheme to match new residents in nearby growing employment areas with empty properties in the Pays de Menat, helping to restrict demands for new development and to renovate vacant and abandoned buildings.

Carmarthenshire, UK: Promotion of tourism through the development of information boards and literature on the theme of the Land of History and Legend, with input from local people.

Source: LEADER II Magazine, various issues 1999–2001.

Food tourism and farmers' markets

A common theme of endogenous rural development is to 'add value' to existing rural landscapes, environments and products. This may involve the 'repackaging' of a rural locality to attract tourists, perhaps by emphasizing local traditions and heritage, as is discussed in more detail in Chapter 12. It can also involve a new approach to agriculture, seeking to promote economic development not through agricultural modernization but rather through an emphasis on traditional food products and on direct sales by farmers and local producers. As Bessière (1998) observes, regional food and gastronomy have become important elements in rural tourism. Rural areas market themselves to tourists through their speciality food products, drawing on classifications such as the *appellation d'origine contrôlée* system in France to define themselves, and with marketing initiatives supported by rural development funds. Moreover, the sites of local food production, such as farms, dairies, cheesemakers,

vineyards and breweries, are additionally marketed as tourist attractions, creating a second income stream (Figure 10.3).

Farmers' markets are an increasingly commonplace component of endogenous rural development, in part because they contribute towards regeneration on three levels. Not only do they contribute to food tourism, but they also help to support locally based small-scale food processing and can increase incomes for farmers by removing the commission of wholesalers and retailers. Based on the model of the 6,000 weekly markets in France, there are now around 3,000 farmers' markets in the United States, whilst in the UK the number of markets has expanded rapidly from the first, in Bath in 1997, to 200 in 2000 and 450 in 2002. Consumer spending at US farmers' markets exceeds $1 billion per year, and in the UK amounted to £166 million in 2001/2 (Holloway and Kneafsey, 2000; NFU, 2002). As Holloway and Kneafsey (2000) describe, the markets appeal to ideas of localism, quality,

Figure 10.3 Tourist leaflet promoting
visits to local food producers in west Wales
Source: Woods, private collection

Holloway and Kneafsey's case study of Stratford Farmers' Market in England suggests that there is a significant degree of turnover in the stalls present at the market from month to month. Secondly, farmers' markets favour certain products over others. Holloway and Kneafsey (2000) record that the most common purchases at Stratford were of vegetables, eggs, apple juice, cheese and honey. Farmers' markets offer less of an opportunity for livestock farmers, who predominate in many more depressed rural areas. Thirdly, and related to the above, there is a clear spatial concentration in the distribution of farmers' markets. The majority of markets in the United States are located in or near metropolitan regions (Figure 10.4), including urban markets such as the Union Square Greenmarket in New York (Figure 10.5), at which producers from neighbouring rural districts sell to city residents. In contrast, the more rural states of Montana and Georgia had just seven farmers' markets apiece in 1998, and Wyoming had six. Similarly, the growth of farmers' markets in the UK has occurred primarily in southern England rather than in the more peripheral rural regions of Wales, Scotland and the north of England. Thus, farmers' markets provide an illustration that, whilst the transfer of ideas is important within bottom-up rural regeneration, endogenous development also means finding appropriate solutions to local problems and recognizing that the same strategies will not work everywhere.

Small Town Regeneration

The regeneration of small towns is a distinct and important challenge in rural development. Small towns are key nodes in the rural economy. In addition to their historic role as service centres for rural districts, the shift of rural employment from land-based activity to industrial and service sector work has made small towns the main sites of employment in rural

authenticity and community, and thus can be simultaneously read as 'alternative spaces' that are challenging the dominance of supermarkets and global agri-food corporations, and 'reactionary or nostalgic spaces' that represent a notion of the rural idyll.

However, three caveats must be attached to the apparent success of farmers' markets. First,

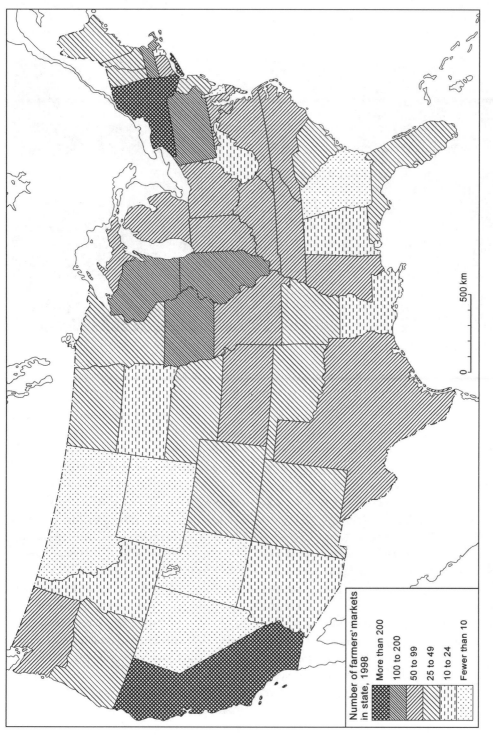

Figure 10.4 Location of farmers' markets in the United States, 1998
Source: Based on Price and Harris, 2000

Number of farmers' markets
in state, 1998

More than 200

100 to 200

50 to 99

25 to 49

10 to 24

Fewer than 10

500 km

0

Figure 10.5 The Union Square Greenmarket in New York City
Source: Woods, private collection

labour markets (see Chapter 5). They have also absorbed a disproportionate share of population growth and new housing development, and are the site of social investment in the countryside (including public services such as education and health, but also support for arts and cultural facilities and events) (Edwards et al., 2003). Yet, many small towns have also experienced considerable economic problems with the closure of manufacturing plants and other traditional employers, and many are less prosperous than their neighbouring rural communities. Shops and services may also have been lost to out-of-town commercial centres or to nearby larger towns and cities (see Chapter 7). As such, a small town may need to be regenerated in itself, but there is also perceived to be a 'trickle-out' effect through which the regeneration of small towns can boost the economy of a wider rural region.

This discourse was articulated in the UK government's 'Rural White Paper' policy document for England in 2000, which devoted an entire chapter to the regeneration of small, or 'market', towns. As the document declared, 'market towns play a critical role in helping rural communities to thrive and in regenerating deprived areas' (MAFF/DETR, 2000, p. 74). Proposals to help small towns become the focus for economic development and centres for meeting people's service needs were subsequently carried forward through a bottom-up approach in the Countryside Agency's 'Market Town Healthcheck' initiative, in which towns are invited to evaluate their own situation and agree 'action plans' for regeneration (Edwards et al., 2003).

Successful regeneration strategies for small towns tend to rely on achieving an appropriate combination of local resources and external assistance. Kenyon and Black (2001), in a handbook for small town regeneration in Australia, argue that successful projects share a number of key ingredients, including: timing, use of community planning processes, enthusiastic local leadership, a positive belief in the town and its future, local entrepreneurship, willingness by locals to contribute financial resources, the smart use of outside resources, new community networks that actively support new ideas and a focus on local youth (see Box 10.5) (see also Herbert-Cheshire, 2003). Many of the Australian examples discussed by Kenyon and Black involve some element of direct economic investment and job creation, but, frequently, the simple refurbishment and renovation of a small town's built environment is regarded as an equal focus for regeneration schemes. The National Main Street Program in the United States supported regeneration in over 200 small towns between 1979 and 1989, including over 650 facade renovations and 600

rehabilitation projects (Lapping et al., 1989). The initiative operated through project managers employed in each of the participating towns to work with stakeholders including merchants, banks, civic groups, local government, the media and residents, in implementing the three dimensions of the programme. These were: first, the diversification of downtown areas, including the conversion of unused upper storeys for residential accommodation and offices, and the recruitment of new stores. Second, the physical renovation of main streets, and especially of historic buildings, together with the adoption of design policies for new buildings. Third, activities to promote small town centres as places to shop, work and live: 'activities may vary from distributing shopping bags with a special downtown logo to publishing a directory of downtown businesses or sponsoring special events such as craft fairs, farmers' markets, and sidewalk sales' (Lapping et al., 1989, p. 293).

Box 10.5 *Small town regeneration – Deloraine, Tasmania*

Deloraine is a small town of 2,100 residents located in the north west of the Australian island of Tasmania. Like many small towns, its economy suffered from the decline of agriculture, whilst the opening of a highway by-pass in 1990 resulted in the closure of 12 businesses. The town has also experienced conflict between local residents and in-migrants pursuing alternative lifestyles. The regeneration of the town involved a number of components, including the formation of a business centre to support local business development and training; the creation of a land-fill and recycling site; community beautification and park projects; sponsoring highway signage to attract visitors to the town; the establishment of an annual Tasmanian Craft Fair, attracting over 30,000 visitors; and a community 'Artwork in Silk' project, creating a portrait of the local area on a 57 square metre hanging that now forms a tourist attraction. These projects have drawn on external funds where appropriate, but have been initiated and led by the local community. Kenyon and Black (2001) identify the presence of a strong sense of belief and expectation, the leadership of local government and a strong focus on young people as key elements in Deloraine's regeneration. The town was named the 'Australian Community of the Year' in 1997 in recognition of its achievements.

For more see Peter Kenyon and Alan Black (eds) (2001) Small Town Renewal: Overview and Case Studies (Barton, Australia). Report for the Rural Industries Research and Development Corporation. Available at: www.rirdc.gov.au/fullreports/hcc.html

The Limits to Rural Development?

Bottom-up endogenous development may have become the fashionable approach for addressing the problems of the rural economy, but it is not free from problems. McDonagh (2001), in a study of rural development in western Ireland, points to 'the frustration of bottom-up development' (p. 128), including difficulties involved in coordinating and representing the diverse interests of local people, and concerns that the 'core groups' that drive initiatives do not necessarily represent the interests of the wider community. Issues of accountability and power, and of the extent to which different sectors of a 'community' are engaged in active participation, have been raised by a number of other commentators (see Edwards et al., 2000, 2003; Storey, 1999) and are discussed further in the next chapter.

There are also signs that the shifting of responsibility for rural development from the state to local communities is creating an uneven geography of regeneration, as some communities are better able to initiate projects or to bid competitively for government funds than others (Edwards et al., 2000; Jones and Little, 2000). Moreover, there are some rural communities that may be near impossible to regenerate once traditional economic activities have declined or disappeared. As Herbert-Cheshire (2000, 2003) observes in an examination of Australian country towns, the strategy not only raises false hopes as towns attempt to follow supposed 'blueprints' for regeneration but fail to reap the benefits, but also suggests that the responsibility for failure rests with the local community itself. It is only a small step from this logic to a rationality that justifies the withdrawal of state aid for development from certain 'uneconomic' localities. This prospect was floated by an Australian economist, Gordon Forth, in 2000, following the logic that 'many of these towns are going to go into on-going decline and the population will not only become smaller but poorer and increasingly disadvantaged' (Gearing and Beh, 2000). Although the comments provoked a fierce reaction at the time, the complete withdrawal of the state from rural development responsibilities in the least favourable regions would not be too much of a divergence from the recent trajectory of rural development policy.

Summary

A new paradigm has emerged in rural development and regeneration in the last quarter-century, that has replaced a previous emphasis on top-down, state-led, development through large-scale infrastructure projects and industrialization, with a bottom-up approach based on endogenous development. The new approach seeks to regenerate rural areas by enhancing and adding value to local resources, both physical and human, according to the priorities and preferences of the local community. As such it has been strongly advocated as both a form of empowerment of rural communities, and a step towards more sustainable economic development. However, bottom-up, or endogenous, rural development is not a panacea for all rural ills. Not all rural localities are equally able to regenerate themselves through the enhancement of their endogenous resources, and not all rural communities are equally equipped to compete successfully for external funding and support. As such, the paradigm shift in rural development can, in fact, be argued to have contributed to the production of a new geography of uneven rural development.

Further Reading

An overview of theory and practice in rural development, with European case studies, is provided by Malcolm Moseley, *Rural Development: Principles and Practice* (Sage, 2003). For more on the EU's LEADER programme see the themed issue of *Sociologia Ruralis* from April 2000, which includes papers with studies of LEADER in Italy, Spain, France, Germany and the UK. Further material on farmers' markets in the UK can be found in Lewis Holloway and Moya Kneafsey, 'Reading the spaces of the farmers' market: a case study from the United Kingdom', *Sociologia Ruralis*, volume 40, pages 285–289 (2000). The question of small town regeneration is discussed from a British perspective by Bill Edwards, Mark Goodwin and Michael Woods, 'Citizenship, community and participation in small towns: a case study of regeneration partnerships', in R. Imrie and M. Raco (eds), *Urban Renaissance: New Labour, Community and Urban Policy* (Policy Press, 2003); and from an Australian perspective by Lynda Herbert-Cheshire, 'Translating policy: power and action in Australia's country towns', *Sociologia Ruralis*, volume 43, pages 454–473 (2003).

Websites

Information about the European Union's programmes for regional development and rural development can be found on the EU's website. Details about Objectives 1 and 2 are at europa.eu.int/comm/regional_policy/index_en.htm; details about LEADER+ are at europa.eu.int/comm/agriculture/rur/leaderplus/index_en.htm. Web resources are also available for farmers' markets in the UK (www.farmersmarkets.net) and the US (www.localharvest.com/farmers-markets), and for the National Main Street Program (www.mainst.org).

11

Rural Governance

Introduction

The structures through which rural areas are governed vary between countries depending on the constitutional framework, the dominant political ideology and historical precedent. The United States and France, for instance, have strong government institutions at the community scale that enjoy considerable autonomy and authority to adopt policies that reflect local circumstances and local opinion. In contrast, local government in New Zealand and the UK has far less autonomy and far fewer responsibilities, and is much more strongly directed in its actions and policies by the national-level central state. These differences are important to students of rural societies, as they will inform decisions about at which scale to examine policy-making and responses to rural change, and also determine the extent to which responses to restructuring are developed from within rural areas or are imposed from outside.

Despite the different administrative structures, however, rural local government has in most countries been subject itself to significant change in recent decades. The nature and timing of these changes will again vary with the national context, but in broad terms, rural government has moved through a transition from a *paternalist* era in the early twentieth century, to a *statist* era in the mid-twentieth century, to a new era of *'governance'* at the turn of the twenty-first century. This transition has both reflected and been part of rural restructuring, and has important implications for the formulation and implementation of rural policy, the regulation of rural societies, economies and environments, and the distribution of power within the countryside. This chapter explores these themes by first briefly detailing the transition and then focusing on the new structure of rural governance, its characteristics and the issues that it raises.

From Paternalism to Governance

Historically, the distribution of power within rural societies has been determined by the control of resources, and particularly by the control of land. In an economy based on primary production, land ownership was the key to economic wealth, and also brought in workers and tenants, who in many cases were dependent on the landowner for both employment and accommodation. Wealth, in turn, enabled landowners to purchase other scarce resources such as transport, and allowed them time to participate in public service and government. Land and wealth also afforded status in accordance with the popular discourses that informally set the rules of rural leadership. And status, time and wealth all permitted access to the exclusive clubs, private parties and social networks through which politicking and patronage took place (Woods, 1997).

In Europe, the structure of land ownership was the legacy of feudalism. The families who owned the majority of the land and formed the leadership class in rural society at the beginning of the twentieth century were predominantly the aristocratic descendants of feudal barons, together with a number of industrial capitalists who had bought land in order to acquire the status and power that it brought. The European settlement of rural North America, Australia and New Zealand, in contrast, was supposed to be a more egalitarian endeavour. Yet, here too, the entrepreneurs who developed the mines and established the largest farms and ranches rapidly came to dominate. As Mattson (1997) observes, in frontier regions

a distinguishing feature of local leadership was the ascribed influence based on wealth and position. Because illiteracy rates were high compared to towns along the coast, men obtained positions of prominence based on their charismatic leadership skills and economic influence. Thus, large landowners, local merchants and land speculators became the magistrates. (p. 127)

Paternalism meant more than just the concentration of power with an economic elite, however. Under paternalism the elite took responsibility for discharging many of the conventional functions of government through their own private channels. It was the landowners and business owners who developed the local infrastructure and led economic development. They provided housing and employment, and the more benevolent endowed schools and hospitals, and supported local charities. The role of the state, including that of local government, was therefore limited.

By the mid-twentieth century, paternalism was becoming more difficult to sustain. The aristocratic elite in Europe began to decline in both number and wealth and started to withdraw from their role in local political leadership (Woods, 1997). Elements of paternalistic culture persisted and communities turned to large farmers to provide leadership and assume the role of 'squire' (Newby et al., 1978). However, neither the new farming elite nor the merchants and professionals who already formed the dominant elite in rural small towns, had the resources to provide the kind of private government that had characterized paternalism. Instead they exercised their power through the machinery of local government, whose offices they dominated (see Box 11.1).

Box 11.1 The changing rural power structure in Somerset, England

In 1906, the 67 members of Somerset County Council, in south-west England, included 26 landowners and at least eight farmers. Of the 22 county aldermen, 15 were signifi-cant landowners. The dominance of this largely aristocratic, land-owning elite was typical of most of rural England at the time, and rested on three sources of power: the control of resources, most notably land; the exercise of patronage and influence through an exclusive network built on kin, hunting and country house parties; and a 'discourse of the country gentleman' which positioned the gentry as the superiors of the rural population and hence as the natural leaders of rural society.

The power of the elite, however, was eroded after the First World War by death duties and recession that prompted the sale of land and led to aristocratic families withdrawing from leadership positions or even leaving the county altogether. Their place was taken by a new elite of small farmers, traders and rural community leaders such as postmasters, clergy and doctors, supported by the twin discourses of the 'agrar-ian community' and the 'organic community' that positioned farmers and visible com-munity figures respectively as the appropriate leaders of rural government. By 1935, the 74 county councillors included 12 small farmers, at least 15 councillors from com-mercial backgrounds, and a reduced group of 17 significant landowners. Farmers dom-inated parish councils and rural district councils, whilst merchants dominated town and borough councils in the county.

During the last quarter of the twentieth century, Somerset again underwent signifi-cant social and economic restructuring, including most notably the in-migration of a large middle class population with little identification with the imagined 'organic com-munity' and motivated by a very different discourse of place in their understanding of rurality. As such the new middle class residents did not feel represented by the exist-ing elites and began to compete themselves for positions in local government. In 1995, only two of the 57 members of Somerset County Council were from the 'landed gentry', four more were farmers but nine were teachers or ex-teachers and ten were or had been employed elsewhere in the public sector. The older elites, however, con-tinued to have a greater presence in appointed positions in local government, includ-ing as magistrates and on the local health authority and National Park boards. As such, the local power structure in Somerset had moved from dominance by a single, closed and exclusive elite at the start of the twentieth century to a fragmented structure with competing mini-elites at the end.

For more see Michael Woods (1997) Discourses of power and rurality: local politics in Somerset in the 20th century. Political Geography, 16, 453–478.

The state, most notably in the shape of local government, thus expanded its activity in rural areas. However, the new statist era was characterized by a fundamental contra-diction. On the one hand, the expansion of elected local government was represented as a

democratization of rural society after the elitism of paternalism. Oliveira Baptista (1995), for example, notes that the introduction of democratic local government in Portugal following the end of the Salazar dictatorship in 1974 'gave citizens the opportunity to oppose those holding economic control over the territories in the management of local areas' (p. 319), and helped to create a basis through which rural decline could be addressed. In countries such as the UK, France and the United States, this apparent democratization involved not just local government, but also the participation of elected farmer representatives on the various bodies responsible for delivering agricultural policy.

Yet, on the other hand, the statist era also involved an unprecedented degree of centralization, and reflected the need of the state to intervene in rural areas in support of capitalist economic activity. As detailed in earlier chapters (see Chapters 4, 9 and 10), this included actions to guarantee agricultural markets and prices; to absorb risk by subsidizing agricultural investment and modernization; to ensure a stable supply of energy and resources through nationalized industries; to protect agricultural land through land use development controls; to promote leisure consumption by regulating for rural leisure use; and to regulate population movement by investing in rural development. To deliver these objectives, new state bodies were created – including agricultural intervention boards, national park and forestry services, conservation authorities, state-owned utility companies and rural development agencies – that operated in rural space alongside the structure of elected local government, but usually without the democratic participation of local people.

The statist era was itself brought to a close by a combination of pressures from within and outside rural space. At one level, the restructuring of rural government has been part of a more extensive process of state restructuring driven by economic and ideological factors. These included the changing requirements of capitalist production; concerns about the growing cost of state welfare provision, public opposition to high rates of taxation, the inefficiency of state-owned enterprises, and the power of public sector trade unions; and the election of 'New Right' governments in the 1980s committed to an ideology of the 'minimal state', empowering individuals, active citizenship and the engagement of business knowledge. At a second level, the processes of social and economic restructuring in rural areas also undermined elements of the statist structure and created a rationale for reform. This has been advanced through five key changes (Woods and Goodwin, 2003):

- The scaling back of state activities in rural government, including deregulation in sectors such as agriculture and transport, the privatization of state-owned agencies and companies, and the engagement of private and voluntary sector organizations in local government functions.
- The shifting of responsibilities from the state to 'active citizens' and the engagement of communities through partnership working on a local scale.
- The greater coordination of rural policy delivery, including the amalgamation of government departments and agencies and formation of partnerships between different tiers and sectors of government.
- The replacement of some specifically rural institutions in favour of regional bodies encompassing both rural and urban areas.
- Reforms to elected local government, including changes to the powers, finances and territories of local councils.

Collectively, these changes have been argued to represent a transition from a system of 'rural government' to one of 'rural governance' (see Box 11.2).

Box 11.2 Key term

Governance: New styles of governing that operate not only through the apparatuses of the sovereign state but also through a range of interconnecting institutions, agencies, partnerships and initiatives in which the boundaries between the public, private and voluntary sectors become blurred. The actors and organizations engaged in governance exhibit differing degrees of stability and longevity, take a variety of forms and operate at a range of scales above, below and coincident with that of the nation-state.

The Characteristics of Rural Governance

The concept of 'governance' was first developed by urban researchers in the 1980s who observed how the authority of elected local government was being compromised by the increasing involvement of the private sector in urban policy-making and delivery, and the establishment of non-elected agencies with responsibility for areas such as economic development (Jessop, 1995; Rhodes, 1996; Stoker, 2000). For a while, this system of 'new local governance' was implicitly treated as an intrinsically urban phenomenon. Yet, almost unnoticed, the same processes were also at work in rural areas. By the mid-1990s, a landscape of rural governance had emerged that included not only the established institutions of government, but also a plethora of partnerships, community initiatives, inter-governmental organizations, business forums and co-funding arrangements, including, for example, LEADER action groups in Europe (see Chapter 10) and watershed management partnerships in the United States. As Goodwin observed,

The signs are that these tangled hierarchies which increasingly govern rural areas in a complex web of interdependence, are now the favoured mechanisms for rural policy formulation and service delivery at each level from the local to the European. Official policy statements, at all levels, emphasize the role of partnerships and networks beyond the formal structures of government. (Goodwin, 1998, p. 6)

The shift from government to governance, however, implies more than just a change in the institutional framework. It also involves a change in the style, rhetoric and discourse of governing. The state is no longer assumed to have a monopoly on governing, but rather there is a blurring of the responsibilities of the state and other sectors. The state is also no longer positioned as the provider of public goods, but is cast as a facilitator that enables communities to govern themselves. Similarly, the legitimacy of governance is perceived to come from the direct participation of citizens and stakeholders in governing activities, rather than from the electoral mandate of traditional government (see also Box 11.3).

Box 11.3 Five propositions about governance

1 Governance refers to a complex set of institutions and actors that are drawn from but also beyond government.
2 Governance identifies the blurring of boundaries and responsibilities for tackling social and economic issues.

> ### Box 11.3 (Continued)
>
> 3 Governance identifies the power dependence involved in the relationship between institutions involved in collective action.
> 4 Governance is about autonomous self-governing networks of actors.
> 5 Governance recognizes the capacity to get things done which does not rest on the power of government to command or use its authority. It sees government as able to use new tools to steer and guide.
>
> *From Gerry Stoker (1996) Governance as theory: five propositions mimeo, quoted in Mark Goodwin (1998) The governance of rural areas: some emerging research issues and agendas. Journal of Rural Studies, 14, 5–12.*

Drawing together these ideas, evidence for the emergence of rural governance can be identified around two key interlocking components: partnership working, and community engagement and active citizenship.

Partnerships

Partnership working is core to the idea of governance and may be manifest in a number of ways. 'Working in partnership' can mean that organizations hold liaison meetings or are involved in consultative forums, that there is co-funding of an initiative, or that two or more organizations are working together on a specific project. At the most concrete level are 'partnership organizations', defined by Edwards et al. (2000) as 'a formal or semi-formal body consisting of two or more partners but with an identifiable financial and administrative structure and an identity distinct from that of its constituent partners, which has been created to combine the resources of its constituent partners to achieve a capacity to act with regard to specified objectives' (pp. 2–3). Arrangements of all these types have proliferated in rural areas, such that, as Edwards et al. note:

Close examination of the organizations operating in any small town or rural district in England or Wales is likely to reveal LEADER groups, Local Agenda 21 groups, training partnerships, community enterprise or development projects, civic fora, and rural development programmes, as well as a plethora of groups focused on marketing, product valorization, sustainable development, transport or tourism – all constituted as some form of partnership, bringing together a range of organizations, often from across the public, private and voluntary sectors. (Edwards et al., 2000, p. 1)

Three types of partnership in particular have gained prominence within rural governance. First, there are strategic partnerships that are aimed at coordinating the policies and initiatives of the various state agencies operating in a rural area, including those operating at different scales or in different sectors. In some cases strategic partnerships also involve other stakeholder groups, such as farmers' unions, business confederations and voluntary sector representatives. The National Rural Development Partnership (NRDP) in the United States is a prime example of a strategic partnership. Formed in 1990, the NRDP is a network of over 40 federal agencies and national organizations with the aim of identifying programme duplication and gaps in service to rural areas, building collaboration and coordination between agencies, disseminating information and representing rural interests in the policy-making process. It operates on two scales with the national partnership, the National Rural Development Council, supported by 36 State Rural Development Councils which are themselves partnerships of state-level agencies and key private and not-for-profit sector stakeholders (Radin et al., 1996).

Secondly, there are delivery partnerships that are formed at a local level to manage the implementation of a particular policy or initiative. Local government will normally be a key partner, but other partners may include the appropriate funding body, the chamber of commerce, local development or enterprise agencies, civic and residents' associations and groups representing specific sections of the community, such as young people. Delivery partnerships are often involved with the implementation of rural development projects;

indeed, partnership working has become a requirement of many rural development programmes. Westholm et al. (1999), for example, surveyed local rural development partnerships in Europe, many of which are the product of EU rural development initiatives (see Chapter 10). They reveal the widespread application of the principle of partnership, but significant variations in the form and structure of partnership organizations that result from the different political contexts and traditions of civil society in the member states. Furthermore, Westholm et al. illustrate that whilst the EU has been a driving force in promoting partnership working in Europe, partnerships have increasingly become a feature of domestic rural development programmes (see Box 11.4). As such, Edwards et al. (2000) recorded that the number of partnerships with a rural regeneration remit operating in three neighbouring counties in the UK increased from fewer than 20 in 1993 to over 140 in 1999 (Figure 11.1).

Box 11.4 Rural Challenge

Rural Challenge was a regeneration initiative run by the Rural Development Commission in England which funded 24 local projects in four annual funding competitions from 1994 to 1997. The competition had two stages, with one selected to go forward from each of 16 counties eligible to participate in each year of the programme, from which six projects were funded annually. As Jones and Little (2000) record, the 18 projects funded in 1994–7 ranged in value from a £1.5 million initiative to provide mobile information, leisure and training facilities for rural youth in Somerset, to a £13 million development of a business park and Eco Tech centre at Swaffham in Norfolk. The competition rules stipulated that all applications must be made by partnerships involving public, private and community sector partners. The strength of partnership working was also part of the criteria used in evaluating the applications, with credit given for bids that included 'the highest proportion of private sector investment' (p. 175), and that involved a broad range of partners. As the programme guidance stated, Rural Challenge aimed 'to stimulate organisations not normally involved in rural regeneration to join in the local partnerships. The bid must show that a wide range of interests has been consulted and that the key partners from the private, public and voluntary sectors, e.g. local employers, the police, schools, colleges, the health authority, with a direct interest in the bid are involved as partners and committed to the proposals' (quoted in Jones and Little, p. 176).

> **Box 11.4 (Continued)**
>
> However, Jones and Little demonstrate that the process of partnership building was far from straightforward. The limited size of the rural private sector and high dependence on small enterprises meant that it was often difficult to enrol private sector partners, and, particularly, private sector finance. Community groups also frequently found that their participation was compromised by a lack of resources. Thus, most partnerships in the programme were dominated by public sector institutions, including the local county and district councils. In a few cases, 'false partnerships' were created with some 'partners' participating in name only to enable the bid to meet the programme criteria.
>
> *For more see Owain Jones and Jo Little (2000) Rural challenge(s): partnership and new rural governance. Journal of Rural Studies, 16, 171–183.*

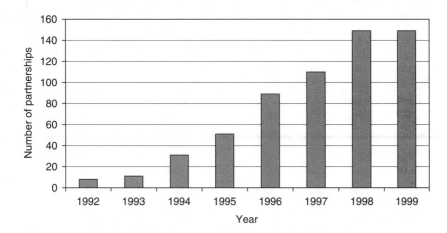

Figure 11.1 Number of rural regeneration partnerships operating in Ceredigion, Powys and Shrophire, England, 1992–1999
Source: Edwards et al., 2000

Delivery partnerships are also increasingly employed in the implementation of environmental management schemes. The involvement of all stakeholders, including government agencies, environmental groups, resource users and local communities, is seen as an important step in creating shared responsibility and building consensus in tackling environment problems. Partnerships for integrated ecosystem and watershed management have been established in number of parts of rural America, particularly in the western United States (Swanson, 2001), and through the Landcare programme in Australia (Lockie, 1999b; Sobels et al., 2001).

Thirdly, consultative partnerships operate on a range of scales as mechanisms for engaging communities in the governing process, as discussed below.

Community engagement and active citizenship

The promotion of community participation as a central tenet of governance has developed hand-in-hand with the rise of community engagement in rural development, as discussed in the previous chapter. Indeed, in both cases, the direct involvement can be seen both as an empowerment of local people and as the passing of responsibility from the state to citizens themselves. This is done through enrolling community groups, such as residents' associations, civic societies and village amenity societies, in partnership organizations, and through the use of surveys, appraisal exercises and public meetings to engage directly with local residents.

Edwards (1998) traces the gradual adoption of community engagement in the government of rural Britain from the early 1980s onwards. It was not until the 1990s, however, that the direct involvement of local communities was positioned as a core principle of rural policy in landmark documents and events such as the British Rural White Papers in 1995/6 (Edwards, 1998; Murdoch, 1997) and the Positive Rural Futures conference in Queensland, Australia, in 1998 (Herbert-Cheshire, 2000). As such, community involvement has become a standard practice not only in rural development and regeneration initiatives (Aigner et al., 2001; Edwards, 1998; Lapping et al., 1989), but also across a range of other areas of rural governance including education (Ribchester and Edwards, 1999), crime prevention (Yarwood and Edwards, 1995), public access to the countryside (Parker, 1999) and housing and homelessness (Cloke et al., 2000).

Yet, engaging 'the community' is far from straightforward in an era when rural communities have become increasingly fragmented (see Chapter 7). In some cases the local council is engaged as the 'representative' of the community, but the wider thrust of community engagement as a practice may be read as an implicit critique of the inclusiveness of elected local government. Instead, partnerships are encouraged to enrol a range of community groups to represent a broader cross-section of the population, with some programmes specifically requiring that partners are included that represent particular groups, such as young people.

Effective community engagement therefore relies on the active participation of members of the community. The promotion of community engagement is hence closely linked to the promotion of *active citizenship* – the discourse that individuals have both a *right* to seek their own solutions to problems and a *responsibility* to be actively involved in doing so (Herbert-Cheshire, 2000; Parker, 2002; Woods 2004b). At a modest level, active citizenship might simply mean voting in local elections or filling in community surveys, but it also means the mobilization of some individuals to become community leaders. This latter dimension is explicitly developed through leadership training schemes such as the 'Building Rural Leaders' programme in Queensland and the W.K. Kellogg Foundation's Rural America Initiative in the United States (see Box 11.5).

Box 11.5   Leadership training in Vermont

Environmental Partnerships in Communities (EPIC) is a regeneration initiative in Vermont, in the north-eastern United States. Founded and run by faculty at the University of Vermont, the initiative aims to assist communities in developing strategies for sustainable rural development. Its work includes a leadership training

---

Box 11.5   (Continued)

programme that involves a ten-week series of evening meetings and weekend retreats. In the workshops, 'participants role play, practice coping with tensions over local issues (such as the siting of a landfill), learn how to access resources, learn how to give a television interview, write a press release, and persuade a reporter to write about their work' (Richardson, 2000, pp. 112–113). The course participants, who are nominated by existing leaders in rural communities, are expected to return to their communities and assume leadership roles themselves.

For more see Jean Richardson (2000) Partnerships in Communities: Reweaving the Fabric of Rural America (Island Press).

Issues Raised by Rural Governance

The 'new' system of rural governance is a still-evolving phenomenon, yet researchers have already begun to raise a number of concerns about its workings and its consequences for the distribution of power in rural societies. There is not room to discuss all of the issues raised in detail here, but six key issues may be flagged up.

First, there are concerns about the *exclusivity* of the structures of rural governance. Partnerships and community engagement may superficially use the language of inclusion, but they can concentrate power in a small group of established organizations and individuals. More marginal sections of the community can find themselves excluded, even if they are the intended focus of the initiative. Cloke et al. (2000), for example, describe how one partnership aimed at tackling homelessness in a rural town excluded homeless people from participation. The working culture of rural governance can also be exclusionary. Little and Jones (2000) argue that the emphasis on competition and private sector engagement within rural governance, and decisions about the types of initiative to support within regeneration programmes such as Rural Challenge, reflect and reinforce particular masculine working practices and values.

Secondly, the *legitimacy* and *accountability* of new governance structures have been questioned. Whereas elected government institutions are accountable to citizens through elections and draw legitimacy from their democratic mandate, the organs of governance such as partnerships draw legitimacy from the breadth of organizations involved and are accountable only to their partner organizations and funders (Edwards et al., 2000). Furthermore, the continuing institutions of traditional, elected, local government must find new ways of maintaining their legitimacy. Welch (2002) demonstrates through case studies of rural local government in Australia and New Zealand that legitimacy is a major concern for those involved, and that to some extent it now derives from the involvement of local councils in partnerships and with other community groups.

Thirdly, the rhetoric of partnership is frequently undermined by the *unequal resources* of different partners. Community sector partners often have nothing to contribute other than their time and opinion, whilst finding private sector inputs in rural areas can be difficult (Edwards et al., 2000; Jones and Little, 2000; Welch, 2002). Accordingly, public sector partners tend to dominate. Community inputs can also be moderated by the need to reach consensus among partners and the principle of collective responsibility (Edwards et al., 2000; Westholm et al., 1999).

Fourthly, partnerships and other initiatives linked to particular programmes or funding competitions may have very *short lifespans*. Partnerships that do achieve greater longevity often spend considerable time and energy on simply securing their existence. As such, the institutional framework of rural governance can be very unstable (Edwards et al., 2000, 2001).

Fifthly, *new territories and scales of rural governance* have been created along with the establishment of new partnerships and, in some places, the restructuring of elected local government. Problems of cooperation can arise between overlapping institutions and partnerships operating over differently defined territories, and the accountability of governance bodies to local people becomes further confused (Edwards et al., 2001; Welch, 2002).

Finally, as the state has dissipated its responsibilities to 'communities', so the notion of universal state provision has been denuded. The provision of certain facilities within a community, or the availability of funds for economic development, may depend upon the capacity of that community to organize appropriate partnerships and compete for resources. Thus, it has been argued, rural governance is creating *geographical unevenness* between 'partnership-rich' and 'partnership-poor' communities that can be strongly pronounced in rural areas (Edwards et al., 2000, 2001).

Summary

The structures through which rural areas are governed have changed significantly over the past century, reflecting both the consequences of rural restructuring and the patterns of state restructuring more broadly. An essentially paternalist system at the start of the twentieth century was gradually replaced by a more centralized and more comprehensive structure of local government through state institutions, which in turn has been reworked during the past two decades as responsibility and authority have been dispersed to a wider network of actors from both within and outside the state in a new system of rural governance. This transition has impacted on the power structure of rural areas, with different elites being favoured at different times. Indeed, despite the democratic framework of elected local government and the language of inclusion in governance, all ways of governing implicitly favour some voices over others and produce a concentration of power in line with the distribution of valued resources. Thus, landowners formed the elite in rural Europe under paternalism, whilst with the new structure of rural governance it is the institutions that provide funding and the managers who sit on partnership boards who enjoy disproportionate influence. The emerging system of rural governance in particular has raised concerns about issues of power and accountability. The difficulty experienced by many community and voluntary sector partners in participating equally with public sector partners, and the absence of a strong rural private sector, have led some commentators to suggest that governance is little more than the continuation of old-style government by a different name (Edwards et al., 2001). Yet, if power has not been transferred under governance, responsibility certainly has been. Working in combination with the emphasis placed on 'bottom–up' endogenous development (Chapter 10), the principles of the new rural governance suggest that responsibility for shaping the future of rural areas has been shifted from the state to communities themselves.

For many communities this shift has been empowering, but, as Herbert-Cheshire (2000), notes, communities 'could be (unfairly) held responsible for any failure to improve their own conditions because they were regarded as deficient in entrepreneurial skills or because they were reluctant to "self change"' (p. 210).

Further Reading

A themed issue of the *Journal of Rural Studies* published in 1998 (volume 14, issue 1), is a good starting point, with a number of papers addressing concern associated with rural governance, including papers by Mark Goodwin on 'The governance of rural areas: some emerging research issues and agendas' (pages 5–12), and Bill Edwards on 'Charting the discourse of community action: perspectives from practice in rural Wales' (pages 63–78). The papers in the journal are primarily focused on the UK, but perspectives from Australia, New Zealand and the United States respectively can be found in Lynda Herbert-Cheshire, 'Contemporary strategies for rural community development in Australia: a governmentality perspective', *Journal of Rural Studies*, volume 16, pages 203–215 (2000); Richard Welch, 'Legitimacy of rural local government in the new governance environment', *Journal of Rural Studies*, volume 18, pages 443–459 (2002); and Beryl Radin et al., *New Governance for Rural America* (University of Kansas Press, 1996).

12

Selling the Countryside

Introduction

One of the most significant elements of rural restructuring has been the transition from an economy based on production to an economy based on consumption (Chapter 5). The consumption-based rural economy is broad-ranging and includes many diverse activities, from financial services through to retailing, but its most visible component is tourism. Accurate statistics for rural tourism and its contribution to the rural economy are difficult to find, particularly at a comparative level. However, the significance of rural tourism can be indicated by a few 'snapshot' facts and figures:

- Of all overnight domestic tourism trips in the UK in 2001, 23 per cent were to countryside or village locations.
- Rural tourist operations in Galway, Republic of Ireland, received 659,000 visitors in 1999, and those in neighbouring County Clare, 310,000 visitors. Rural tourism operators in Galway received revenues of Irish £101 million, and those in Clare, Irish £122 million.
- The 81,000 visitors to rural Cochise County, Arizona, spent close to $1 million dollars in 1994.
- Over 200,000 overnight stays were made by tourists in rural areas of Andalucia, Spain, in 2002. The number of rural tourism companies in the region increased by 50 per cent between 1999 and 2002.
- There are between 15,000 and 18,000 rural tourism-related businesses in New Zealand.

Whilst these figures provide a useful glimpse at the contribution of tourism in particular rural localities, it should also be noted that the type and significance of rural tourism varies considerably between different regions. Some areas, such as the national parks of North America, the European Alps, the English Lake District and the Scottish Highlands, have a long history of tourism extending back to the nineteenth century. In other, more traditional agricultural regions, tourism on a noticeable scale is a relatively recent development. Furthermore, the types of activities engaged in by day-trippers to recreational sites in the urban–rural fringe will be very different from those of 'adventure tourists' in the more remote regions of North America, Australia from New Zealand.

The growth of rural tourism reflects both a general expansion of tourism of all types, and a shift in popularity from 'traditional' seaside resort holidays to a broader range of tourist experiences, which Walmsley (2003) identifies with the rise of a lifestyle-led and leisure-oriented society. These societal trends have created opportunities that rural localities have attempted to respond to. In many regions, tourism has been promoted as part of endogenous rural development strategies (see Chapter 10), such that Walmsley observes that tourism has come to be seen as a panacea for declining rural communities. Tourism has also been promoted as part of farm diversification (Chapter 4). Shaw and Williams (2002) estimate that 20 per cent of farms in the UK were involved in tourism in some form in 1990, and this figure is now likely to be notably higher. Nearly two-thirds of farm tour operators in New York State expanded their business between 1986 and 1991 (Hilchey, 1993), whilst six farm accommodation marketing cooperatives in the East of England Tourist Board region collectively generate an annual gross income of £1.6 million.

Rural tourism, therefore, embraces a wide range of activities. Butler (1998) distinguishes between 'traditional activities' such as driving, walking, visiting historic sites, picnicking, sightseeing and fishing, and 'new activities' including snowmobiling, mountain biking, off-road vehicles and endurance sports. A more useful classification, however, might be between those tourism activities that are located in rural areas but are not distinctively rural in character, and those activities that actively engage with the rural landscape, environment, culture or traditions. The former might include theme parks and self-contained holiday centres such as Center Parcs, as well, arguably, as many 'activity-based' holidays and courses. The latter category, meanwhile, variously includes mountain walking, farm holidays, 'traditional' craft attractions, and more embodied, thrill-based, forms of engagement with nature through adventure tourism. What connects this second group is the centrality of 'the rural' to the 'tourist gaze' (Urry, 2002). By talking about the 'tourist gaze', Urry focuses attention on tourism as a process of seeing, experiencing, understanding and representing place as different from the everyday and mundane. The tourist gaze hence has a transformative effect. Tourists transform the rural place that is gazed upon to meet their preconceptions and expectations; and rural places are themselves packaged, marketed and even in some cases physically changed, in anticipation of the tourist gaze and in order to direct and exploit it.

The Commodification of the Rural

The importance of the search for experience that is intrinsic to the 'tourist gaze' suggests that a further step can be taken beyond stating that the countryside has moved from a production-based economy to a consumption-based economy, to proposing that the rural economy has changed from one based on exploiting the physical environment to one based on exploiting the aesthetic appeal of the countryside. As a consequence, the relative value of different aspects of the rural environment has changed. Rural land is increasingly valued less for its productive potential, and more for the opportunities that it offers for tourism and other forms of aesthetic consumption, such as use as a film set. In other words, the countryside has become a commodity (see Box 12.1), to be 'bought' and 'sold' through the consumptive practices of tourism, property investment by in-migrants, the marketing of rural crafts and products, and the use of rural images to sell other products.

Box 12.1   Key terms

Commodity: An object that is produced for the purpose of being exchanged (that is, bought and sold).

Commodification: When the value at which an object can be sold (its 'exchange value') exceeds its 'use value'. In other words, the object is valued for some cultural or aesthetic reason above its actual usefulness. When an object is commodified it becomes removed ('abstracted') from its use and from discussions about the need for it.

Cloke (1992) demonstrates that the commodification of the countryside has resulted from multiple factors, including not only the declining economic fortunes of agriculture and forestry compared with the growing significance of tourism, but also the rise of a society in which brands and signs and symbols are fundamental to the way in which we understand the world, and the pressures placed on companies to extract maximum value from their assets. Thus a piece of rural land might be exploited simultaneously for its production value and its exchange value (see for example the section below on the marketing of hydro-power stations, reservoirs and forests as tourist attractions). As such, commodification is part and parcel of the capitalist economy in rural areas and, as Cloke describes, has occurred in a number of different ways, including:

the countryside as an exclusive place to be lived in; rural communities as a context to be bought and sold; rural lifestyles which can be colonized; icons of rural culture which can be crafted, packed and marketed; rural landscapes with a new range of potential from 'pay-as-you-enter' national parks to sites for the theme park explosion; rural production ranging from newly commodified food to the output of industrial plants whose potential or actual pollutive externalities have driven them from more urban localities. (Cloke, 1992, p. 293)

In common with all commodities, the countryside is packaged and marketed in a way designed to appeal to the largest number of potential customers. In a rural context, the landscapes, environments, traditions and practices that have greatest exchange value are those that conform most closely to the ideal of the rural idyll. Therefore, the marketing of rural places frequently means repacking and *re*-presenting rural areas and features to emphasize characteristics associated with the rural idyll (see Chapter 1).

Through a study of publicity material for rural tourist attractions in Wales and south-west England, Cloke (1993) identifies five recurring themes that illustrate this emphasis. First, the rural *landscape* as a setting is commonly referred to, particularly by attractions that are disconnected from agriculture such as the Oakwood Theme Park in Wales, which notes its location in 'eighty acres of beautiful Pembrokeshire countryside' (p. 62). Secondly, *nature* is emphasized through reference to animals and plants that form part of the visitor experience, as at the 'Milky Way and North Devon Bird of Prey Centre', which promises, 'a fun, hands-on farming experience. Milk a cow! Bottle feed and cuddle the baby animals!' (p. 62). Thirdly, *history* is strongly presented as an important component in the social construction of the countryside, as is, fourthly, the *family*. Finally, the availability of rural craft products and country 'fayre' is promoted, 'in almost all cases the commodification of the countryside as represented in the brochures creates the impression that certain craft items and types of refreshment are somehow part of the total packaging of a countryside experience' (p. 63).

Similar themes are identified by Hopkins (1998) in an examination of the 'symbolic countryside' constructed by promotional material for rural tourist attractions in southern Ontario, Canada. Nature is again often emphasized, including references to specific animals, as are the amenity value of the environment and themes of family, community and history. Hopkins notes that these messages are conveyed through the publicity material in a number of ways, both textual and visual.

These include the logos of attractions, most of which symbolize nature or the environment in some form, such as this logo for a family camping ground:

> Another forest animal – this time a bright-eyed, smiling 'teddy' bear – is depicted standing upright roasting a hot dog on the end of a stick over a campfire. Myths of 'innocence', 'childhood', 'nourishment' and 'nature domesticated' are symbolized once again ... Still, there are other myths connoted here by the campfire: 'romantic', 'socializing', 'summertime', 'time immortal' and 'wilderness'. This is a visually simple logo which effectively captures the natural joys, perhaps the primal pleasures, of camping in secluded woodland. (Hopkins, 1998, p. 76)

The remainder of this chapter is written around a series of case studies that illustrate five prominent elements in the commodification of the countryside: the marketing of rural production sites as tourist attractions; the repackaging of rural heritage; the promotion of 'fictional' rural landscapes; the rural as a site for extreme experiences through adventure tourism; and the use of the rural as a 'brand' to sell goods and products to urban consumers.

Rural Production Sites as Tourist Attractions

The shift from an economy based on production to one based on consumption has taken place not just at the level of the countryside as a whole, but also in the business practices of individual enterprises. Many farms have branched into tourism in order to diversify their sources of income away from a dependence on agriculture. Most commonly this involves the provision of on-farm accommodation, including bed and breakfast rooms, self-catering cottages and camping sites. Farm

Figure 12.1 Publicity leaflets for 'farm park' attractions in England
Source: Woods, private collection

shops, nature trails, horse-riding facilities and fishing lakes have also been experimented with, but in all these cases the farm tends to remain primarily a working agricultural enterprise. A further level of abstraction has been reached, however, by a number of farms that have reinvented themselves as 'farm parks', replacing the landscape and practices of a working farm with those that reflect idyllized popular images of farming (Figure 12.1). The 'Farmer Giles' attraction in Wiltshire, England, for example, is described in its publicity leaflet as a safe environment where families 'enjoy themselves whilst learning about farming methods of past and present. You can cuddle, groom, bottle or hand feed a variety of animals in the many paddocks and enclosures'.

In the same way that farm parks play to the importance of sanitized images of farming in the rural idyll, forests have also been commodified because their landscape has a high value within popular discourses of the rural. Until recently, forests were planted and managed as commercial enterprises for the production of timber, with very limited public access. Increasingly, however, forest managers have realized that forests have not only a use value in terms of the harvested wood, but also an exchange value that can be exploited through recreational activity. In the UK, for example, Forest Enterprise – the semi-commercial body that is now responsible for managing Britain's state-owned woodland – actively promotes its properties as tourist attractions and has developed facilities including visitor centres, picnic sites, waymarked walks, art installations and mountain bike trails.

Significantly, Forest Enterprise's promotion of its forests combines the common themes of nature, landscape, tranquillity and wilderness with an educational component that attempts

Figure 12.2 Interpretative noticeboards in the Seymour Demonstration Forest, Vancouver
Source: Woods, private collection

to inform the public about nature conservation, forest management and the timber trade. This educational agenda is also a prominent feature of the Seymour Demonstration Forest on the northern fringe of Vancouver, Canada. Originally a closed part of the Greater Vancouver Water District's watershed lands, the forest was opened in 1987 'to provide educational and recreational opportunities to the public' (publicity leaflet). Its attractions include a 1.6 kilometre 'Integrated Resource Management Trail' with interpretative brochures and signs (Figure 12.2), that 'provides a good overview of the cycle of forest management' (leaflet).

The Commodification of Rural Heritage

The examples above translated the rural into a commodity by stressing the motifs of nature,

environment and landscape that are core elements in social constructions of the rural idyll. A further feature of the 'rural idyll' is nostalgia and the sense that the countryside has been less changed and corrupted by modernity than the city. It is this belief that is appealed to by sites that seek to commodify rural heritage in attracting tourists. These include places where the commemoration of the past has been employed as a strategy for regeneration following the collapse of the traditional economic base, of which the most notable example is Chemainus on Vancouver Island in Canada. When the town's sawmill closed in 1983 with the loss of 654 jobs, the various regeneration schemes attempted included a small project to paint five murals depicting scenes from local history. The

Figure 12.3   Murals depicting rural heritage in Chemainus, Vancouver Island
Source: Woods, private collection

unexpected response to the project stimulated a much larger initiative to create a 'Festival of the Murals', with 32 murals and six sculptures completed by 1992, all portraying representations of the district's agrarian and pioneer heritage (Figure 12.3). The murals became a significant tourist attraction, with over a quarter of a million visitors each year (Barnes and Hayter, 1992), and the initiative has been copied by other towns, such as Sheffield in Tasmania (Walmsley, 2003).

The Chemainus murals necessarily involve a distanced consumption of representations of the past. Other ventures, however, have attempted to 'reconstruct' an historic rural in order to give visitors the opportunity to directly 'experience' the past. Wilson (1992) discusses four examples of this in southern Appalachia, one of the most deprived rural regions of the United States, which Wilson suggests provides 'a sterling example of a geography in crisis, for its

old mountains and river valleys have in the past sixty years witnessed the collision of a distinct and isolated regional culture with the relentless project of modern development' (p. 206). Of the four sites, two – Dollywood and Heritage USA – are overtly commercial enterprises in which 'rural heritage' is part of the package sold to visitors. Dollywood, at Pigeon Forge, Tennessee, is a theme park owned by and themed around the country singer Dolly Parton, the publicity material for which invites visitors to 'experience the only place in America where the tradition and pride of the Smokies [Mountains] are brought to life through old-time crafts, breathtaking scenery, food, fun and friendly folks, and Dolly's kind of upbeat music!' (Dollywood brochure). As Wilson describes, the mountain heritage is reproduced in Dollywood through material representation – of buildings, looms, farm implements, banjos, corn-cob pipes, washboards and Bibles – and symbolically portrayed on the T-shirts, mugs and posters on sale in the gift shop. Beyond the materiality, Wilson argues that Dollywood also commodifies the idea of the wholesome, simplistic, rural home:

> People lived like this once; it was simple out here in the bush, but it was wholesome. The fact that Dolly doesn't live this way any more seems irrelevant. It is the very artificiality that fascinates … Dollywood makes sure that our relation to what might or might not be left of mountain culture is full of irony. (Wilson, 1992, p. 211)

The moral agenda that is implicit in Dollywood's representation of rural heritage was explicit at Heritage USA, a now-closed Christian real-estate development near Charlotte, North Carolina. Heritage USA employed tradition to guide and justify a moral vision for the present, which has a material manifestation in the landscape of the site. Yet, Wilson comments, 'it's hard to describe Heritage USA without using quotation marks every few words. There is a "steam" train, an "old" farm with "log" houses, a "Main Street" mall with "Georgian" architecture. Inside most of the buildings we are enveloped by air conditioning, perfume, and discreet music "from yesterday and today"' (p. 214). This included Farmland USA, described in the publicity material as 'a glimpse into the country life of the 19th century', but which embodied the ambiguous presentation of time and place in its stereotyped collection of a Victorian farmhouse and barn, a petting zoo, horse and carriage rides, a windmill, a rustic workshop and a country chapel.

In contrast, the Museum of Appalachia at Norris, Tennessee, markets itself as 'the most authentic and complete replica of pioneer life in the world'. The open-air site contains over twenty 'authentic' cabins and other structures, relocated from different parts of the region, but

these have been artificially arranged and furnished in the museum to convey 'the "lived-in" look, striving for, above all else authenticity' (museum brochure) (see Figure 12.4). As with Dollywood, the material representations on display are there to signify a deeper message, articulated in the museum founder's statement that 'the true breed of diminishing mountain folk of Southern Appalachia are among the most admirable people in the world' (museum brochure). Yet, Wilson points to the selectivity of the representation:

> The stories are meant to take visitors back to an earlier day: the years of victory over wilderness and savages. I look but can't find documentation of the subsequent years of loss: loss of land to erosion, to inundation by reservoirs, to poverty and all the displacements of modern life. There are no traces here of what happened to these places, to their disappearance in the mid twentieth century. Nor is there any sense of a culture that predated the white one. So we're left wondering how our present connects to these tools and buildings. (Wilson, 1992, p. 207)

The final site, Cades Cove, has perhaps the strongest claim to authenticity. It was once a lived-in community, but the population was moved out following the creation of the Great Smoky Mountains National Park in 1934 (see also Box 13.1). Most buildings were demolished, but some were left standing as a representation of the former way of life and now stand in isolation as stops on a circular driving tour – empty cabins and barns, a school, a church, a blacksmith shop and a mill (Figure 12.5). The structures are unfurnished and there are no attendants or interpretative boards. For Wilson, this absence of additional materials makes Cades Cove more authentic than the other sites, but this conclusion is open to contest. As empty structures, reasonably well maintained by

Figure 12.4 Representation of a mountain cabin interior at the Museum of Appalachia
Source: Woods, private collection

Figure 12.5 Remnants of the former community at Cades Cove in the Great Smoky
Mountains
Source: Woods, private collection

the National Park Service, the buildings of Cades Cove convey little about the people who once lived there, the practices they acted out and the hardships they faced. Instead, the buildings are neutral vessels into which visitors can pour their own idyllized perceptions of historic mountain life without being challenged by counternarratives.

The sites discussed by Wilson are marketed as representations not just of rural heritage but, more specifically, of the mountain heritage of rural Appalachia. The spatial reference is important as it positions rural heritage sites and the commodities associated with them as symbols of regional distinctiveness against a background of globalization and homogenization (see Chapter 3). This use of rural heritage as an expression of local identity can also be observed elsewhere. In the Darlana region of central Sweden numerous villages have a preserved heritage site, or *hembygdsgård*, usually focused on an abandoned farm. These sites, which Crang (1999) demonstrates are as abstracted from the reality of past rural life as the historic parks of Appalachia, are cherished as representations of the rural foundation of Darlana culture, which in turn is regarded as the iconic culture of Swedishness.

Fictional Rural Landscapes

Nostalgic ideas of rural heritage provide one framework for the construction of the tourist gaze on the countryside, but the tourist gaze is also informed by fictional representations of rural life and landscape in film, television programmes and literature. Indeed, the use of rural places as locations for filming is in itself a form of commodification, providing an additional source of income for landowners – a CD database of some 400 British farms offering 'quiet, traditional and spectacular' locations for filming was launched at the Cannes Film Festival in 1999. As well as location fees, rural places used as the sets for successful films or television programmes can also look forward to increased tourism as devotees travel to see the places represented in 'real life'. Mordue (1999), for example, discusses the example of Goathland, a village of 450 residents in the North Yorkshire Moors of England that since 1991 has been the filming location for a popular British television drama, *Heartbeat*. As a result of the programme, tourist visits to Goathland have increased from around 200,000 a year, to over 1.2 million. The series depicts the life of a country policeman in the fictional village of Aidensfield in the 1960s. Thus, whilst many visitors to Goathland may have been attracted by the landscape shown in the programme, others will also be searching for the romanticized, simple and slow-paced rural life that the programme represents.

The tourist gaze that is fixed on rural localities that provide the settings for fictional tales commonly blurs the distinction between the real and fictional landscape and, over time, this can come to have a transformative effect on the physical environments of the 'real' localities. In Goathland, for example, elements of the film set for *Heartbeat* have been retained in the permanent villagescape, most notably the frontage of the village shop, and become focal points for tourists providing a connection with the fictional place they are searching for. A more developed case is found in Cavendish on Prince Edward Island in Canada, which is the setting for L.M. Montgomery's popular children's novel *Anne of Green Gables*. The novel, published in 1908, told the story of an orphan adopted by a farmer and his wife and the rural childhood she enjoys. Hence, as with *Heartbeat* and Goathland, the visitors to Cavendish are in part searching for a romanticized rural past. As Squire (1992) observes, Montgomery's particularly evocative imagery helped to create a prototype of the Canadian pastoral idyll (see Box 12.2).

Box 12.2   Anne of Green Gables

A huge cherry-tree grew outside, so close that its boughs tapped against the house, and it was so thick-set with blossoms that hardly a leaf was to be seen. On both sides of the house was a big orchard, one of apple trees and one of cherry trees, and show-ered with blossoms ... In the garden below were lilac trees purple with flowers, and their dizzily sweet fragrance drifted up to the window on the morning wind.

Below the garden a green field lush with clover sloped down to the hollow where the brook ran and where scores of white birches grew, upspringing airily out of an undergrowth suggestive of delightful possibilities in ferns and mosses and woodsy things generally.

Extract from L.M. Montgomery, 1968 edition, Anne of Green Gables, Toronto: McGraw-Hill Ryerson p. 33–34.

Although Cavendish has become identified as the model for the community of Avonlea in the novel, some aspects of the fictional land-scape are imagined or amended, such that the real geography of Cavendish and the fictional geography of Avonlea do not directly corre-spond. Faced with a disjuncture between the two, literary tourists to Cavendish opt to prioritize the fictional account, as does, signi-ficantly, the Parks Canada agency in its management of the 'Green Gables' site in the Prince Edward Island National Park. As Squire notes:

> Parks Canada's interpretative policy at Green Gables recognizes that historical authenticity must sometimes be com-promised with literary accuracy. Site redevelopment has been guided by details from the novels, and only if these sources proved inconclusive would 'infor-mation about the actual farm that existed on the site', or a comparable nineteenth century farmstead be used. (Squire, 1992, p. 143)

Thus, Squire argues, whilst elements of Prince Edward Island were transferred by Montgomery into her novels, the tourist industry has inverted the process, 'giving that which was fictional a factual identity through a number of tourist attractions' (p. 143).

Embodied Experiences of Rural Adventure

Rural environments are increasingly becom-ing the location for tourist experiences that seek to go beyond the conventional practices of holidaymakers in a quest for adventure. This includes not only participation in tradi-tional 'outdoor pursuits' such as canoeing, trekking and ski-ing, but also in a range of newer adventure tourism experiences includ-ing jet-boating, bungy jumping, snowboard-ing and canyoning. These activities require a different form of tourist engagement with the rural environment than that involved in the more traditional 'sightseeing' activities described in the sections above. As Cloke and Perkins (1998) argue, adventure tourism goes beyond the metaphor of the 'tourist gaze' to a more embodied experience based on 'being, doing, touching *and* seeing' (p. 189).

Adventure tourism has become a signifi-cant recreational – and economic – activity in a number of rural regions, including the

Rocky Mountains, British Columbia, New England, California and, perhaps most notably, the South Island of New Zealand. Estimates suggest that each year some 150,000–200,000 visitors to New Zealand go jet-boating, and 50,000–100,000 participate in bungy jumping, climbing/caving, mountain biking and rafting (Cloke and Perkins, 1998; Swarbrooke et al., 2003). The centre for many of these activities is Queenstown in the interior of the island, a town that has been a winter sports resort since the 1950s but which has seen its economy and population rapidly expand with the boom in adventure tourism (Cater and Smith, 2003). As Cloke and Perkins observe, the natural setting and the opportunities for adventure are drawn together in the commodification of Queenstown: 'adventure and excitement set amongst scenic natural landscapes characterizes the tourist experience and thereby fuels the place-myths of social spatialization' (p. 201). Projected onto individual tourist activities, the location of adventure experiences in a deep rural natural environment is important in two respects. First, it promises to take tourists 'off the beaten track' to places that are accessible only in adventurous ways:

> Combine skiing, rafting, mountain biking and kayaking with unspoilt scenery, excellent food and Kiwi hospitality. See the attractions for which this country is famous, then head off the beaten track to encounter the Aotearoa that only the locals know. (Alpine Excellence brochure, quoted by Cloke and Perkins, 1998, p. 202)

Secondly, adventure tourism is presented as an embodied experience that involves overcoming the challenges of nature. The adventure 'involves exploration of uncharted territory; experiencing the danger and adrenaline rush of past explorers; traveling the untravelable; seeing the unseeable; generally pitting adventurousness, personal bravery, and technological expertise against natural barriers – and winning' (p. 204).

In these ways, adventure tourism also contributes to a commodification of rural places, but does so by reproducing a different but equally historic social construct of the rural not as a pastoral idyll, but as a wilderness and a place of adventure.

The Rural as a Marketing Device

All the previous examples have involved the commodification of the rural for consumption practices that have taken place within rural space. However, as a commodity, 'rurality' has a mobility that enables it to be attached to other commodities that are bought and sold in an urban environment. Moreover, the value of these other commodities is enhanced because of their association with the perceived qualities of rurality. An obvious example of this is the use of 'rural' brands and symbols in the marketing of premium food and craft products. Hinrichs (1996), for example, discusses the construction of Vermont as a distinctive rural place that subsequently allows the tagging of products with the 'Made in Vermont' label to imply 'certain standards of quality, clearly linked to the positive aspects of landscape, tradition and place reinforced by more general promotion of Vermont' (p. 269).

However, the products that are marketed with a rural association need not necessarily have a rural origin. They need only to signify a lifestyle that corresponds with urban aspirations for middle class rural culture. As Thrift (1989) notes, 'the countryside and heritage have met and blended with consumer culture. Countryside and heritage *sell* products, and in turn these products strengthen the hold of these traditions' (p. 30; original emphasis).

Among the most notable examples of this are clothes, including waxed Barbour jackets and Gore-tex outdoor wear, and cars – particularly four-wheel drive and sports utility vehicles – where rural imagery is frequently used to suggest a particular macho, masculine, engagement with the rural that promises the conquest of nature and wilderness. One advertisement for Land Rover in the late 1990s, for example, showed the car perched on a hillside with a people-less moorland landscape stretching out behind it and the strapline, 'Sunday, all this could be yours' – explicitly positioning the car as the key to weekend consumption of the rural idyll by urban dwellers.

Summary

The commodification of the countryside is part of the ongoing economic restructuring of rural areas. As traditional, production-based economic activities have declined so the 'use value' of rural environments and landscapes has begun to be exceeded by the 'exchange value'. Packaged to conform to popular social constructions of rurality, the countryside as a commodity has many buyers. These include not only tourists, but also in-migrants, relocating businesses, film production companies, adventure seekers, recreationists, consumers of premium rural food and craft products, and urban dwellers who wear Gore-tex clothing, drive SUVs and install country-style kitchens.

However, the process of commodification changes rural places and generates conflict. In order to market the countryside as commodity, representations of the rural are fixed that belie the dynamism and diversity of rural society and space. Moreover, as marketing images are selected to correspond with the pre-existing expectations of consumers, the representations employed frequently owe more to myth than to the everyday lived experience of the place concerned, drawing on nostalgic ideas of the rural idyll or on references to film, television or literature. As such, conflicts can emerge over the way in which a particular place is represented – and over the consequences of commodification. Large-scale tourism can create social, economic and environmental problems, including traffic congestion, footpath erosion, increased property prices, an over-dependence on seasonal employment and the tailoring of shops and services towards tourists' rather than residents' needs. Local residents may also feel that they are losing control over the identity of the place, and those involved in more traditional economic sectors such as farming may find that their interests are constrained by those of tourism or other consumptive practices. For example, once a rural landscape is valued more for its aesthetic appeal than for its productive potential, the conservation of the visual appearance of the landscape has more economic weight than agricultural modernization practices that might alter it, such as the removal of hedges. Rural conflicts are explored further in Chapter 14, whilst the next chapter examines the broader issue of countryside conservation.

Further Reading

The concept of the commodification of the countryside is most comprehensively introduced by Paul Cloke in Sue Glyptis's edited collection, *Leisure and the Environment* (Belhaven, 1993), a book that is unfortunately not widely available. Briefer accounts are given in many of the other papers cited in this chapter. The case studies discussed in this chapter can be followed up in the books and papers in which they were originally published. For more on the representation of rural heritage in Appalachia see Alexander Wilson, *The Culture of Nature: North American Landscape from Disney to the Exxon Valdez* (Blackwell, 1992); for more on TV-inspired tourism in the North Yorkshire Moors, see Tom Mordue, 'Heartbeat country: conflicting values, coinciding visions', *Environment and Planning A*, volume 31, pages 629–646 (1999); and for more on Anne of Green Gables and Prince Edward Island see Sheelagh Squire's chapter 'Ways of seeing, ways of being: literature, place and tourism in L.M. Montgomery's Prince Edward Island', in P. Simpson-Housley and G. Norcliffe (eds), *A Few Acres of Snow: Literary and Artistic Images of Canada* (Dundurn Press, 1992). Adventure tourism in New Zealand is discussed further by Paul Cloke and Harvey Perkins in 'Cracking the canyon with the awesome foursome: representations of adventure tourism in New Zealand', *Environment and Planning D: Society and Space*, volume 16, pages 185–218 (1998), and by Carl Cater and Louise Smith in 'New country visions: adventurous bodies in rural tourism', in P. Cloke (ed.), *Country Visions* (Pearson, 2003).

Websites

A number of the tourist attractions mentioned in this chapter have their own websites, which also convey their particular representation of rurality. These include:

Farmer Giles Working Farm Park	www.farmergiles.co.uk
Rays Farm	www.virtual-shropshire.co.uk/rays-farm/
Umberslade Children's Farm	www.umbersladefarm.co.uk
Seymour Demonstration Forest	www.gvrd.bc.ca/LSCR/
Chemainus murals	www.chemainus.com
Dollywood	www.dollywood.com
Museum of Appalachia	www.museumofappalachia.com
Cades Cove	www.cadescove.net/auto_tour.htm
Green Gables Park	www.annesociety.org/anne/
Queenstown, New Zealand	www.queenstown-nz.co.nz

13

Protecting the Countryside

Introduction

The protection of the rural environment has been an important challenge for campaigners and governments alike for over a hundred and fifty years. As early as the mid-nineteenth century, American writers, including Ralph Waldo Emerson and Henry Thoreau, were advocating the need to protect the spectacular natural 'wilderness' of North America from the impact of settlement, cultivation and development. Similarly, in Britain, writers in the Romantic movement, including William Wordsworth, John Ruskin and William Morris, promoted an appreciation of the aesthetic value of the countryside that led eventually to the formation in 1895 of the National Trust, a charity that acquired valued landscapes and historic sites to preserve them for the public interest, thus initiating a practice of rural conservation through private philanthropy. The role of the state in the protection of the rural environment, meanwhile, was established by the creation of the first national park, Yellowstone, in 1872, and the pioneering work of Gifford Pinchot as the first head of the US Forestry Service, founded in 1909, in which he developed a utilitarian model of conservation, combining the protection of the environment with the stewardship of economic resources.

The early advocates of countryside preservation tended to be motivated by a concern for the aesthetic value of pastoral or wilderness landscapes, often informed by religious beliefs or by perceptions about the significance of such landscapes to national identity and culture (Bunce, 1994; Green, 1996). Pinchot differed from this approach in emphasizing the material benefits of conservation to the rural economy, seeking maximum sustained yields from agriculture, forestry and fishing through the controlled exploitation and management of biological resources. More recently, the rise of the environmental movement has contributed to a further 'greening' of rural policy, based on scientific analysis of damage to the rural environment (see Chapter 8), sometimes combined with ethical motivations rooted in a 'deep ecology' philosophy (Green, 1996). These different rationales for the protection of the rural environment led to different objectives for environmental initiatives. Aesthetic motivations are associated with *preservation* initiatives aimed at maintaining a rural landscape in a relatively unchanged state. Utilitarian motivations based on material benefits, in contrast, support *conservation* projects, which implies stewardship, managed change and the avoidance of over-exploitation. Whilst both these approaches essentially follow a modernist compartmentalizing of nature and culture (see Chapter 3) in implying that nature can be protected alongside the broader development of rural space (either by designating specific 'protected landscapes' or by managing resource exploitation), the recent wave of environmentalism has more fundamentally challenged rural policy by demanding that the environmental impact should be given priority consideration in all policy areas – for example, by creating a bias against new road building which was once regarded as a staple element in rural development.

The different motivations for the protection of the countryside also reflect the importance placed on different threats to the rural environment at different times and in different places. To the early preservationist movement, the threat came from industrialization and the spread of urban sprawl. Yet, as Chapter 8 detailed, environmental change in the countryside over the course of the past century has been caused by a much wider range of factors, including indigenous development in rural areas themselves and the impact of modern agricultural practices.

As such, efforts to protect the rural environment consist of a number of different strategies, aimed at addressing different problems and following different rationales. This chapter discusses three such approaches. First, the designation of 'protected areas' within rural areas, in which land use and management is tightly controlled. Secondly, the use of land use planning policies to regulate the development of the countryside more generally. And thirdly, the promotion of agri-environmental schemes to reduce the detrimental impact of modern agriculture and to encourage conservation through farming. Finally, the chapter shifts its attention from landscape to animals, examining initiatives to preserve rare breeds of livestock and to reintroduce extinct native wildlife.

Protected Areas

The principle behind protected areas is that there are particular rural landscapes or rural environmental sites that are of such aesthetic, cultural or scientific importance that they warrant specific protection from detrimental human activity. The designation of protected areas hence seeks to preserve the most valued natural features of the rural environment whilst permitting the development of the wider countryside. The best known type of protected area are national parks, but these are in fact just one level of designation. As classified by the World Conservation Union (also known as IUCN), protected areas range from scientific reserves which are very strictly managed for scientific purposes with little human access, to 'managed resource protected areas' in which resource exploitation takes place but is managed for sustainable use (Table 13.1). As well as variations in the level of regulation and protection, protected areas also differ in their size – ranging from small nature reserves up to national parks covering several thousand square kilometres – and in the degree of human activity that is permitted. The most stringent protected areas are uninhabited and access may be tightly restricted but other types

Table 13.1   IUCN classification of protected areas

	Type	Notes
I	Strict Nature Reserve/Wilderness	Very strictly managed for science or wilderness protection
II	National Parks	Largely uninhabited, managed for ecosystem protection and recreation
III	Natural Monument	Managed for conservation of specific features
IV	Habitat/Species Management Area	Areas of conservation through management intervention
V	Protected Landscape or Seascape	Aim at balance between humans and nature
VI	Managed Resource Protected Area	Managed for the sustainable use of natural ecosystems

Source: IUCN website www.iucn.org

of protected areas, mostly notably level V 'protected landscapes' in the IUCN classification, are inhabited and must balance the interests of the local population and nature conservation.

National parks

The world's first national park was established at Yellowstone, Wyoming, in 1872. The designation of the park followed decades of lobbying by a group who were less immediately motivated by environmental concerns than by the conviction that the United States needed to demonstrate that it possessed natural wonders to rival those of Europe and that such places should be the property of the people not sites of private profit (Runte, 1997; Sellars, 1997). In 1864, concerns that the outstanding landscape of the Yosemite Valley in California might be abused by private entrepreneurs had led President Abraham Lincoln to cede ownership and responsibility for the area to the State of California, thus establishing a prototype national park (Yosemite was later designated as a full national park in 1890). At Yellowstone it was the discovery of a remarkable array of geysers, waterfalls, canyons and artefacts that suggested the remnants of an earlier lost civilization that marked it out as a place in which the United States could demonstrate its commitment to protecting its

cultural heritage (Runte, 1997). It was not until later that national parks became appreciated as sites of wilderness preservation and the fact that Yellowstone happened to offer protection to a large expanse of wilderness makes it an accidental role model.

Indeed, two other precedents set at Yellowstone were also accidental – the inclusion of an expansive territory (legislators were unsure that all of Yellowstone's treasures had been discovered) and ownership by the US federal government (there was no state government in Wyoming at the time). These accidental origins notwithstanding, Yellowstone became the model for future national parks: they would be focused on outstanding natural or cultural phenomena, cover extensive territories, be entirely publicly owned, uninhabited, free from commercial exploitation and managed by the government on behalf of the nation. Whilst the designation of further national parks in the United States occurred only slowly, the idea quickly spread to the dominions of the British Empire, with national parks established in Australia (1879), Canada (1885) and New Zealand (1887) – all on the Yellowstone model.

The model did not, however, translate easily back to Europe, where there were few areas of extensive uninhabited countryside left, but where concerns mounted during the early twentieth century about the threat to the rural landscape from urbanization. When national parks were established in the mid-twentieth century, one of two compromise approaches was adopted. Countries such as Ireland, Italy and Switzerland remained true to the principles of the Yellowstone model (level II in the IUCN classification), that national parks are publicly owned, uninhabited and strictly managed, but were restricted to designating relatively small areas of land as national parks. In the UK and Germany, in contrast, larger territories were designated as national parks, but these included privately owned land and were inhabited, and offered a much lesser degree of environmental protection (level V in the IUCN classification). France innovated a mixed approach, with a core national park conforming to IUCN level II standards, surrounded by a peripheral zone that is inhabited and more akin to the UK model. Consequently, whilst the term 'national park' is used around the world, its meaning varies significantly between countries (see Table 13.2).

These differences can be illustrated by examining in more detail the national park systems of the United States and the UK. By 2003, 56 national parks had been designated in the United States, many of which had been originally awarded the less protected status of 'national monument' before later elevation. The expansion of the national park system occurred in four phases. First, there was a continued designation of parks in the relative wilderness regions of the western states following the precedent of Yellowstone and Yosemite, including Grand Canyon (1919), Sequoia (1890) and Rocky Mountain (1915), but also smaller sites such as Hot Springs, Arkansas (1921), designed to protect thermal springs, and Wind Cave, South Dakota (1903). Secondly, travellers from the eastern states to

the west were inspired to campaign for the creation of national parks in the east, most notably in the Great Smoky Mountains (see Box 13.1). The establishment of such parks in the already settled eastern US proved more complex, involving difficult land purchase negotiations and the resettlement of communities, and relatively few national parks have been designated east of the Mississippi (Figure 13.1). Thirdly, the Alaska Lands Act of 1980 facilitated the creation of seven new national parks in the state and the enlargement and renaming of the Mount McKinley nation park (originally established in 1917) (Runte, 1997). The Alaskan national parks encompass vast areas of both scenic and scientific importance, collectively covering an area larger than England. Finally, a handful of new parks have been designated since 1990 – Death Valley and Joshua Tree in California (both created by the Desert Protection Act of 1994), Black Canyon of the Gunisson in Colorado (1999), Cuyahoga Valley in Ohio (2000), Great Sand Dunes in Colorado (2000) and Congaree in South Carolina (2003).

The management strategies adopted by the National Park Service have also evolved in line with contemporary concerns and conservation practice. In the 1920s, for example, the Park Service eradicated grey wolves from Yellowstone national park as part of a predator control strategy, only to reintroduce them in 1995 (Sellars, 1997). Challenges also come from the need to balance conservation and recreation, the public use of national parks having been part of their rationale from the outset even if it was decades before mass tourism reached Yosemite and Yellowstone. The provision of camping grounds, visitor centres and walking trails is an important part of the National Park Service's work, but the demand for recreation has begun to clash with conservation interests, particularly in the most untouched parks, such as the Alaskan

Table 13.2 Comparison of national parks in eight countries, 2003

	Number of parks	First established	Latest to be established	Largest park by area (km²)	Smallest park by area (km²)	Average area (km²)	Population	Ownership of land	Management/ governance
United States	56	Yellowstone 1872	Congaree 2003	Gates of Arctic 34,287	Hot Springs 22	3,917	Uninhabited	Public	National Parks Service
Canada	42	Banff 1885	Sirmilik 1999[a]	Wood Buffalo 44,840	St Lawrence Islands 8.7	5,344	Most parks uninhabited. Some small towns and nomadic populations	Public	Parks Canada – government agency
United Kingdom	13	Peak District 1951	Cairngorms 2003[b]	Cairngorms 3,800	The Broads 303	1,407	All parks are inhabited. Range 2,200 – 43,000	Mostly private (c.75%).	Appointed National Park Authorities
Ireland	6	Killarney 1932	Ballycroy 1998	Wicklow Mountains 159	The Burren 16.7	99	Uninhabited	Public	Dúchas – state heritage agency
France	7	La Vanoise 1963	La Guadeloupe 1989	Les Cévennes (913 core; 3,214 total area)	Les îles de Port Cros 37 (total area)	Core: 530 total area: 3,068	Uninhabited in core; populated in periphery	Mixed	Appointed Council of Administration
Germany	13	Bayrischer Wald 1970	Hanich 1997	Schleswig– Holsteinisches Wattenmeer 4,440	Jasmund 30	731	Limited local population	Mixed	Responsibility of local government
Australia	516[c]	Royal (NSW) 1879	30 new national parks established in Western Australia 2002–4	Kakadu (NT), 13,000	The Palms (QL) 0.12	500	Uninhabited	Public	Responsibility of state authorities
New Zealand	14	Tongariro 1887	Rakiura 2001	Fiordland 12,570	Abel Tasman 225	2,204	Uninhabited	Public	By Department of Conservation

[a] Gulf Islands National Park proposed 2003.
[b] South Downs and New Forest National Parks proposed 2003.
[c] Not including proposed new parks 2003.

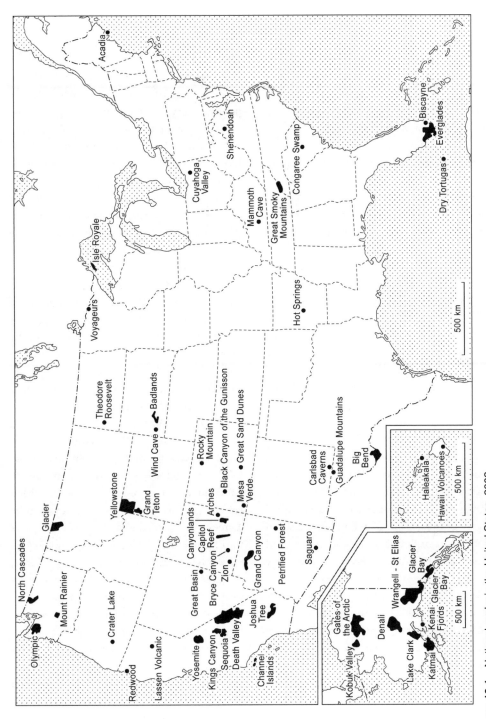

Figure 13.1 National parks in the United States, 2003

national parks. Similarly, whilst commercially valuable land was historically excluded from national parks, the creation of the large parks in Alaska was fiercely contested by business interests who identified the potential for oil extraction and mining, and conflicts have persisted over the development of roads and pipelines across the parks and over access for hunting and commercial fishing.

In the UK the groundwork for the creation of national parks was provided by two government reports during the Second World War – the Scott Report on Land Utilisation in Rural Areas (1942), which identified the need to protect valued rural landscapes from urban and industrial development, and the Dower Report on National Parks in England and Wales (1945), which proposed the framework for the establishment of national parks. These were followed by the National Parks and Access to the Countryside Act 1949, eventually leading to the designation of the first parks, the Peak District and the Lake District, in April and May 1951 respectively. Between 1951 and 1957 a further eight national parks were established, mainly in upland areas of western and northern England and Wales, although several were of close proximity to major urban centres (notably the Peak District, Yorkshire Dales and Northumberland) (Figure 13.2).

Both the Scott and the Dower reports had perceived urbanization and industrialization to be the major threats to the countryside, and accordingly the primary function of the new national parks was to impose strict controls on land use development. Significantly, farming was seen to be part of the conservation process, such that no specific controls were placed on agricultural practice in the national parks. As MacEwen and MacEwen (1982) have observed, this proved to be 'the fatal contradiction' (p. 10) of the system, as modern, productivist, farming practices dramatically

changed the landscape and damaged wildlife habitats (see Chapters 4 and 8), sparking conflicts in a number of national parks, particular Exmoor where the conversion of heather moorland to agricultural grassland led to a special inquiry (Lowe et al., 1986; Winter, 1996). Similarly, mining, hydro-electric power, reservoirs, military training and the construction of highways were all theoretically permitted in national parks, although the Dower Report had recommended that 'these should be permitted only upon clear proof of requirement in the national interest and that no satisfactory alternative site could be found' (Williams, 1985, p. 360). In practice, the failure of the national park authorities to prevent large-scale military training in the Brecon Beacons, Northumberland and Dartmoor national parks, the routing of the Okehampton by-pass through part of the Dartmoor park, or the construction of the Trawsfynydd nuclear power station in the Snowdonia national park (see Box 13.1), are all cited by critics as evidence of the weakness of the British system (MacEwen and MacEwen, 1982).

The designation of long-settled and cultivated areas as national parks prohibited the UK from following the American model, either in resettling residents or in acquiring land in national parks for public ownership. Three-quarters of the land in national parks in England and Wales is privately owned, 40 per cent as farmland. Among other things, the dominance of private landownership has shaped the nature of the recreational use of national parks. Although they have become major tourist attractions, public access has, until the Countryside and Rights of Way Act 2000, been restricted to designated 'rights of way' footpaths and to land owned and opened by philanthropic organizations such as the National Trust. Despite the respect for private

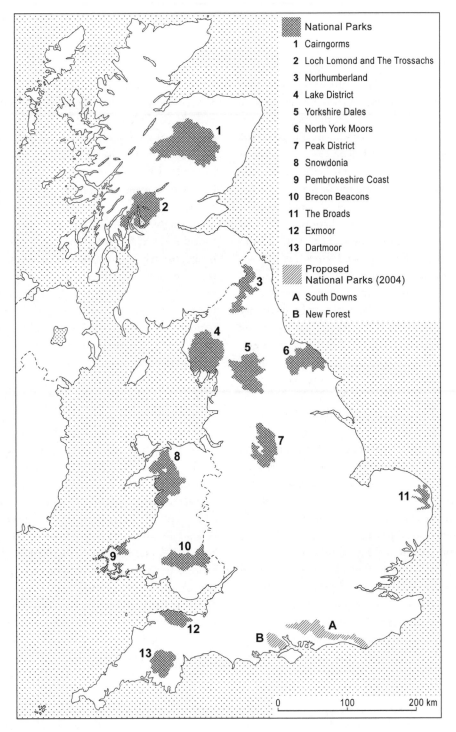

Figure 13.2 Established and proposed national parks in Great Britain, 2003

property, landowners strongly resisted the creation of many of the national parks. Landowner opposition precluded any move to establish national parks in Scotland during the 1950s, whilst successful landowner opposition to a proposed Cambrian Mountains national park in mid-Wales brought the programme of national park designations in England and Wales to an abrupt halt in the 1960s. It was not until the 1990s that further national parks were created, including the formal recognition of the Broads as a national park, the establishment of the first two national parks in Scotland, and proposals for two national parks in southern England (see Figure 13.2).

National parks in the United States and the UK conform to very different models, but there are two issues that cross-cut both systems, and those of other countries. The first is the balance between national and local interests. The very notion of a 'national park' implies that the management of the site should transcend local interests and be a national responsibility. Thus, in the United States, Canada and New Zealand, management is undertaken by a national government agency with little local input. Whilst most parks in these countries are uninhabited, there are neighbouring communities with particular economic and cultural interests in the parks, and in a number of parks there are small first nation populations whose welfare and cultural interests (notably hunting) may conflict with conservation priorities. In countries with more substantially populated national parks there is greater local involvement in park governance. National parks in Germany are managed through the normal local government system, whilst in the UK and France governance is undertaken by independent boards on which elected local politicians are represented alongside the appointees of national government. This does not, however, prevent protests from local residents that national park authorities are undemocratic or

that local economic interests and property rights are over-ridden by outside appointees.

The second cross-cutting issue is recreational use. As rural tourism has grown (see Chapter 12), national parks have emerged as key sites through which the countryside is consumed, particularly through activities that seek some kind of 'outdoor adventure' or 'reconnection with nature'. Traditionally these have included camping, hiking and sightseeing by car (the main visitor activities in British national parks according to a 1994 survey), but increasingly they also include 'adventure tourism', such as long-distance treks and helicopter trips. The majority of visits are made to national parks within relatively easy reach of urban centres, or with an established history of tourism, but increasingly even remote parks such as the Denali and Gates of Arctic parks in Alaska are becoming tourist destinations. Only the most inaccessible national parks are immune – the extremely remote Ivvavik national park in arctic Canada received just 170 visitors in 1995. Recreation has hence become a major part of the mission of national parks, with great economic importance for nearby communities. Indeed, the creation of new national parks in Australia, the UK and Canada during the past 20 years has been driven as much by anticipated economic benefits as by conservation. However, tourism has created pressures on the environment of national parks, through pollution, erosion and the demand for the development of roads, car parks and facilities. The desire to balance recreation and conservation has led the Federation of Nature and National Parks of Europe to promote sustainable tourism, encouraging activities such as walking, climbing, cycling, photography, school visits and nature camps, whilst discouraging large hotels, leisure parks and holiday villages, large group visits, skiing and the use of motorized boats and off-road vehicles.

Box 13.1   Comparison – *the Great Smoky Mountains and Snowdonia national parks*

The differences between national parks in the United States and the UK can be explored further by comparing two parks of similar size – the Great Smoky Mountains national park on the border of North Carolina and Tennessee (2,110 km^2) and Snowdonia national park in north-west Wales (2,142 km^2). The Great Smoky Mountains national park was established in 1934 following a campaign initiated by Mrs Willis P. Davis of nearby Knoxville, who had been inspired by the national parks of the western states. The Smoky Mountains (the 'Great' was added by the National Park Commission), however, were already settled and cultivated, with a significant logging operation. Creating the park meant purchasing land from private owners, with funds largely generated by public appeals and by a $5 million donation from the Rockefeller family. Some owners resisted selling and an unusual compromise allowed a number of families in Cades Cove (now a tourist attraction – see Chapter 12) to remain on a life-time lease after the establishment of the park. Generally, however, residents were resettled (ironically, European settlement had been enabled by the earlier resettlement of the Cherokee nation in the 1830s).

Today the landscape of the Great Smoky Mountains is dominated by forests which are home to 130 species of trees, 4,000 other plant species, and wildlife including the black bear, elk and red wolves – reintroduced by the National Park Service in 1991. It is recognized as an International Biosphere Reserve and a World Heritage Site. The national park is crossed by just one major road (built as part of the national park's establishment) and much of the area is accessible only by unpaved roads or walking trails. It is, however, the most-visited national park in the United States, with over 9 million visits each year, mostly between June and October. Facilities operated by the National Park Service include nine camping grounds, five riding stables, picnic areas, nature trails, three visitor centres and a lookout tower on the park's highest peak, Clingman's Dome (2,023 m). Environmental problems include water pollution and plant and animal disease, as well as air pollution that drifts in from as far as Cleveland, Ohio and Birmingham, Alabama. Average visibility at the Newfoundland Gap viewpoint has declined from 145 km (90 miles) to 35 km (22 miles) in 50 years and damage to plants has been recorded from sulphate and ozone pollution.

The Snowdonia national park was the third to be created in the UK, in October 1951. Covering a predominantly upland landscape, it includes the first and third highest summits in Wales – Snowdon (1084 m) and Cader Idris (892 m). Parts of the park are designated as a World Biosphere Site and as a World Heritage Site. Like all UK national parks, Snowdonia is populated and mostly privately owned. Private land constitutes 69.9 per cent of Snowdonia, with 15.8 per cent owned by the Forestry Commission, 8.9 per cent by the National Trust charity, 1.7 per cent by the Countryside Council for Wales, and just 1.2 per cent by the National Park Authority itself. The population of 26,267 is largely concentrated in towns and villages, including Dolgellau and Bala, with development tightly restricted outside these settlements. The former slate-mining town of Blaenau Ffestiniog is excluded from, but entirely encircled by, the national park.

The creation of the national park was opposed by landowners and local resistance forced proposals for an independent planning board on the model used in the Lake District and the Peak District to be dropped and responsibility to be vested in a joint committee of the elected county councils. The national park authority eventually gained independent status in 1995 and presently comprises a mix of local government representatives and members appointed by the Welsh Assembly. The authority has a budget of around £5 million and employs 120 staff.

(Continued)

Box 13.1   (Continued)

Agricultural land uses predominate. Nearly 45 per cent of the national park is open country, mostly grazed moorland, to which the public has recently gained a right of access. A further 31 per cent is enclosed farmland (again mostly grazing) and 15 per cent is forested. Agriculture remains significant to the local economy but agricultural modernization has been one of the key environmental challenges faced by the park – addressed in recent years through agri-environmental schemes. A more controversial source of local employment was the Trawsfynydd nuclear power station, the construction of which in the heart of the park in 1959–65 was regarded by environmentalists as an indictment of the park's ability to control development. The power station closed in 1993 and is undergoing decommissioning.

Snowdonia is the third most visited national park in Britain, with around 10 million visitor days annually. Just under half of visits are day trips and recreation in the national park successfully combines traditional activities such as sightseeing with more adventurous 'outdoor pursuits'. Both types of activity, however, have put pressures on the park's environment. In particular, problems of traffic congestion and associated pollution prompted the mooting of a 'congestion charge' in 2003 which would charge visitors to enter the national park by car or private coach and encourage use of public transport.

For more see www.great.smoky.mountains.national-park.com/info.htm and www.snowdonia-npa.gov.uk.

Other protected areas

National parks represent just one fraction of the matrix of protected areas that have been designated to promote countryside conservation. In the UK, for example, national parks are complemented by Areas of Outstanding Natural Beauty, National Scenic Areas, Sites of Special Scientific Interest and Nature Reserves (Table 13.3), whilst Australia's national parks comprise only around 43 per cent of the 604,000 km² of designated protected areas in the country. The objectives of these other protected areas generally differ from those of national parks in one of three ways. First, some designations apply similar conditions to those of a national park to a single natural feature or to a specific type of landscape. The US National Parks Service, for example, is also responsible for 70 national monuments focused on specific natural or heritage sites, ten national seashores, four national lakeshores, six national rivers and nine national wild and scenic rivers, as well as national historical parks, national battlefields, national

memorials and national historic sites that are protected for their historical and cultural significance, and national recreation areas, national scenic trails and national parkways which have a stronger recreational emphasis. Secondly, protected areas may recognize areas of scenic or cultural importance but afford a lesser level of protection or regulation than national parks. Areas of Outstanding Natural Beauty (AONB) in England and Wales, for example, are officially considered to be of equal importance to national parks, and involve specific measures to protect natural features, but until recently did not have independent management structures and are not charged with the same conservation and recreation functions as national parks (Green, 1996; Winter, 1996). Thirdly, some types of protected area have a stronger scientific rationale than national parks in their protection of habitats and wildlife. Sites of Special Scientific Interest (SSSI) in the UK, for instance, protect vulnerable habitats ranging from individual wildlife colonies to whole ecosystems such as woodlands or bogs. SSSIs

Table 13.3   Protected areas in the UK, 2002

Designation	No.	Total area (km²)	Notes
Area of Outstanding Natural Beauty (AONB)	50	24,087	England, Wales and Northern Ireland only. Require conservation of flora, fauna and landscape features
National Scenic Area	40	10,018	Scotland only. Equivalent to AONBs
Site of Special Scientific Interest (SSSI)	6,578	22,863	Scientifically important sites with regulation of certain activities
National Nature Reserve	396	2,405	Sites designated for the protection and study of flora and fauna. Managed by national conservation agencies
Local Nature Reserve	807	455	Sites of nature preservation managed by local authorities and conservation trusts

Source: Whitaker's Almanack 2003

receive specific protection from development in the planning system, but also regulate agricultural usage. Landowners are restricted in the activities that they can undertake and are required to notify conservation authorities of operations that may impact on the site. In practice, however, the reliance on voluntary cooperation by landowners means that the SSSI regulations can be difficult to enforce and around a quarter of SSSIs suffered some damage between 1982 and 1989 (Winter, 1996). In the United States, additional protection for wildlife and sensitive habitats has been introduced through the designation of national preserves, wildlife refuges and wilderness areas, many of which overlap with national parks. These areas should in theory experience minimal human activity, including prohibitions on the use of any motorized means of access or equipment in wilderness areas (Runte, 1997), but the threats that can still arise from the clash of conservation and economic interests were demonstrated in 2001 when the US administration floated plans to permit drilling for oil in the Alaska Wildlife Refuge.

Land Use Planning and Development Control

Protected areas help to conserve the most highly valued rural environments, but cover only a small proportion of rural space. Outside these areas, the character, appearance and environment of the 'everyday countryside' are threatened by land use change and development (see Chapter 8). Attempts to counter these threats by regulating rural land use and development have been introduced in many countries through systems of land use planning. Land use planning is not just about development control, and may be used proactively to encourage economic development and to provide infrastructure, but it is the development control function that is generally most significant in a rural context (see Cloke, 1988; Hall, 2002; Lapping et al., 1989). Yet there are also significant variations between countries in the form and scope of the planning system and in the degree of protection that is afforded to the rural environment. In Western Europe the development of rural locations tends to be subject to quite strict controls and regulation within a comprehensive national planning framework. In the United States and Australia, in contrast, the regulation of land use outside protected areas is more liberal and such development controls as exist tend to be initiated at a local level with no overarching national strategy. Accordingly, this section compares the highly regulated planning system in Britain with the more fragmented approach to development control in the United States.

The planning system in England and Wales

The introduction of the modern planning system in England and Wales by the 1947 Town and Country Planning Act was a victory for rural preservationists who had campaigned for the protection of the rural landscape and environment from urban and industrial development. As Hall (2002) observes, the 1947 Act effectively nationalized the development rights of land in England and Wales, such that the state – in the form of local planning authorities – decides which land may and may not be developed and that landowners do not have a right to develop their land without obtaining permission from the planning authority. This enables the state to exercise control over what is built where and to protect areas from development.

The system operates in a top-down manner. Planning law is formulated at a national level and guidance on its interpretation is issued by the planning ministers in England and Wales. Additional guidance may be issued on a regional basis, and regional planning conferences made up of local authorities in the region have a role in agreeing quotas for housing development in England. The national and regional policies inform the production by county councils of structure plans that outline the framework for land use and development control in a locality, and by district councils of local plans that designate particular plots of land for development on a community-by-community basis (see Murdoch and Abram, 2002; and Murdoch and Marsden, 1994 for an example of Buckinghamshire). Where the county and district councils have been merged into a single tier of local government (as in Wales), a single development plan is produced that combines the purposes

of the structure plan and the local plan. National park authorities are also responsible for producing specific plans for their territories. Once agreed, these plans form the regulations against which applications to develop land are assessed. A landowner or builder wishing to develop a piece of land (or alter an existing building) must apply to the local planning authority (usually the district council) for planning permission. This is granted only if the land is within an area designated for that type of development and if the proposal meets other criteria relating, for example, to the proposed building materials or to the safety of vehicular access.

The separation of urban and rural space has been a fundamental principle of the British planning system from the beginning (Murdoch and Lowe, 2003). This was most notably enforced through the creation of 'greenbelts' around metropolitan areas in which there is a strong presumption against any development. The first greenbelt, around London, was designated in 1947 and later extended such that it now forms a ring up to 80 kilometres (30 miles) wide (Hall, 2002). Further greenbelts were subsequently established around the other major British cities and conurbations. They have proved to be highly effective in meeting their original objective of limiting urban sprawl as well as in protecting agricultural land and in providing areas for countryside recreation close to urban populations. However, greenbelts have also contributed to counterurbanization (see Chapter 6) by restricting the opportunities for suburban expansion and thus encouraging migrants from the city to 'leapfrog' into rural areas beyond the ring (Murdoch and Marsden, 1994). Such is the pressure that has accumulated on these adjacent rural areas that in the late 1990s proposals were mooted to permit new housing development within greenbelts in Bradford, Newcastle and Hertfordshire (see Chapter 14). Furthermore, restrictions on development within greenbelts have had the consequence of inflating property prices, effectively turning them into exclusive middle class enclaves (Murdoch and Marsden, 1994).

The separation of rural and urban space is also practised in the local planning policies of broadly rural regions, as new development tends to be concentrated in small towns and large villages. Different planning authorities have adopted different strategies for achieving this. Most commonly, structure plans have employed a 'key settlements' policy of identifying particular towns and villages that are earmarked for expansion whilst development in other settlements is tightly restricted (Cloke, 1983; Cloke and Little, 1990). Other authorities have policies of market town concentration or of the severe restraint of development across the whole district (Cloke and Little, 1990). In all cases, however, local plans draw 'development envelopes' around towns and villages to which new build is generally constrained, thus prohibiting building in open countryside.

Overall, the planning system in England and Wales has been successful in regulating the development of rural land in a rational and systematic manner, in generally preventing random and unsightly structures from blighting the rural landscape, in preserving the rural character of small villages, in containing urban growth, in safeguarding agricultural land, and in protecting environmentally sensitive sites. Yet, it can also be criticized on a number of grounds. First, planning policies are more concerned with *where* things are built than with *what* is built. Even low-impact developments crafted into the landscape and using natural materials will be prohibited in open countryside, whilst new buildings within development envelopes are usually not obliged to conform to local architectural styles or building materials. Secondly, agriculture is largely exempt from the planning system. Agricultural buildings may be erected outside development envelopes without requiring planning permission and planning authorities have no powers to regulate farming practices such as the removal of hedgerows, which can dramatically change the appearance of a landscape. Thirdly, the planning system has been less effective at controlling large-scale infrastructure developments in rural space. Proposals for new roads, power stations, airports and similar projects generate considerable opposition and tend to be decided by a public inquiry, yet they are rarely rejected. Public inquiries are expensive and can take a number of years to complete, which led the government to propose new procedures in 2002 aimed at speeding up the planning process for large projects, but which rural campaign groups have argued will result in even less protection for the countryside.

Fourthly, the planning process favours the interests of the middle classes. The production of structure plans and local plans is in theory subject to consultation and democratic accountability. However, the groups that have the greatest influence in this process are the middle classes who dominate local government in rural areas and are most able to mobilize resources to lobby planners and to make representations in the appropriate technical language. The consequence is that development tends to be restricted in the more exclusive, middle class villages, and instead concentrated in towns and villages with a more mixed population. This tendency becomes self-reproducing as shortage of supply forces up property prices in the villages where development is most restricted, limiting the range of people who can afford to purchase housing in those communities. Murdoch and Marsden (1994) describe this as the production of a middle class space, as they describe for Buckinghamshire, to the north-west of London:

> The result is a delightful stretch of coun-
> tryside in the midst of an urban region.
> With Milton Keynes to the north, green
> belt and London to the south, the strug-
> gle to hold onto the rural in Aylesbury
> Vale is by no means easy. But the social
> make-up of the place means that a for-
> midable array of actors can usually be
> assembled to orchestrate opposition to
> unwelcome development. As its posi-
> tional status grows, the area will become
> even more attractive to those would-be
> residents who are trapped on the
> 'outside'. Thus, competition for resour-
> ces, notably housing, will continue to
> increase, making it more and more diffi-
> cult for those on low incomes to either
> stay in, or move to, such areas. The
> middle-class complexion of the locality is
> thus assured. (Murdoch and Marsden,
> 1994, p. 229)

Finally, the top-down nature of the planning system means that local level planning
policies are essentially responsive to wider
trends. As will be discussed in the next chapter,
this became markedly evident in the late
1990s when local authorities were asked to
produce plans that would accommodate a
projected 2.2 million new dwellings in rural
England, sparking widespread conflict.

Development control in North America

There is no comprehensive national framework for development control in either the
United States or Canada. Responsibility for
land use planning rests with states and
provinces and with local government and the
fragmentation of authority between different
agencies and authorities has severely restricted the effectiveness of the planning system.
Attempts to engender a more integrated
approach to planning have been made through

the establishment of planning commissions
for a number of urban regions, including
New York, Calgary and Edmonton (Hall, 2002).
Regional plans are not just concerned with
development control, but many have included
measures to preserve agricultural land or open
land for recreation. However, such plans tend
to be advisory rather than statutory and do
not in themselves offer protection for rural
space.

Although planning boards do exist within
American local government, they are generally
under-resourced and lack powers of enforcement. In the United States land use is more
significantly regulated through the zoning
process, which operates in parallel to but sep-
arately from the planning process, administered by separate zoning authorities and
enforced using laws concerned with public
health (Hall, 2002). Zoning designates different areas of land for different types of use,
such as housing, industry, commerce and so
on. However, the use of zoning to protect
open rural land has been limited. Only three
states – Hawaii, Oregon and Wisconsin – have
introduced state-wide zoning laws that permit land to be zoned for exclusive agricultural
use (Lapping et al., 1989; Rome, 2001).
Additionally, some municipalities have zoned
agricultural land as very large building lots,
thus reducing the impact of any development
(Hall, 2002). Yet, the zoning system as a whole
is considered to be only semi-efficient in regulating development, with developers usually
able to get their way eventually, particularly as
the right of landowners to develop their land
is perceived to be protected by the US constitution (Hall, 2002; Rome, 2001).

In the absence of an equivalent of the
British Town and Country Planning Act,
which nationalized development rights,
authorities in the United States wishing to
control development have introduced
schemes to purchase the development rights

Table 13.4 Farmland preserved by 46 easement programmes in 15 US states

	Scale of programme(s)	Farmland preserved Hectares	Farmland preserved Acres	Cost of programme ($m)
Maryland	County and local	105,019	259,307	464.6+*
Pennsylvania	County and local	60,286	148,861	394.0
Vermont	State-wide	40,763	100,651	56.8
California	County and local	34,189	84,418	102.4
Delaware	State-wide	26,478	65,377	69.5
Massachusetts	State-wide	21,384	52,800	135.0
Colorado	County and local	20,589	50,788	75.1
New Jersey	County and local	20,153	49,761	254.3
Connecticut	State-wide	11,684	28,850	84.2
Washington	County and local	6,693	16,527	62.1
New York	County and local	3,669	9,060	68.3
Virginia	Local	2,570	6,346	13.5
Wisconsin	Local	836	2,064	3.38
Michigan	Local	752	1,856	6.0
North Carolina	County	508	1,255	2.6

*Figures not available for two programmes in Maryland.
Source: Sokolow and Zurbrugg, 2003

of agricultural land. Farmland preservation schemes are employed by a number of states, counties and municipalities, mostly in the north-eastern United States, and have collectively protected over 730,000 hectares (1.8 million acres) at a cost of more than $2 billion (Sokolow and Zurbrugg, 2003). The most extensive coverage is in Maryland, Pennsylvania and Vermont, although financially the largest programmes are administered by Howard County, Maryland, which has spent $193 million on acquiring development rights, and the state of Massachusetts, which has spent $135.9 million, with other significant initiatives elsewhere in Maryland and in Pennsylvania, California and Vermont (Table 13.4). The payments made to landowners, known as 'easements', are generally for the difference between the value of the land as farmland and its value as land for development, and average around $810 per hectare ($2,000 per acre), although the amount can vary significantly depending on location. Funding is drawn from local taxes, bond issues and grants from federal, state and local governments. Federal funding as part of the

Farmland Protection Program was introduced in 1996 and has contributed to a more than three-fold increase in the amount of farmland protected between 1996 and 2003. A further $1 billion of federal funds were committed to the programme by the Farm Security and Rural Investment Act, 2002.

Elsewhere, tax incentives have been employed to encourage landowners to keep farmland in agricultural use. Schemes of this kind operate in every US state and Canadian province, including notably California and New York, where landowners are offered tax concessions in return for agreeing to maintain land as farmland for a specified period of time (Beesley, 1999; Hall, 2002), and Michigan, Wisconsin and Alberta, where lower taxes are offered to farm operators (Beesley, 1999). Agricultural districting has also been developed as an alternative incentive-driven approach, involving the collective, voluntary agreement of farmers to maintain agricultural land uses within a defined agricultural district in return for benefits including tax deferrals and exemptions from nuisance laws (Beesley, 1999; Lapping et al., 1989). Although over a third of farmland in New York State was included in agricultural districts in the 1970s, the approach has proved to be less effective in rural areas with immediate development pressures than in areas where the prospect of development is less critical (Lapping et al., 1989).

A final approach to controlling development on rural land in North America has been for public authorities to purchase land itself and manage it for public use. This is the model used in US national parks but it has also been employed on a smaller scale by county and municipal authorities to protect vulnerable land from urban development. San Francisco, for example, established a *de facto* greenbelt around the city in the 1960s and 1970s by this method, but by the 1970s many local authorities had found that property prices had become too high to permit large-scale acquisitions of undeveloped land (Rome, 2001). Private and community land trusts have also purchased open land to protect it from development, with more than 900 land trusts managing over 810,000 hectares (2 million acres) of farmland in the late 1980s (Beesley, 1999).

Overall, assessments of the effectiveness of development control strategies in North America have been mixed. Individual programmes have had a significant local impact by safeguarding land that would otherwise have been developed, but the majority of undeveloped rural land remains unprotected. The variety of approaches in operation, the costs involved in many of them and the lack of statutory enforcement powers have all militated against the emergence of comprehensive, integrated planning and weakened the degree of protection that can be maintained.

Agri-environmental Schemes

The primary objective of countryside conservation policy during much of the twentieth century was to protect the rural environment from urban-style development. Whilst the strictest forms of protected landscapes, such as national parks in North America, Australia and New Zealand, have also prohibited the farming of the land, more generally agriculture was regarded either as not part of the problem, or even as an ally of conservation. Development control policies in particular have been driven as much by a desire to protect land for agriculture as for its environmental value. Since the 1950s, however, it has been widely acknowledged that modern farming practices are also damaging to the rural environment (see Chapter 7), and that agri-environmental programmes are also needed that seek to change the conduct of agriculture.

As agri-environmental schemes involve the extensification of farming, they have fortuitously combined with pressures to reduce agricultural production in the post-productivist transition (see Chapter 4). Agri-environmental policies were first introduced in the European Union as part of attempts to reform the Common Agricultural Policy in the 1980s, and in the United States following lobbying by conservation groups in advance of the 1985 Farm Security Act. Having gained purchase during the wave of popular environmental awareness in the 1980s, agri-environmental schemes have been extended and reinforced in subsequent reforms of agricultural policy (Potter, 1998; Swanson, 1993; Winter, 1996).

Agri-environmental schemes may be targeted at any of a wide range of objectives, including reducing chemical use, pollution control, converting arable land to grassland, tackling soil erosion, reducing livestock densities, encouraging organic farming and maintaining and replanting woodland (see also Box 13.2). Some schemes are specifically tailored to addressing one of these objectives, whilst others aim to develop a more integrated approach within a particular designated area. One of the first schemes to be introduced in

Britain, for example, involved the designation of 28 Environmentally Sensitive Areas (ESAs), collectively comprising 16,889 km², selected on the basis of their environmental significance and the potential for a promotion of traditional farming practices to prevent further damage to the environment (Winter, 1996). Farmers in ESAs are eligible to receive an acreage-based payment in return for signing up to a management agreement that typically includes restrictions on fertilizer use and stock densities, prohibitions on herbicide and pesticide use and on the installation of new draining and fencing, and commitments to maintain landscape features such as hedges, ditches, woods, walls and barns.

Box 13.2 Re-foresting the countryside

The natural state of much of rural Europe is woodland, yet centuries of cultivation have cleared the forests to allow for agriculture and, later, urbanization. However, in recent decades initiatives have been introduced to increase native woodland cover by planting surplus farmland with trees. In England, where the proportion of forested land had fallen from 15 per cent in 1086 to just 4.8 per cent in 1890 before recovering with the planting of industrial coniferous forests in the early twentieth century, policies to replant broadleaved woodlands were introduced in the 1980s. The Farm Woodland Scheme (later the Farm Woodland Premium Scheme – FWPS), operated as part of the European Union's Agri-Environment Programme, paid up to £195 per hectare to farmers to replant broadleaved woodland. In recognition of the long timescale involved in broadleaf afforestation, payments were guaranteed for 40 years for oak and beech, 30 years for other broadleaves and 20 years for other woodland (Mather, 1998). In 1995–6 a total of £3.6 million in grants was paid under the FWPS, with a further £16.1 million paid under the Woodland Grant Scheme administered by the Forestry Commission.

Afforestation has also been promoted through large-scale projects including the National Forest, designated in 1991 in the former rural coalfield area of the English Midlands and similar schemes in Lancashire and around Bristol. The National Forest project aimed to plant 30 million trees over one-third of its 500 km² area through a mixed strategy of land purchases by Forest Enterprise (the commercial arm of the Forestry Commission) and the Woodland Trust (a charity), the development of community woodlands and incentives to encourage voluntary planting on farmland, including support from the FWPS (Cloke et al., 1996). As well as environmental enhancement, projects such as the National Forest also aim to create spaces for recreation and to stimulate tourism and economic regeneration.

A similar scheme in the Republic of Ireland – which has also experienced substantial historic deforestation – celebrated the Millennium by planting the People's Millennium Forests. Using funds from private sponsorship, the project planted a tree for every household in Ireland at 16 sites around the country. Each household was given a certificate with details of the location of 'its' tree, which they could visit using the grid maps at the woodland sites (Figure 13.3).

Both the English National Forest and the People's Millennium Forests projects have attempted to involve local people in the reforestation of the countryside. However, as Cloke et al. (1996) demonstrate, public attitudes to forests are mixed and draw on deep-rooted cultural associations. Whilst some people perceive forests to be 'a living, breathing, peaceful place in which humans and wildlife can cohabit in tranquility and

(Continued)

Box 13.2 (Continued)

happiness' (p. 569), others identify forests with fear and being 'closed in' and 'overpowered' by trees. Similarly varied representations of forests – and variable usages of woodland as, alternately, places of refuge and places beyond surveillance – are also found in communities close to more established woodlands and commercial forests (Marsden et al., 2003).

Although some scepticism has been expressed about the achievements of the Farm Woodland Scheme (see Mather, 1998), the combination of agri-environmental programme payments, large-scale forestry projects and natural growth on abandoned farmland had increased woodland cover in England to 8.4 per cent in 2000.

For more on the National Forest and public perceptions of woodland see Paul Cloke, Paul Milbourne and Chris Thomas (1996) The English National Forest: local reactions to plans for renegotiated nature–society relations in the countryside. Transactions of the Institute of British Geographers, 21, 552–571. For more on rural forestry in general see the chapter by Alexander Mather in Brian Ilbery (ed.) (1998) The Geography of Rural Change (Longman). More information on the English National Forest is also available at www.nationalforest.org and on the Irish People's Millennium Forests at www.millenniumforests.com

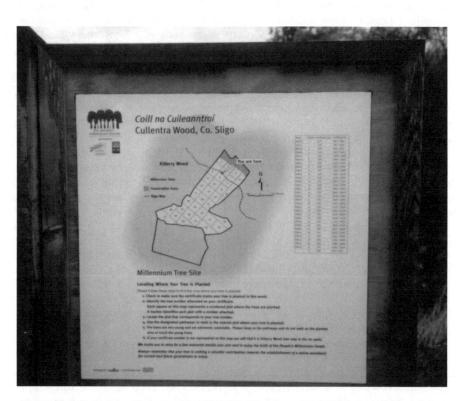

Figure 13.3 Map of tree planting in the People's Millennium Forest at Cullentra Wood, County Sligo, Ireland
Source: Woods, private collection

Whilst the ESA programme in the UK targeted a broad range of conservation issues, the Conservation Reserve Program (CRP) introduced by the 1985 Farm Bill in the United States was initially focused on the single problem of soil erosion. Under the programme, farmers were paid around $60 per hectare to protect land highly prone to erosion, for example by planting trees or grassland (Potter, 1998; Swanson, 1993). Like many agri-environmental schemes, the rationale behind the CRP was as much about production control as about environmental gain and critics have claimed that during the first wave of the CRP the USDA was keener on the former than the latter and had targeted the programme at land where the greatest reductions in production could be achieved, not at the land most at risk from erosion (Potter, 1998). None the less, by 1992 a total of 14.5 million hectares (11 per cent of all US cropland) had been enrolled into the CRP and an estimated 22 per cent reduction in soil erosion – 700 million tons per year – had been achieved (Potter, 1998). Moreover, in 1991 the scope of the CRP was extended to include filter strips to watercourses, wellhead protection areas and state water quality and conservation areas, whilst two parallel schemes, the Wetlands Reserve Program and the Agricultural Water Quality Protection Program, were also introduced (Green, 1996).

The CRP and its offspring correspond with most initiatives in the EU agri-environmental programme in using financial incentives to promote conservation through farming. In contrast, the Conservation Compliance policy introduced in the United States at the same time as the CRP employed the threat of negative sanctions. Conservation Compliance required farmers to implement an approved conservation plan for their farms that would demonstrate that their operations did not seriously contribute to soil erosion (Potter, 1998; Swanson, 1993). Farmers not implementing a plan by 1 January 1995 forfeited their eligibility for federal subsidies, including commodity payments, crop insurance and disaster relief funds. Some states, such as Iowa, have taken the negative sanctions approach further by drafting strict laws to protect against soil erosion and water pollution, with the threat of prosecution for farmers who fail to comply (Simon, 2002).

Production control is not the only side-benefit of agri-environmental schemes. The Landcare programme in Australia combines conservation with the promotion of public participation in environmental management. Landowners and farm businesses are encouraged to participate with other community actors in Landcare groups to address problems of land degradation through the principles of self-help, cooperation and localized action (Lockie, 1999a, 1999b). The *Tir Cymen* scheme in Wales, meanwhile, and its successor, Tir Gofal, have positioned agri-environmental action as part of broader rural development goals. By paying farmers who enter into agreements to maintain environmental features on their land, the schemes aimed to support the sustainability of traditional forms of agriculture in the rural economy and to create new employment opportunities. Overall, it was calculated that Tir Cymen generated 29 days of environmental work per year in its three pilot areas, which translated into longer periods of employment for casual workers on farms (Banks and Marsden, 2000).

Agri-environmental schemes are based on the voluntary participation of farmers and hence their effectiveness depends upon farmers' evaluations of the incentives and benefits promised by agri-environmental schemes against the costs and effort involved, including the loss of productivity. Although the uptake of some early schemes was modest, and there remain significant geographical variations, agri-environmental schemes are now implemented by a significant minority of farms. Around 20 per cent of farmland in the European Union was estimated to have been enrolled into schemes in the EU's agri-environmental programme by 2000, and 30 per cent of farms in Australia participate in the Landcare programme (Juntti and Potter, 2002; Lockie, 1999a). Participation is greater among younger farmers and more entrepreneurial farmers, and the principal factors cited by farmers as reasons for participation are varied. However, whilst Wilson and Hart's (2001) study of two agri-environmental schemes in the UK found that around half of participants said that they would not continue with conservation measures if the scheme's funding was withdrawn, they also reported indications of a shift in attitude towards more environmental thinking among participating farmers. As such, Wilson and Hart reiterated the findings of previous studies in suggesting that participants in agri-environmental groups can be classified as 'active adopters' motivated by environmental imperatives and 'passive adopters' for whom financial incentives are important and where little change is required to existing farm management practice, whilst non-participants can be divided into 'conditional non-adopters' who might join if the conditions of the scheme were different, and 'resistant non-adopters' who are strongly opposed to participation (Wilson and Hart, 2001).

Mixed conclusions have also been reached about the effectiveness of agri-environmental schemes in protecting the rural environment. Specific objectives are often met, but not necessarily with a wider impact. Winter (1996) notes concerns about a 'halo effect' whereby farmers entering into agri-environmental agreements with respect to one element of their activities (for example, stocking on rough grazing land) compensate by intensifying production elsewhere, whilst Lockie (1999b)

argues that Landcare in Australia actually contributes to the intensification of farming. More positively, studies in the UK have suggested that agri-environmental schemes have contributed to a revival in populations of butterflies, insects and birds.

Animals and the Rural Environment

Initiatives to protect the rural environment have largely been concerned with the preservation of current landscapes and habitats against further degradation and with relatively modest projects to reverse the decline of plant or wildlife populations or to restore neglected landscape features. In only a minority of cases has action been taken on a more ambitious scale to reconstruct past rural environments. These include the retreat from agricultural cultivation in a number of national parks and afforestation schemes, as well as initiatives to reintroduce once native animal species. Animals are important both to natural environmental systems and to discursive representations of the countryside. As such, the protection of wildlife has been a key objective in many conservation initiatives, with action focused on habitat protection informed by scientific knowledge and supported by popular sympathy for the vulnerability of high-recognition fauna such as birds, butterflies and brown bears. Notably, less popular attention has been directed to the significance of livestock to the countryside and to the homogenization of livestock through selective breeding and the predominance of the most productive varieties under productivist agriculture (Yarwood and Evans, 2000). Although conservation projects have been launched to protect rare breeds of cattle, sheep, pigs and so on, such schemes have little public profile compared with wildlife preservation (Evans and Yarwood, 2000).

More controversial are initiatives to reintroduce animal species that had been consciously eradicated or reduced in number by a previous generation because of their perceived incompatibility with modern farming. Perhaps the most ambitious scheme of this kind is the 'Buffalo Commons' project in the Great Plains of the United States to restore the native population of bison, which was virtually eradicated in the nineteenth century, partly for sport, partly for the trade in hide, but also partly because the bison's extinction was seen as a precondition for the establishment of commercial cattle ranching (Manning, 1997). In recent decades, bison herds have been re-established in a number of small pockets on the plains, but the Buffalo Commons idea would see bison return to roam across a vast region from Montana and North Dakota south to Texas. As envisaged by its proponents, Frank and Deborah Popper, the initiative is based on a projection of continuing depopulation and agricultural decline on the plains, which would create the opportunity for cultivated farmland to be restored to an unfenced grassland common. As Popper and Popper (1999) describe, the 'Buffalo Commons' is a metaphor as much as a detailed proposal, representing an idea of a new regional economy based on tourism, hunting, bison meat and leather and the exploitation of native plants (see also Manning, 1997; Popper and Popper, 1987). Yet, as they observe,

the metaphor has developed its own life, reproduced in different ways by the many proponents of the initiative in the region, but also by its opponents, including landowners and ranchers who fear the loss of property rights and the marginalization of their commercial activity.

Further opposition again is generated by schemes to reintroduce animals that are perceived to be a direct pest to agriculture. Opposition from landowners was blamed for the stalling of plans to reintroduce the beaver to Scotland, and projects to reintroduce large carnivores such as wolves have provoked fierce resistance in a number of localities. Brownlow (2000), for example, describes how proposals to reintroduce the grey wolf to the Adirondack Mountains of New York State clashed with local cultural constructions of the wolf as 'vermin', out of place in the settled countryside, where it preyed on livestock and more valued wildlife such as deer. Moreover, the proposals were represented as an attempt by urban-based conservationists to impose their ideological values on the rural environment. This resonates with the idea of the 'globalization of values' discussed in Chapter 3, in which an increasingly globalized environmental ideology has promoted conservation standards that conflict with local lay knowledge of nature and rurality.

Summary

The protection of the natural environment has become a major influence on the way in which rural space is managed. This greening of rural policy has been a response to a number of factors associated with rural restructuring. First, there has been growing recognition of the damage caused to the rural environment by modern agriculture and by urbanization and physical development (see Chapter 7), which together with the broader spread of environmentalism, has generated public support for countryside conservation. Secondly, as agricultural overproduction has become a major policy problem (see Chapter 4), the rationale provided by conservation interests for reducing the intensity of agriculture has become increasingly attractive to policy-makers. Thirdly, the economic restructuring of rural space has involved the commodification of the countryside as a space of consumption more than a space of production (see Chapter 12), such that the protection of aesthetically valued landscapes makes more economic sense than the degradation of the environment through resource exploitation. These various imperatives have coalesced to form a coalition of interests that collectively have achieved the greening of rural policy.

Within the scope of countryside protection, however, there are a wide range of different schemes and initiatives, addressing many specific problems, that have adopted different strategies and approaches. Some rely on voluntary participation, others have compulsory measures backed by law. Some use financial and other positive incentives to encourage conservation, others employ negative sanctions against those who fail to comply. Some work with farmers, landowners and other traditional users of rural space, others seek a more fundamental change in the use of rural land. Even within the environmental lobby there are differences about objectives and methods. Although 'conservation' is frequently used as an overarching term for environmental protection, it in fact implies an acceptance of moderate change that is absent from the more dramatic goal of 'preservation'. Both conservation and preservation, meanwhile, take the present state of the rural environment as a starting point, and thus differ from schemes that aim to reconstruct past environments.

Furthermore, any initiative to protect the rural environment might still encounter opposition from farmers, landowners, developers, hunters, logging companies, oil and mineral exploiters, and other commercial operations, who may contend that their economic interests, their welfare and their rights are being subordinated to concerns for plants and animals. The scientific and philosophical rationale of environmental projects may be contested, with local lay discourses of nature asserted in defence of the status quo. Resistance may also be organized around the perceived imposition of alien, urban-based, environmental values on rural people. As such, environmental issues have proved to be fertile ground for the emergence of rural conflicts, as is discussed in the next chapter.

Further Reading

Fairly comprehensive accounts of measures to protect the rural environment, including protected landscapes and agri-environmental schemes, are provided from a predominantly British perspective by Bryn Green in *Countryside Conservation* (Spon, 1996) and Michael Winter in *Rural Politics: Policies for Agriculture, Forestry and the Environment* (Routledge, 1996). The story of national parks in the United States is authoratively told by Alfred Runte in *National Parks: The American Experience* (University of Nebraska Press, 1997). Adam Rome, in *The Bulldozer in the Countryside* (Cambridge University Press, 2001), discusses the history of attempts to restrict urban expansion into the American countryside. Development control strategies and farmland preservation schemes are also discussed by a number of contributors to Owen Furuseth and Mark Lapping's edited volume *Contested Countryside: The Rural Urban Fringe in North America* (Ashgate, 1999). For more on the Australian Landcare programme see work by Stewart Lockie, whilst the work of Clive Potter explores many aspects of agri-environmental policy in Europe. Recommended readings include S. Lockie, 'The state, rural environments and globalisation: "action at a distance" via the Australian Landcare program', *Environment and Planning A*, volume 31, pages 597–611(1999); C. Potter, 'Conserving nature: agri-environmental policy development and change', in B. Ilbery (ed.), *The Geography of Rural Change* (Addison Wesley Longman, 1998); and C. Morris and C. Potter, 'Recruiting the new conservationists: farmers' adoption of agri-environmental schemes in the UK', *Journal of Rural Studies*, volume 11, pages 51–63 (1995).

Websites

More information on national parks can be found on a number of official national websites, including the British Council for National Parks (www.cnp.org.uk), British Association of National Park Authorities (www.anpa.gov.uk), Parks Canada (www.pc.gc.ca), the Irish National Heritage Service (www.duchas.ie/en/NaturalHeritage/NationalParks), the New Zealand Department of Conservation (www.doc.govt.nz) and the United States National Park Service (www.nps.gov), as well as on a good unofficial site for American national parks (www.us-national-parks.net). The website of the UK Department for the Environment, Food and Rural Affairs (www.defra.gov.uk) provides details of both planning policy and agri-environmental schemes in Britain, whilst the website of the American Farmland Trust (www.farmland.org) includes information on both farmland preservation programmes and agri-environmental schemes in the United States. For more on the Australian Landcare programme see the websites of Landcare Australia (www.landcareaustralia.com.au) and the National Landcare Program (www.landcare.gov.au).

14

Rural Conflicts

Introduction

Social and economic restructuring has turned the countryside into a far more complex space than it once was. In the past the economic dominance of agriculture and other resource-exploitation industries and the relative stability of rural communities meant that hegemonic discourses could represent the rural as a homogeneous space and that such homogenizing representations were taken as the basis for rural policy and the organization of rural life. For example, the powerful identification of the rural with agriculture meant that agricultural interests were prioritized in rural policy (see Chapter 9) and mainstream rural life was organized around farming. The processes of restructuring, however, have exploded such simple representations (Mormont, 1990). There are now many different representations of the rural mapped over the same physical space, informed by different social constructions of rurality (see Chapter 1) and by different economic and ideological interests. In some cases, different representations of rural space can co-exist, but often the implications that follow for the management of rural space prove to be incompatible. There is, for example, an inherent contradiction between a representation of the rural as a working community in which employment needs to be provided by the exploitation of natural resources, and a representation of the rural as a pleasant place to live, whose attraction rests on the absence of industry and the preservation of the landscape; or between a representation of a field as a piece of land to be grazed as a part of agricultural production and a representation of a field as a habitat of rare plants and insects which must be protected. Tensions of this kind have given rise to what Mormont (1990) describes as 'the symbolic battle over rurality' (p. 35), in which a multitude of conflicts have erupted about the legitimacy or appropriateness of different developments, initiatives and policies in 'rural' space.

Rural conflicts range from disputes over relatively parochial matters such as noise and smells from farms, access to footpaths and the provision of street-lighting, to protests against developments of new housing, industrial sites, roads, waste dumps and power stations, to debates over the designation and management of protected areas, agricultural practices and the regulation of 'rural' activities such as hunting. In the wake of social and economic restructuring, rural conflicts tended to emerge first on a local scale – the level at which everyday life was most directly impinged upon. However, in many cases the conflicts involved not just local actors, but individuals, pressure groups, companies and agencies located outside the immediate rural area. As such there has been an 'up-scaling' of rural conflicts, as campaigners have been forced to engage in local, regional and national politics in attempts to change policy decisions (see for examples Murdoch and Marsden, 1995; Woods, 1998b, 1998c). At the same time, rural campaigners have also mobilized in response to perceived threats to rural communities, landscapes and culture from national government policy initiatives to reform agriculture, introduce new conservation measures, restructure public services, regulate hunting and promote public access to the countryside (Woods, 2003a).

In this way, rural issues have since the 1980s moved from the margins to the mainstream of political debate in a number of countries. There has, of course, always been a degree of political debate about rural policy, but as Chapter 9 discussed, much of this was traditionally channelled into relatively closed, private, policy networks. Such rural protests as did from time to time occur tended to concern issues of private property interests, environmental protection, or sector-specific disputes, most notably with regard to agricultural policy (see Box 14.1). The transformation that has taken place over the past two decades is that such 'rural politics' have been replaced by a new 'politics of the rural' in which the very meaning and regulation of rural space is the defining issue (Woods, 2003a). Or, as Mormont put it:

> if what could be termed a rural question exists it no longer concerns issues of agriculture or of a particular aspect of living conditions in a rural environment, but questions concerning the specific functions of rural space and the type of development to encourage within it. (Mormont, 1987, p. 562)

This chapter examines three types of rural conflict that are typical of the new 'politics of the rural'. The first concerns the development of rural space and conflict between planning rationales that promote the need for development and concerns about environmental impact and the loss of 'rural character'. The second case study is about conflict over the use of natural resources in rural space, and the balance between agricultural and conservation interests. The third conflict is about the perceived threat to a 'rural way of life' from attempts to prohibit or regulate the hunting of wild animals. The chapter concludes by discussing the emergence of a broader 'rural movement' comprising groups concerned with campaigning across a range of different issues in order to defend or promote particular representations of rural identity.

Box 14.1   Farmers' protests

The political mobilization of farmers has played an important part in shaping the historical trajectory of rural politics and policy. The unionization of farmers in the late nineteenth and early twentieth centuries helped to reinforce the position of agriculture at the heart of rural policy. Yet, even as farm unions were incorporated into agricultural policy communities (see Chapter 9), protests and demonstrations by farmers continued in many countries, either to exert further pressure on politicians, or as an expression of discontent by dissident farm groups with the way in which agricultural interests were represented by the mainstream unions. The latter motivation was behind protests by the American Agricultural Movement (AAM) in the 1970s. A loose alliance of small farmers, the AAM organized two 'tractorcade' protests in Washington, DC to demand increases in price support for agricultural commodities and action to tackle farm debt. The first tractorcade, in January 1978, brought 3,000 farmers to the capital, whilst the second, in February 1979, caused traffic congestion with a 40 km (25 mile) long line of tractors (Stock, 1996).

Discontent with the mainstream farm unions also fuelled periodic protests by militant farmers in France from the 1950s onwards, usually directed at trade policies and proposed reforms of the European Union's Common Agricultural Policy (CAP) which threatened to cut farm incomes. Blockades of roads, railways and ports, mass demonstrations, graffiti and the hijacking of lorries carrying imported meat have all formed part of French farmers' protests, as has occasional violence (Naylor, 1994). The tradition is most visibly continued by the *Confédération Paysanne* (see also Box 3.3), though with a more progressive, anti-globalization spin.

In the 1990s, falling farm prices and the collapse of traditional agricultural policy communities provoked the adoption of protest tactics by farmers in the UK, Ireland and Australia. The first protests in the UK targeted imports from Ireland, with an impromptu blockade of ferry ports in the winter of 1997–8. Subsequent protests coordinated by the radical grassroots Farmers for Action group have been directed at supermarkets, food processing plants, dairies and creameries as recession spread across agricultural sectors (Woods, 2004a). Most notoriously, farmers joined with hauliers in September 2000 to blockade oil refineries and fuel depots as part of a Europe-wide series of protests against the level of fuel taxes.

For more on farmers' protests in the US, France and the UK, respectively, see Catherine McNicol Stock (1996) Rural Radicals (Cornell University Press); Eric Naylor (1994) Unionism, peasant protest and the reform of French agriculture. Journal of Rural Studies, 10, 263–273; and Michael Woods (2004) Politics and protest in the contemporary countryside, in L. Holloway and M. Kneafsey (eds), Geographies of Rural Societies and Cultures (Ashgate).

Contesting Development in the Countryside

The development of the built environment has become a frequent focus for rural conflicts for a number of reasons. Building projects involve highly visible changes in the rural landscape that potentially affect a number of people in the surrounding area and also present a clear, tangible object around which protest can be mobilized. They also raise issues that are significant for several different discourses of rurality.

From the perspective of government officials and planners, for example, the location of large-scale developments in the countryside has been conventionally regarded as both acceptable and appropriate. The availability of land and the relatively sparse population have made rural areas attractive locations for large-scale or noxious developments such as power stations, airports and waste dumps that would be unacceptable in densely populated urban areas. Similarly, it was the received wisdom that rural spaces needed to be traversed by highways and railways connecting major cities; that the construction of reservoirs and dams was part of a rural economy based on resource exploitation; and that housing and industrial developments were necessary as part of regional development strategies.

Support for development has also come from rural local government and rural businesses for whom the development of infrastructure is a necessary part of the modernization of rural areas. Thus, new houses are needed to accommodate in-migrants and to replace sub-standard existing housing; industrial plants and tourism sites are needed to create employment as agriculture declines; and new roads, railways and airports are needed to ease the economic disadvantages of peripherality.

These pro-development discourses dominated in rural policy until the closing years of the twentieth century, since when they have been challenged by anti-development campaigners. Protests against developments have often mobilized concerns about the environmental impact, including damage to sensitive landscapes and habitats; but they have also been motivated by concerns about the aesthetic quality of rural space, as represented by the ideal of the 'rural idyll' (see Chapter 1). Proposals to develop rural land have therefore been opposed on the grounds that the new structures would disfigure the landscape, disturb the tranquillity of the countryside with noise or light pollution, compromise the 'rural character' of small settlements, or introduce land uses that are perceived to be 'urban' or, at least, 'non-rural'.

The emergence of conflicts organized along these lines has been frequently associated with counterurbanization and, as such, there is a temptation to represent them as conflicts between locals and in-migrants. Spain (1993), for example, identifies conflict over development in Lancaster County, Virginia, as a struggle between 'come-heres', or in-migrants, who place a higher value on environmental quality and preservation and longer resident 'been-heres', who are more accommodating of growth that brings economic benefits. Similar distinctions have been recorded as voiced by local politicians in rural southern England (Woods, 1998b).

However, closer examination reveals that the situation is often more complex than the local/incomer dichotomy suggests. Many in-migrants do, as Spain observes, have greater access to resources than longer-term residents and are more able to mobilize politically. But this is usually a reflection of the class composition of counterurbanization, and conflicts that are purely expressed in terms of differential resources might be more accurately portrayed as class conflicts than as local/incomer conflicts. In-migrants do have a particular motivation to oppose developments if they perceive that their financial and emotional investment in a rural location is under threat, but this motivation may be shared by longer-term residents who have made property investments or who have strong emotional attachments to place. There are also significant cleavages within both the in-migrant and locally raised communities. For example, the introduction of street-lighting may be proposed by one group of in-migrants used to illuminated urban streets on safety grounds, but opposed by others who regard the absence of artificial lighting as part of the rural character of a community. It is hence more useful to think of attitudes towards developments as stemming from different discourses of rurality that may cut across categories of class, length of residence, age and so on, and of conflicts over development as involving ad hoc coalitions of actors motivated by a range of different rationales.

Housing development in rural Britain

The complexities involved in conflicts over development in rural space are demonstrated by the example of conflicts over the development of new housing in rural areas of the UK. New housing development is controversial in all developed countries because it implies an increase in population with spin-off requirements for more infrastructure, and is easily regarded as 'urbanization' (see Chapters 9 and 13). As detailed in Chapter 13, as a result of such concerns, new housing development in rural Britain is regulated through the planning process which determines both the quantity and the location of new building. Periodic plans are produced by democratically elected local councils through a process that involves consultation with stakeholders and the public (see Murdoch and Abram, 2002; Murdoch and Marsden, 1994). Campaign groups representing both developers and conservationists have sought to influence the outcomes of this process through representations to public inquiries and lobbying of decision-makers, but up to the 1990s there was a general consensus that outside greenbelts and national parks a degree of new housing development was necessary and could be accommodated without significantly impairing the rural environment.

The preparation of new plans in the mid-1990s, however, proved much more controversial. Population projections based on the trend of counterurbanization and anticipated changes in social behaviour estimated that properties for an additional 4.4 million new households would be required in England between 1991 and 2006, and that half would need to be built on undeveloped land in rural areas. These figures were agreed by the national government, and allocated between counties by regional planning conferences, before they attracted public attention as county councils sought to build them into local structure plans. The multi-scalar conflicts that followed are illustrated by the case of Somerset, a rural county in south-west England, which was allocated a quota of 50,000 new houses between 1991 and 2016 (Woods, 1998b).

The conflict over the proposed housing development progressed on three scales. First, a campaign was launched at the county council to challenge the target of 50,000 new houses. The campaigners, who included both locals and in-migrants, conservation organizations and environmental groups, emphasized concerns about the environmental impact and the loss of rural character. The local branch chair of the then Council for the Protection of Rural England (now the Campaign to Protect Rural England – CPRE), for example, told a local newspaper that 'the way of rural village life which has slowly evolved over the centuries is going to be wiped away at a stroke' (quoted by Woods, 1998b, p. 20). Such representations reflected typically in-migrant and middle class discourses of rurality, but the campaign was also supported by councillors aligned with local working class interests who opposed building houses for in-migrants when there was a shortage of affordable housing for local people.

Secondly, the debate about medium-term housing development highlighted a number of localized conflicts in the county where protesters were opposing more immediate plans for housing developments. In these cases, opposition was mobilized around the perceived threat of the developments to views of rural landscapes, to village character and to natural habitats. At least one correspondent to the local newspaper was explicit about their motivation of defending their investment in the rural idyll:

> We object to any development at all. We came here from an estate in Taunton [Somerset's largest town] for the views and the privacy and that didn't come cheap. (Quoted by Woods, 1998b, p. 20)

Within Somerset, the number of pro-development representations was limited. Those prepared to speak publicly suggested that development was needed in order to create housing for local people, to prevent economic opportunities being lost to other parts of the country and simply in the name of 'progress'. This minority did, however, have the inertia of the planning system on their side. Despite the popular campaign there was in effect very little that the county council could do to change the plans without a significant change of policy at a national level. Thus, thirdly, the campaign in Somerset fed alongside others into a national campaign orchestrated by the CPRE against the target of 2.2 million new houses across rural England as a whole, which portrayed the plans as the destruction of the English countryside. Only this national-scale campaign met with any success, eventually achieving a modest reduction in the amount of new build allocated to rural sites.

Rural Resource Conflicts

Rural resource conflicts have a long history, but the increasing emphasis placed on nature conservation combined with the declining economic significance of resource-exploitation industries, including agriculture and forestry,

means that the appropriate use of resources has returned as an explosive political issue that brings different representations of rural space into conflict. Flashpoints for resource-related conflicts include the management of agricultural land and forestry. The Great Bear Rainforest along the coast of British Columbia, for example, has been the focus of conflict between logging companies and conservationists over the impact of clear-cutting operations on the natural ecosystem of the region, which the Wilderness Committee campaign group represents starkly as a choice between commercial forestry and wildlife.

One of the highest profile rural resource conflicts in recent years, however, is located further down the west coast of North America, in the Klamath Basin of southern Oregon and northern California. The region is naturally high desert, but bisected by the Klamath River and associated wetlands. In 1905 the Bureau of Reclamation embarked on an ambitious project to support agriculture in the basin by reclaiming wetlands and irrigating desert land through a constructed network that now includes seven dams, 45 pumping stations, nearly 300 km (185 miles) of canals and 830 km (516 miles) of irrigation channels (LaDuke, 2002). Almost 100,000 hectares of farm and ranchland normally rely on water from the project, diverting around 25 per cent of the river's mean annual flow. The region has been heralded as a model of watershed restoration and biodiversity preservation, but tensions between the different users of the limited water supply – including farmers and ranchers, the indigenous Klamath tribe and the basin's wildlife – are deep-seated (Doremus and Tarlock, 2003).

In the summer of 2001, however, a severe drought dramatically reduced the water level in Upper Klamath Lake, threatening stocks of suckerfish and Coho salmon. Under the provisions of the Endangered Species Act, project managers closed the headgates of the irrigation system, maintaining the flow of water to the lake but cutting the supply for irrigation by 90 per cent. As crops failed, farm incomes plummeted and several faced bankruptcy. In response, farmers, ranchers and supporters mounted a campaign of civil disobedience, including an unauthorized attempt to reopen the headgates by flag-waving protesters on 4th July.

An above-average snowmelt helped to avoid a repeat of the crisis in 2002, but the conflict between farmers and conservationists continued, with farmers resentful that their interests had been placed below those of fish. Tensions have also emerged within the farming community, between those who want to sell up and others who believe that any further decline in farming would threaten a support economy of tractor dealers, fertilizer suppliers and seed distributors, thus undermining a whole rural community.

Hunting and the Rural Way of Life

In Chapter 3 it was observed that one of the most significant processes of globalization to impact on rural areas was the 'globalization of values'. In particular, the late twentieth century witnessed the global spread and popularization of a set of values concerning human interactions with nature, which introduced new standards for environmental protection and promoted ideas of animal rights. These values – which are founded on a mixture of environmental philosophy, green ideology, scientific representation and lay knowledge of benign nature – are often at variance with the understandings of nature that form part of traditional rural folk culture and knowledge. As such, conflicts have developed as governments and agencies have attempted to introduce new laws and regulatory frameworks based on the new, globalized, environmental values that relate to traditional rural activities. These are manifest,

for example, in conflicts over animal welfare in farming, including the use of battery pens and the transport of live animals (Buller and Morris, 2003). However, the issue that has generated the most notable conflict is hunting.

Hunting is a highly symbolic activity for many traditional discourses of rurality. It represents a rural way of life that is closely entwined with nature, in which humans must pit their wits against nature to survive, but are ultimately able to exert their power and control over nature. Yet, from the perspective of the global values of animal rights, hunting is represented as a cruel and barbaric activity. The growing influence of the latter position within society as a whole has led to the adoption of new laws and measures aimed at regulating or prohibiting hunting in a number of countries including France, Belgium and, most controversially, the UK.

The debate in the UK has focused on the hunting of wild mammals, including foxes and deer, with hounds. This quintessentially British form of hunting is undertaken by organized groups, with specially trained hounds chasing the scent of the quarry, followed by the hunters, usually mounted on horseback. Hunts, and the rituals associated with them, have been an important part of rural culture since at least the early nineteenth century and were traditionally both an element in the maintenance of elitist rural power structures (Woods, 1997) and a focal point for community activity (Cox et al., 1994; Cox and Winter, 1997). Although hunting remains a significant activity for particular sections of the rural population, and within particular localities (Milbourne, 2003a, 2003b), its presence within the British countryside as a whole has been diluted by in-migration, the decline of agriculture and changes in social attitudes.

Opposition to hunting also has a long history in the UK, fuelled by a combination of concern for animal rights and class politics.

From 1945 onwards a number of attempts were made to introduce legislation to ban the hunting of wild mammals with hounds, gaining momentum during the 1980s and 1990s. In 1997 the general election produced both a Labour government with a manifesto commitment to holding a free vote on hunting, and a perceived anti-hunting majority in the House of Commons. A Bill to outlaw hunting was quickly introduced, but fell foul of parliamentary procedure. Other attempts followed, but the strength of opposition mobilized by the pro-hunting lobby tempered the enthusiasm of the government for a ban and led to a series of stalemates and stand-offs. Demands for more objective evidence resulted in the establishment of an independent Commission into the likely impact of a ban on the rural economy and society (the Burns Inquiry), whilst the newly devolved parliament in Scotland pre-empted the rest of the UK by legislating for a ban in December 2001.

Whereas the anti-hunting argument is mainly grounded in animal rights, pro-hunting campaigners have advanced their case by identifying hunting with rurality, such that an attack on hunting becomes an attack on the rural. There are three elements to this strategy. First, hunting is represented as being core to the rural way of life. A ban on hunting, it is argued, would cost jobs, remove a focal point of community life and increase social exclusion (see Woods, 1998c). Secondly, the scientific and ethical basis of a ban are challenged by the assertion of rural representations of nature which hold that hunting is entirely natural, that hunted animals do not suffer unnecessary cruelty, and that hunting is needed as a form of pest control for farmers (Woods, 2000). To ban hunting would therefore be to ignore the wisdom of rural society. Thirdly, the right of an urban-centred society to impose its values on the countryside is contested. Thus the conflict over hunting is

recast as a conflict between the rural and the urban in which the rural is portrayed as the underdog, defending civil liberties and rights.

The association of hunting and rurality was reproduced in the tactics adopted by pro-hunting campaigners. Judging that the issue of hunting alone would not attract sufficient public support to halt legislation, campaigners formed the Countryside Alliance pressure group and positioned hunting as just one area of rural concern in which, they argued, the rural way of life was under attack from a misguided urban government. This message was conveyed through a range of protest events, including three mass demonstrations in London – the Countryside Rally (July 1997), the Countryside March (March 1998) and the Liberty and Livelihood March (September 2002). The publicity for the demonstrations emphasized a number of issues, including agricultural recession, housing development and the closure of rural services, but the defence of hunting remained the core motivation for both the organizers and the majority of participants (Woods, 2004a). References to hunting predominated on the banners and placards carried by marchers, whilst many slogans played up the notions of a rural–urban conflict and of the countryside being under siege (Figure 14.1).

The strategy of the Countryside Alliance has been to represent the British countryside rising up as one in support of hunting. However, the realities of the restructured countryside are far more complex. Even within rural areas there is a significant minority opposed to hunting and discourses of rurality are also drawn on to support a ban. Hunting, for example, is portrayed as an infringement of the rural as a space of nature, and as an activity that is offensive to other users of rural space such as walkers and picnickers (Woods, 1998c, 2000). As such, conflict over hunting is as much an intra-rural conflict as a conflict between rural and urban society.

Figure 14.1 Hunting associated with civil liberties in the 2002 Liberty and Livelihood
March in London
Source: Woods, private collection

Summary

The case studies discussed in this chapter provide an indicative illustration of the types of issue
around which rural conflicts have developed in recent years. The arguments mobilized, the
actors involved and the precise way in which the meaning and regulation of rurality is
contested in each case will depend on the context. However, there are common themes and
regularities that emerge across different specific conflicts and which reflect the influence of
broader discourses of rurality. The Countryside Alliance in Britain, for example, has
demonstrated the potential for a range of issues to be connected together under an umbrella
theme of 'defending' the rural from urban interference. Yet, the representation of the rural that
is mobilized and defended by the Countryside Alliance is based on a very particular discourse
that associates the rural with agriculture, private landownership, homogeneous communities
and 'traditional' activities such as hunting. This is a discourse that is in itself exclusionary and
which is not subscribed to by many contemporary residents of the British countryside. It is
hence little surprise that adherents to this discourse will find themselves engaged in numerous
conflicts as its assumptions are challenged in the more complex rural world created by
restructuring. Moreover, there are other campaign groups that have also sought to make
connections between rural conflicts, but from very different perspectives. The *Confédération
paysanne* in France, for example, has made connections between the economic interests of
small farmers, environmental issues and counter-globalization (Woods, 2004a). The Rural

Coalition in the United States, meanwhile, draws similar links between sustainable farming, environmental protection, social justice issues, indigenous and minority rights and community development. A progressive politics of rurality of this type has also been pioneered by groups seeking to follow alternative rural lifestyles, such as low-impact housing developments, as discussed in Chapter 21.

Collectively, these various rural campaign groups together with the multitude of small-scale, informal protests focused on localized conflicts, represent a substantial political mobilization of individuals around the issue of rural identity. Increasingly, these groups are taking the form of a new social movement (Woods, 2003a). Like all social movements, the emergent rural movement is a loosely structured collection of autonomous groups, with no organizing centre, identified leaders or even a coherent ideology, and is united only by the centrality placed on rural identity. Yet there is even disagreement about what rurality means, with at least three strands identifiable (Woods, 2003a):

- *Reactive ruralism* involving the mobilization of a self-defined 'traditional' rural population in defence of purportedly historic, natural and agrarian-centred rural 'ways of life'.
- *Progressive ruralism* that opposes activities and developments that conflict with a discourse of a simple, close-to-nature, localized and self-sufficient rural society.
- *Aspirational ruralism*, which involves the mobilization of in-migrant and like-minded actors to defend their fiscal and emotional investment in rural localities by seeking to promote initiatives that further the realization of an imagined 'rural idyll', and resisting developments that threaten or distract from this imagined ideal rural.

Each strand of ruralism may provide motivation for campaigners to mobilize against perceived external threats to rurality. In some cases, alliances and coalitions are built between groups representing the different strands, for example to protest against the construction of new roads. But, these different strands of ruralism also inform the disagreements around which conflicts within the rural develop, providing the opposing sides in Mormont's 'symbolic battle over rurality'.

Further Reading

For more on the emergence of a politics of the rural and of the rural movement, see Marc Mormont, 'The emergence of rural struggles and their ideological effects', *International Journal of Urban and Regional Research*, volume 7, pages 559–575 (1987) and Michael Woods, 'Deconstructing rural protest: the emergence of a new social movement', *Journal of Rural Studies*, volume 19, pages 309–325 (2003) – two papers published 16 years apart at different stages of the politicization of the rural. A number of journal papers and book chapters provide more information on various aspects of the hunting debate in Britain, including two papers by Paul Milbourne: 'The complexities of hunting in rural England and Wales', *Sociologia Ruralis*, volume 43, pages 289–308 (2003b) and 'Hunting ruralities: nature, society and culture in "hunt countries" of England and Wales', *Journal of Rural Studies*, volume 19, pages 157–171 (2003a) – both of which report research undertaken for the Burns Inquiry on hunting. Michael Woods, 'Researching rural conflicts: hunting, local politics and actor-networks', *Journal of Rural Studies*, volume 14, pages 321–340 (1998), examines an attempt to ban stag-hunting in part of southern England, whilst Woods's contribution 'Fantastic Mr Fox? Representing animals in the hunting debate', in C. Philo and C. Wilbert (eds), *Animal Spaces, Beastly Places* (Routledge, 2000) analyses the use of language and imagery in the hunting debate. For more on the case study of the politics of housing development in rural England see M. Woods, 'Advocating rurality? The repositioning of rural local government', *Journal of Rural Studies*, volume 14, pages13–26 (1998). For more on the conflict over water regulation in the Klamath region, see H. Doremus and A.D. Tarlock, 'Fish, farms, and the clash of cultures in the Klamath basin', *Ecology Law Quarterly*, volume 30, pages 279–350 (2003).

Websites

There are a large number of websites with information relating to the conflicts discussed in this chapter, many of them maintained by campaign groups on either side of the debates. For more on housing development in Britain see the websites of the Campaign to Protect Rural England (www.cpre.org.uk) and the House Builders' Federation (www.hbf.co.uk). For the Klamath conflict see www.klamathbasinincrisis.org for a pro-farmer perspective, and the Klamath Basin Coalition (www.klamathbasin.info) for the conservationist argument. The main pro-hunting organization in Britain is the Countryside Alliance (www.countryside-alliance.org), whilst the main anti-hunting group is the League Against Cruel Sports (www.league.uk.com). The Wildlife Network (hot.virtual-pc.com/wildnet/wildnet.shtml) advocates a 'middle way' of regulated hunting. The report of the government-commissioned inquiry into the impact of banning hunting with hounds on the rural economy and society, together with research and supporting material, can also be found on the web at www.huntinginquiry.gov.uk

Part 4

EXPERIENCES OF RURAL RESTRUCTURING

15

Changing Rural Lifestyles

Introduction

The preceding chapters in this book have discussed the processes of social and economic change that have impacted on rural areas over the past century, and the responses that have been adopted by communities, governments and other policy-makers. Inevitably, much of the discussion has focused on structural changes, institutions and policies. This final part of the book shifts attention to the people who live and work in the countryside and their experiences of rural restructuring and its consequences. As noted in Chapter 3, Hoggart and Paniagua (2001) have argued that rural restructuring involves qualitative as well as quantitative change. By examining the changing nature of rural lifestyles, and by listening to people's own personal narratives of those changes, evidence can be found for the qualitative aspects of rural restructuring to complement the quantitative evidence that has been described in a number of earlier chapters.

The contrast between the lifestyles of rural people today and those a century ago is stark. In the early twentieth century, rural lifestyles were characterized by insularity, a lack of technological appliances, a strong social hierarchy and moral framework for community life, and a deep involvement in agricultural work and connection to the natural world. Humphries and Hopwood (2000), for example, relate the memories of residents of rural England in the 1920s and 1930s, in which hard work and isolation feature prominently:

> I'd work on Saturdays, evenings, doing all the menial tasks on a farm at a very early age, much earlier than most other boys had to. Even as a small boy of five years old I would have to go out into the fields at certain seasons. I remember my father would be digging potatoes with a fork, my mother would be picking them up and I would be behind with a big basket – as big as myself – picking up what we called 'the chats', which were small potatoes for the pigs. (Albert Gillett, rural child, quoted in Humphries and Hopwood, 2000, pp. 34–35)

> We were absolutely isolated. We had no other farm round us for at least three to four miles. And the nearest building, Bigland Hall, was about a mile and a half from us. Oh, we had no

news of the outside world because there was no wireless [radio] and we never had a newspaper; we would have had to walk two and a half miles to pick one up and we couldn't afford one in any case. I just lived from day to day by what my husband came and told me. He'd tell me who had died, who'd bought a farm and who was moving: little titbits of news. (Marian Atkinson, farmer's wife, quoted in Humphries and Hopwood, 2000, p. 130)

Elements of this lifestyle have been romanticized as part of the rural idyll myth (see Chapter 1). For those who lived them, however, they were lives of poverty, ill-health and limited opportunities. The modernization of rural society was for many rural people an emancipation. The story of rural restructuring is hence a complex one that cannot be represented as wholly positive or negative. Such mixed experiences and emotions are also evident in narratives of more recent rural change. The case studies that follow recount personal stories of change from two different communities – the farming community of New Zealand in the context of the liberalization of agriculture in the 1980s and 1990s; and a village in southern England in the context of counterurbanization and gentrification. In each case the stories convey a profound sense of change that is associated with a deep connection to place and to rural identity, yet they also reveal the contingency in individuals' attitudes to change and in the responses that they adopt.

Farmers' Tales of Agricultural Restructuring in New Zealand

The New Zealand farming community experienced one of the sharpest and most severe episodes of economic restructuring of any rural industry following reforms introduced by the New Zealand government in the mid-1980s to deregulate agriculture, removing subsidies and forcing farmers to compete in an unfettered free market (see Chapter 9 for more details). As Sarah Johnsen (2003) observes, most accounts of the deregulation and subsequent restructuring of the farming sector have been written from a macro-level economic perspective based on national-scale statistical evidence. In contrast, Johnsen explores farm families' own narratives of the experience of restructuring through a case study of Waihemo, South Island. She found that the farmers would initially reiterate the conventionally received rhetoric that deregulation had been a 'good thing' that had strengthened New Zealand agriculture. Yet when asked

about their own personal experiences, farmers frequently told a different story that emphasized the 'pain' involved and the unequal ability of farm families to respond to the challenge.

Important elements in the painfulness of the experience for farmers who struggled in the new economic circumstances included the sense of absorption in farming as a tradition and a way of life and of attachment to a particular rural place:

Tradition keeps us here. And just familiarity. It's really difficult to ... To leave farming would be really stepping outside our comfort zone in some ways. It's really hard to make a break with what you know and launch out into the unknown although at times I wish we could do it. (Male farmer quoted by Johnsen, 2003, p. 140)

I remember one year prices were particularly bad. And it was hardly worth it on an economic basis ... It was worth it for the health and wellbeing of your sheep, but it was hardly worth it economically to drench them, to shear them. Because there was just no return for your money. By the time you paid your wages and the prices of your inputs and that. It was just crazy. It's very saddening, disheartening, to work hard all year on a place you care about and make a loss. (Female farmer quoted by Johnsen, 2003, p. 141)

For many farmers the commitment to place had its own narrative, a story about how the farm had been built up through hard work. Thus livelihood, lifestyle, life history and location were all entwined together:

We actually got the loans and bought the whole thing lock, stock and barrel. So we had that big investment. We weren't like a lot of other families where provision had been made for the son to come in. All our savings went into it and we bought it before we got married ... At that stage we put so much into it that it was ours. We put our money into it and did all the upgrades of fences and things. And the girls [our daughters] are closely aligned with the farm too I suppose. They always came out with us ... And I suppose there were nice feelings associated with it. To walk away from it ... I suppose I'll have to do it one day. But I think I'd find it quite hard to do. (Female farmer, quoted by Johnsen, 2003, p. 142)

The loss of a farm, or an inability to farm land to its 'full potential', often produced feelings of disappointment, failure and guilt:

Personally I've been disappointed that we haven't been able to maintain the farm to the standard ... You know, some farms are just, they're lovely. They're like showpieces, aren't they? I would like to have been able to put more into our farm

than we did. It's always nice to leave the land in a better condition than when you found it for the next generation. It's nice to see a place being farmed to its full potential. We just haven't been able to do that. (Farmer, quoted by Johnsen, 2003, p. 144)

Such feelings were informed by the official discourse which implied that 'good farmers' prospered in the deregulated agricultural system, whilst 'bad farmers' floundered. However, Johnsen found that farmers themselves rejected this simplistic reading, pointing instead to the variability of impacts at a farm level. Rather, Johnsen argues, farmers' experiences and responses were mediated at the farm enterprise, household and property level by factors including the level of farm enterprise debt, the division of labour in the household and the stage in the family lifecycle, and the size and land quality of the property; as well as by the gender, knowledge and experience, values and attitudes, goals and sense of place of the individual farmer and by the wider context including the local biophysical conditions, characteristics of the local economy and local farm culture.

Villagers' Tales of Community Change in Southern England

The village of 'Childerley' (a pseudonym) is typical of many in southern England. Michael Bell, who spent six months living in the village for an ethnographic study in the early 1990s, observes that it does not have any particularly outstanding views or noteworthy buildings, but nor has it been subjected to substantial new house building (Bell, 1994). As discussed in Chapter 1, Childerley is small and historic enough to be understood by residents as 'rural', and for many residents the rurality of the village is reinforced by a perception that it offers a connection to a past way of life.

Yet, as Bell also notes, Childerley, like many communities on the edge of London's commuter belt, has undergone considerable social change over the past four decades as new, wealthier migrants have moved in (see also Chapter 6). The narratives of rural life told by the villagers to Bell are hence strongly coloured by a sense of change in the village that has both physical and social manifestations:

> It's become sterile. The sort of nitty-gritty has gone out of it. Now everything is being preserved and washed and painted, but the, what I mean, the heart of it has gone. The buildings have been preserved but the character has been lost.
>
> The big problem is that the community spirit is gone, or at least much declined. There's no longer any common purpose, no common goal. That's what is needed to hold a place together. I can't really see where it's all leading to.(Childerley residents quoted by Bell, 1994, pp. 95–96)

The idea of loss of community spirit is an important feature of the villagers' stories, providing a device for describing the way in which the pattern of social interaction has changed from inward-looking, collective activity within the parish, to more expansive, outward-looking and individualistic lifestyles. As Bell reports, for many older residents this change is recounted with a sense of regret:

> I preferred the old village life. It was real friendly, like one big family. We always used to go out visiting. You didn't need a television. You just went and asked someone. If anything happened, if someone was sick or something, everyone knew it soon enough. When I was a girl, you made your own entertainment. You read, the women sewed, we played cards. We would listen to the wireless.

> You were more self-sufficient. Everyone knew each other. Now your neighbours don't know you. It's not the same anymore. (Long-term resident quoted by Bell, 1994, p. 98)

Even newer in-migrants reproduced the rhetoric of the loss of community, with some acknowledging that the perceived decline of community interaction was a result of the different lifestyle that they follow compared with traditional rural lifestyles:

> I think it has something to do with our characters … I suppose some people make friends with their next door neighbours because it's their next door neighbours. I think we tend to only make friends with them if they were similar sorts of people to ourselves anyway. (Management consultant and newcomer, quoted by Bell, 1994, p. 98)

However, changes in lifestyle were also attributed to economic and social changes within the established village, notably the decline in agricultural manual labour and the weakening of paternalistic class structures. Even though memories of the old era tended to emphasize the hardship of life, some older villagers such as a former estate worker quoted by Bell, looked back on the period with nostalgia, arguing that 'it *was* better' (Bell, 1994, p. 116, emphasis in the original).

Neglected Rural Geographies

The stories told by the individuals quoted by Johnsen and Bell are highly personal and are shaped by the particular characteristics, circumstances and experiences of the people involved. They are 'situated knowledges', constructed from particular personal positions and perspectives (Hanson, 1992). The same observation can be made about the stories told by rural researchers in the academic books and articles that we write. We approach the rural from particular social and educational backgrounds and bring with us particular interests, biases and preconceptions that inform the research that we do and the analysis we undertake. It is no coincidence that in the mainstream of rural studies throughout the twentieth century a bunch of rural researchers who were predominantly white, middle class, middle aged men concentrated almost exclusively on those aspects of rural activity that also involved white, middle class, middle aged men – farming, industry, resource exploitation, policy-making and planning.

This point was forcibly made by Chris Philo in a paper published in the *Journal of Rural Studies* in 1992. The paper, 'Neglected rural geographies', was itself inspired by Philo's reading of a book, *The Child in the Country*, by the environmental writer Colin Ward (Ward, 1990). As Philo (1992) describes, Ward investigates the condition and experiences of children in the British countryside, providing an insight into rural life that was virtually absent from academic rural research at the time (see Chapter 17 for later work on rural children). For Philo, the neglect of children in rural studies highlighted by Ward's book revealed a wider neglect of 'other' rural experiences in mainstream rural research which led to a troubling misrepresentation of the rural experience:

> there remains a danger of portraying British rural people (or at least the ones that seem to be important in shaping and feeling the locality) as all being 'Mr Averages': as being men in employment, earning enough to live, white and probably English, straight and somehow without sexuality, able in body and sound in mind, and devoid of any other quirks of (say) religious belief or political affiliation. (Philo, 1992, p. 200)

In response, Philo issued a challenge to rural geographers and allied researchers to take seriously the 'others' who also occupy rural space:

> why should rural geographers not investigate the social relations of health and illness or of ability and disability, and in so doing inquire more specifically into the geographies implicated in the 'otherness' of sickness, physical disability and mental disability as spun out in rural surroundings? And why should rural geographers not reflect upon the social relations of sexuality, and why do they not consider the possibility that their (as it were) equivalent to the gay and lesbian 'ghettos' and networks described by urban geographers is actually the lack of such phenomena because tightly knit rural communities are such unforgiving sites for the expression of alternative sexualities? And why too should they not think about a multitude of other 'others': gypsies and travellers of all sorts, 'New Age hippies' and companion seekers of 'alternative lifestyles', homeless people and tramps, all of whom trace out complicated geographies both across real 'rural space' and in the spaces of their own imaginations. (Philo, 1992, p. 202)

The challenge resulted in a wave of research during the 1990s that sought to recognize and engage with the diversity of experience in the countryside, much of which employed qualitative methodologies that aimed, as Milbourne (1997a) comments quoting Duncan and Ley (1993), to decentre the 'privileged sites from which representations emanate' (Duncan and Ley, 1993, p. 2) and to enable a 'polyphony of voices' (p. 8) from the rural (see also Cloke and Little, 1997).

Some of these studies are discussed in the following chapters, providing an insight into the rural geographies of children, the elderly, migrant workers, indigenous peoples, ethnic minorities, gay and lesbian communities and travellers, among others. There is little doubt that research of this nature has contributed to a fuller, more sensitive, understanding of contemporary rural life, but it has not been immune from criticism. Little (1999), for instance, raises concerns about the lack of theoretical discussion of the terms 'the other' and 'the same', arguing that,

> Too many studies have rather glibly labelled groups or individuals as 'other' with seemingly little recognition of the power relations and processes of transgression involved in such categorization. Studies of the rural other cannot and should not be undertaken without some reference to the basis of a particular form of othering; why are certain identities othered, who gains or benefits from such positioning and who are those who are 'the same'. (Little, 1999, p. 438)

To do this, Little suggests that rural research needs to engage with the role of broader configurations of power such as racism, patriarchy and homophobia. Similarly, Little critiques studies of the rural other for their static treatment of group and individual identity, underrepresenting the extent to which identities are uncertain and subject to change. Thus, there is further for research on 'neglected rural geographies' to go if it is to be more than simply a form of 'academic tourism' through the experiences of a catalogue of uncritically defined 'othered' groups.

Gender and the Rural

The most numerous 'other' group to be marginalized in conventional mainstream rural studies were women. Practically the only attention afforded to women in traditional rural research was in a few sociological studies of farm households in which women appeared only in a preconceived supporting role undertaking duties of household management or childcare, or secondary economic activity (this stereotype has persisted in many media representations of farm women – see Morris and Evans, 2001). As Whatmore et al. (1994) observed, the farm 'family' was 'treated as an organic entity accessed through, and represented by, a single individual – the farmer, or head of household, both masculine-defined terms' (p. 3), contributing directly to the invisibility of women in academic accounts of rural life.

Despite the popularization of feminist theory in human geography and related social sciences in the 1980s, its application in rural studies remained limited. Feminist perspectives were introduced into studies of farm women (Gasson, 1980, 1992; Sachs, 1983, 1991; Whatmore, 1990, 1991), labour markets (Little, 1991), community life (Middleton, 1986; Stebbing, 1984), and environmental activism (Sachs, 1994), but these remained, in Friedland's (1991) phrase, 'a fugitive literature' (p. 315). Such studies did, however, demonstrate the necessity of building gender into analysis of rural change, as noted by Whatmore et al:

> The interests contesting rural restructuring build on and, in turn, reshape gender relations; empowering and disempowering women (and men) in different ways in particular localities, complicated by their intersection with other axes of social power relations, notably class, 'race' and ethnicity. (Whatmore et al., 1994, p. 2)

Little and Austin (1996) explored one aspect of this, by examining through a case study of the village of East Harptree, in south-west England, how the ideal of the rural idyll impacts on the everyday life of women. In particular, they focused on the importance placed in the rural idyll on 'community' and 'the family'. The sense of community was strongly expressed by women in the village, especially in-migrants, but maintaining a functioning community also placed expectations on women, with one resident observing that,

> Lots of the organization is done by women, but not exclusively. Men get involved with village football. During the week the women run the village. (Quoted by Little and Austin, 1996, p. 108)

Similarly, the belief that the countryside is a 'better' environment in which to raise a family was also commonly stated by women in the village, even if the practicalities of looking after and providing transport for children created difficulties, especially for women in full-time employment. As such, Little and Austin noted the high proportion of women who had effectively become full-time mothers and that several women were conscious that their lives and identities in the village had become oriented around their role as mothers. By the same token, there was awareness that women without children could feel excluded from community activities:

> It would be a bit isolating here without kids. Everything is organised through kids … Children give you a legitimate presence in the village. (Young mother quoted by Little and Austin, 1996, p. 106)

Thus, Little and Austin argue that aspects of the rural idyll serve to reinforce traditional gender relations and roles, including motherhood and the centrality of women in the community. As they conclude, 'those aspects of the rural way of life most highly valued by women appear to be those that offer them least opportunity to make choices (for example, about employment or domestic responsibilities) outside their conventional roles' (p. 110).

Little and Austin's study is also important in marking a transition from work that focused on the structural dimensions of gender difference in rural societies to work concerned with exploring how gender identities and ruralities can be mutually constituted. This latter aspect has become more significant as research on gender and the rural has increased since the mid-1990s. One recent review of gender research in rural studies, for instance, noted not only that gender research has begun to flourish, but that the rural itself has started to stimulate new perspectives on gender, as 'meanings associated with rural places and cultures provide new insights into the study of sexual identity, for example, and the relationship between gender identity and the body' (Little and Panelli, 2003, p. 286). An appreciation of gender dimensions is increasingly demonstrated in research across a diverse range of rural topics. As such, there is consciously no specific chapter on gender in this book, but rather in addition to this short introductory section, gender is also discussed in later chapters in the context of youth lifestyles and sexualities, employment and alternative rural lifestyles.

Summary

Understanding the nature and dynamics of the contemporary countryside requires not just a knowledge of structural changes and their statistical expression, and of institutional and policy responses, but also an appreciation of how rural restructuring has been experienced by people living and working in rural areas and how rural lifestyles themselves have changed. The chapters in this final part of the book concentrate on the experiences of rural restructuring. They focus on some of the key aspects of rural life – the quality of rural housing and health, and fear of crime in rural areas (Chapter 16); the lifestyles of children, young people and the elderly in rural communities (Chapter 17); employment and working life (Chapter 18); poverty, deprivation and homelessness (Chapter 19); the situation of ethnic minorities and indigenous communities in rural areas (Chapter 20); and attempts to pursue 'alternative' rural lifestyles outside the mainstream (Chapter 21). In order to establish the context, these chapters do describe structural changes, citing statistical evidence where appropriate, and discuss policy, but significantly space is also given to the voices of rural people themselves talking about their experiences.

Further Reading

More on the two case studies of farmers in New Zealand and of the English village of Childerley respectively can be found in Sarah Johnsen, 'Contingency revealed: New Zealand farmers' experiences of agricultural restructuring', *Sociologia Ruralis*, volume 43, pages 128–153 (2003) and Michael Bell, *Childerley: Nature and Morality in a Country Village* (University of Chicago Press, 1994). Chris Philo's paper 'Neglected rural geographies: a review', in the *Journal of Rural Studies*, volume 8, pages 193–207 (1992), is still essential reading for students of rural society, whilst discussions of the broadening of concern in rural research to embrace 'othered' groups can be found in the introductions to the edited volumes by Paul Cloke and Jo Little, *Contested Countryside Cultures* (Routledge, 1997) and by Paul Milbourne, *Revealing Rural 'Others': Representation, Power and Identity in the British Countryside* (Pinter, 1997). For more on gender and the rural, see recent writing by Jo Little and co-authors, including: J. Little, *Gender and Rural Geography* (Prentice Hall, 2002); J. Little and P. Austin, 'Women and the rural idyll', *Journal of Rural Studies*, volume 12, pages 101–111 (1996); J. Little and R. Panelli, 'Gender research in rural geography', *Gender, Place and Culture*, volume 10, pages 281–289 (2003).

16

Living in the Countryside: Housing, Health and Crime

Introduction

Is life better in the countryside? Many people apparently think so. A British opinion poll in the late 1990s found that 71 per cent of respondents believed that the quality of life is better in rural areas, and that 66 per cent would live in the countryside if there were no barriers to doing so (Cabinet Office, 2000). Similarly, 59 per cent of Canadians living in urban centres told a 1989 survey that they would prefer to live somewhere 'more rural', whereas 85 per cent of rural farm residents were content with their current location (Bollman and Briggs, 1992). Behind the preferences for rural living lie a comparison of stereotypical images of the city and the countryside. Significantly these include perceptions about the relative quality of housing and health and the level of crime – staple factors in the quality of life. Typically, an image of picturesque, spacious rural housing, set in a pleasant, healthy and pollution-free environment with an absence of crime, is contrasted with images of crowded, sub-standard or monotonous urban housing in a polluted and unhealthy environment in which crime is rife and the streets are unsafe. This chapter critiques this simplistic representation by exploring in turn the actual conditions of rural housing, health and healthcare, and crime levels.

Rural Housing

One of the most powerful elements of the rural idyll myth is the image of the rose-covered cottage as the ideal country property. However, few rural residents live in such homes, and the reality of rural housing is far more complex than the stereotype suggests. Jones and Tonts (2003) quote five features of rural housing described by earlier research for the Australian Housing Research Council.

First, the standard of housing is generally lower in smaller towns than larger settlements. Secondly, the cost of construction in rural areas is higher than in urban areas, although land prices are often less. Thirdly, there is a greater availability of cheaper, older, housing in rural areas, but properties may be poorer quality and less easily accessible than their urban equivalents. Fourthly, the costs of housing maintenance are significantly

Table 16.1 Characteristics of housing owned by 'permanent' residents and 'converters' in two rural Canadian localities

	Rideau Lakes (Ontario)		Cultus Lake (B Columbia)	
	Permanent (%)	Converter (%)	Permanent (%)	Converter (%)
Lot size >10000 sq ft	75.8	80.5	75.7	5.0
House size <1500 sq ft	39.0	48.4	45.9	81.8
Single-storey houses	39.5	62.6	57.8	62.2
Three rooms or less	2.4	10.5	10.5	22.2
Seven rooms or more	40.0	30.3	31.6	8.9
Two or more bathrooms	52.0	63.2	63.2	44.4

Source: After Halseth and Rosenberg, 1995

higher in rural areas. Finally, supply of rental accommodation in rural areas is often limited.

At the same time, Halseth and Rosenberg (1995) warn against generalizing about the nature of rural housing. With particular reference to rural Canada, they argue that there are significant regional differences in the prevailing type of housing and the dynamics of rural housing markets. Locations within what they label the 'rural-recreational countryside' around urban centres are often characterized by small-lot, space-intensive developments that are easily distinguished from the more expansive properties in surrounding rural areas. Moreover, their case study research suggests that there are notable differences in size and facilities of housing between different localities and between the properties of different types of owners within the same locality. For example, in the Cultus Lake area of British Columbia, the houses owned by 'converters' who had recently converted seasonal homes into permanent residences were generally smaller with fewer rooms and facilities than those owned by more permanent residents (Table 16.1).

The importance of local factors in shaping the dynamics of rural housing is also illustrated by Jones and Tonts's (2003) case study of Narrogin, Western Australia. A town of 4,500 population, located 190 km south-east of Perth, Narrogin developed as a railway junction and is now a regional service and administrative centre. The local housing stock is dominated by owner-occupied properties (64 per cent), but also includes both private rented (22 per cent) and public rented (10 per cent) accommodation – the latter including property constructed by the state housing authority and the state-owned railway company in the 1950s and 1960s, as well as more recent developments. Some older public housing had been sold to tenants, and much of what remains is considered to be in a deteriorating condition. The population of Narrogin is highly mobile and includes a significant group of 'spiralists', young public sector workers who spend two to three years in Narrogin before moving to more senior positions. As Jones and Tonts report, 'spiralists' tend not to buy property but depend on private rented accommodation, creating a pool of demand for a limited supply that has inflated rents and marginalized lower income local residents into poorer-quality property. At the same time, Narrogin, like many small towns, has drawn in employment, services and population from the surrounding rural area, increasing the value of property in the town significantly. The double exclusion of low income households from both the private rented and owner-occupied housing markets is particularly discriminating against the indigenous Noongar community, which has been gradually absorbed into mainstream housing provision in the past forty years, but which remains heavily concentrated in poor quality public housing. All these trends combined have, Jones and Tonts (2003) note, produced reduced demand for the traditional 'three-bedroom detached home on a quarter acre block' (p. 57), and increased demand for a wider range of housing options, particularly rental stock. As such, they argue, there is now a mismatch between the characteristics of the population and the type of housing stock in Narrogin.

Although the papers by Halseth and Rosenberg and by Jones and Tonts are derived from specific case studies, the observations that they make about the complexity of rural housing have a far wider currency. In particular, there are three key dimensions that emerge from these studies that are central to any understanding of rural housing: housing quality, affordability and the availability of public rented housing.

The quality of rural housing

The promotion of large, well-maintained and expensive rural real estate by agents marketing to middle class in-migrants disguises the persistence of poor housing conditions as the experience of many rural residents. The problem of rural housing in rural areas results from a number of factors. First, much rural housing is older than urban housing and its condition has simply deteriorated over time. As rural areas did not experience slum clearance programmes in the mid-twentieth century to the same extent as urban areas, older sub-standard housing was often not demolished but still remains. Secondly, a lot of older rural housing was originally tied to jobs in agriculture or other dominant local industries. As employment in these sectors fell and rural areas experienced depopulation, large amounts of housing fell into disuse and disrepair. Thirdly,

the relative remoteness and isolation of many rural communities limited the opportunity for rural housing to be connected to infrastructure that was taken for granted in urban areas, including mains electricity, water and sewerage. Fourthly, old rural properties have become valued for aesthetic and heritage reasons and measures have been taken to preserve their appearance and integrity. As such, planning regulations in countries such as the UK can make it difficult to demolish, extend or carry out improvements to old rural properties that would change the outward appearance, especially in national parks and village conservation areas (see also Chapter 13).

Significant improvements have been made, however, to the quality of rural housing overall. In the 1940s, one in nine houses in rural Britain were deemed to be 'unfit for occupation', and one in three 'in need of repair'. By the 1980s, the proportion considered 'unfit' had fallen to around one in 20 (Robinson, 1992), although some commentators have argued that the number of sub-standard properties increased again in the 1980s (Rogers, 1987). Similarly, the number of sub-standard housing units in the rural United States was reduced from over 3 million in 1970 to 1.8 million in 1997 (Furuseth, 1998). However, sub-standard housing in the US is defined as either lacking complete plumbing, or being overcrowded, with more than 1.1 person per room – the actual condition of the building is not considered. If it were, the proportion of sub-standard housing is likely to be significantly higher. Research in four areas of rural Wales in the 1990s, for instance, found that 12.4 per cent of households were in properties with structural defects including dampness, leaking roofs, loose brickwork and plasterwork and problems with doors and windows (Cloke et al., 1997).

There are notable geographical variations in the condition of rural housing. Over a

Table 16.2 Households without basic amenities in four areas of rural Wales

Households without:	Betws-y-Coed (%)	Devil's Bridge (%)	Tanat Valley (%)	Teifi Valley (%)	All (%)
Mains electricity	0.0	1.8	3.2	0.8	1.4
Mains gas	68.0	99.1	97.6	98.0	90.7
Mains water	3.2	16.4	10.8	3.9	8.6
Mains drainage	9.2	59.5	22.0	14.7	26.3
Sole use of flush WC	2.0	3.1	2.1	1.2	2.1
Sink and cold water tap	0.4	0.9	1.7	1.2	1.0
Running hot water	2.0	4.5	3.0	1.6	2.6
Fixed bath or shower	3.2	4.1	3.4	1.6	3.0
Gas or electric cooker	1.6	1.4	3.4	2.3	2.2
Central heating	27.6	26.2	30.7	13.7	24.4
Households with structural defects	6.8	19.6	12.1	12.0	12.4

Source: After Cloke et al., 1997

quarter of all sub-standard rural housing in the United States in 1990 was concentrated in the three states of Alaska, Arizona and New Mexico – states which also have above-average levels of rural poverty (Furuseth, 1998). At a more local level, Cloke et al. (1997) report variations in the proportion of households without basic amenities in four case study areas in rural Wales (Table 16.2). All four case studies were in districts with above-average levels of sub-standard housing. Poor housing conditions are also more prevalent among particular social groups and types of housing. Elderly and low income groups are more likely to live in sub-standard housing, and rented property is more likely to be of poor quality than owner-occupied property.

Fitchen (1991) observes that local authorities are often reluctant to condemn sub-standard housing because of the lack of affordable alternatives. Affordability and housing quality are closely related problems. Poor quality housing gets used because it is cheap, yet residents frequently have little capacity to improve the property. Cheap, sub-standard housing is also attractive, however, to in-migrant middle class buyers, looking to purchase property

for renovation. Thus, the quality of rural housing has been improved in part through gentrification (see Chapter 6), but the consequence has been to increase the value of the property beyond the reach of local lower income households, exasperating problems of the availability of affordable housing.

Affordability of rural housing

Despite the trend of counterurbanization during the late twentieth century, the construction of new housing generally progressed at a slower rate in rural areas than in urban areas. The result has been to increase pressure on existing housing stocks, inflating property prices and rents. Middle class in-migrants are naturally better placed to compete in such pressurized rural housing markets than local residents in low wage employment, thus making the shortage of 'affordable' housing one of the key issues for rural policy. Over 60 per cent of rural households in Canada have reported that housing affordability is a problem (Furuseth, 1998). In the United States, meanwhile, 70 per cent of poor rural households spend in excess of 30 per cent of their

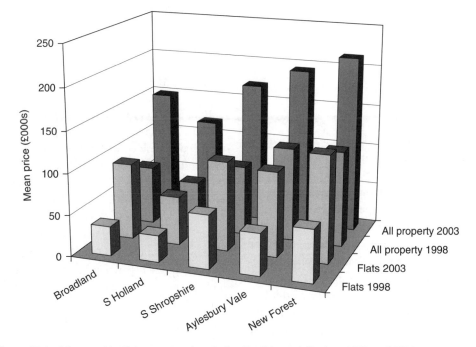

Figure 16.1 Mean residential property prices in five English rural districts, 1998 and 2003
Source: Based on data from the Land Registry

gross income on housing costs, compared with 24 per cent of rural households as a whole (Whitener, 1997). When difficulty with meeting housing costs is added to inadequate amenities and overcrowding, more than a quarter of rural households in the US are judged to have major housing problems.

Some of the most dramatic increases in rural property prices have been experienced in the UK. Fuelled by a combination of significant urban to rural migration, tight planning restrictions on new house building and the liberalization of mortgage lending, property prices in rural areas increased exponentially from the 1980s onwards. Between 1998 and 2003, average residential property prices in most rural districts increased by at least 70 per cent, with the greatest increases in more peripheral areas, such as Broadland in Norfolk, South Holland in Lincolnshire and South Shropshire (Figure 16.1). Although

the rate of increase was slower in areas of longer-term intensive in-migration, such as Aylesbury Vale and the New Forest, average prices in these localities had reached £200,000 by 2003. As Figure 16.1 also shows, property at the lower end of the market attracted some of the sharpest increases, with the average price of flats and maisonettes in rural districts more than doubling in the five years. This rate of inflation has not been matched by average income, thus exacerbating the problem of affordability. In most of southern England average property prices are now at least 4.5 times the average annual household income, and are more than eight times the average income in a number of high-demand localities including Exmoor, South Devon, the Cotswolds and parts of Sussex as well as in parts of the Lake District. The affordability gap is particularly severe for young people on lower incomes. Research by Wilcox (2003) identified 11 districts in rural southern England where the price of starter homes was more than 4.76 times the average annual income of households aged under 40 – a greater margin than in the most expensive parts of London.

In the United States, rural property prices have been inflated by increased demand generated by counterurbanization combined with limited supply due to stricter land use regulations and building ordinances and the reluctance of developers to build low cost housing. However, Fitchen (1991) noted that rising costs were particularly acute for rental property, thus hitting lower income households most severely. She reports that in one rural New York county in 1989 two-thirds of welfare recipients were paying rents that exceeded their shelter allowance by more than $100 per month. One response has been the rapid expansion of mobile home residences as a cheaper alternative. The number of mobile homes in the rural US increased by 61 per cent during the 1980s such that by 1990 they housed 16.5 per cent of the rural population (Furuseth, 1998). In 17 states, mainly in the south and west, mobile homes comprised more than a fifth of the rural housing stock. Yet, Fitchen (1991) observes that the costs of mobile homes have also increased markedly and involve numerous hidden payments, for example for electricity and kerosene. Thus, she quotes one rural resident discussing the problem of housing affordability:

My oldest son, he's got a construction job, in excavation work. There's a lot of work available now in constructing new cabins and houses. He's very good at it. But he has to move from where he's living, and he can't find a place he can afford. The influx from the city has driven up the price of housing something awful. We're trying to help him buy land for a trailer, but even that has gone sky-high.

Just recently you could get land for a trailer for only $200 or $300. Now it's up to $1,000 – just for a trailer space. (Rural New York resident, quoted by Fitchen, 1991, p. 106)

Initiatives to provide affordable housing in the UK have included planning regulations that require builders to include a provision for low cost housing in new developments and 'homebuy' schemes that provide assistance to households with low incomes or in key occupations to purchase property. However, whilst these strategies reflect the preference for owner-occupation, many lower income rural households are dependent on rented property and the problem of housing affordability in rural Britain has been intensified by the dwindling social housing rental stock.

Rural social housing

Britain, in common with a number of other countries, responded to the challenge of poor quality rural housing and the decline of tied rural accommodation in the mid-twentieth century with an extensive programme of building public or social housing for rent by local authorities. Although rural authorities were reportedly more modest in their provision of 'council housing' than urban counterparts and an immediate stigma was often introduced by physically separating council housing from private stock in locations on the edge of villages, the programme did supply reasonable, affordable, housing for lower income rural households. During the 1980s and 1990s, however, around a third of rural council house tenants purchased their homes under 'right-to-buy' legislation introduced by the Conservative government (Hoggart, 1995). Councils were not permitted to invest the receipts in new building and wider restrictions on local government finance meant that little additional housing was constructed to replace sold stock. Instead, primary responsibility for social housing provision has shifted to independent housing associations, but concerns have also been expressed about the level of new building by housing associations and their effectiveness in rural areas (Milbourne, 1998). These factors combined have substantially reduced the stock of social rented housing in rural areas in the UK. In 1999 there were 684,000 social housing units available in rural England (around 14 per cent of the total housing stock) compared with 711,000 units in 1990 (Cloke et al., 2002). As such, the privatization of social housing in the UK and elsewhere (see Jones and Tonts, 2003, for example, on Australia), has reduced the options available to lower income rural residents and contributed to an increase in rural homelessness, as discussed in Chapter 19.

Rural Health

A second element of the rural idyll myth is that rural life is healthier than city life. The statistical evidence in support of this assertion is mixed. Figures from the UK and Canada suggest that there are lower mortality rates and longer life expectancies in rural areas (Cabinet Office, 2000; Wilkins, 1992); but other evidence from North America, the UK and Australia indicates that there are higher accident rates in rural areas and greater prevalence of chronic ill-health (Gesler and Ricketts, 1992; Gray and Lawrence, 2001; Senior et al., 2000; Wilkins, 1992). In rural Wales, for example, Cloke et al. (1997) found that 44 per cent of residents reported serious health problems compared with 17 per cent for Wales as a whole. The apparent poor health of rural residents reflects a number of factors, including the relatively elderly profile of the rural population (reinforced by retirement migration) and the relatively poorer health of indigenous communities in rural Australia, New Zealand and North America. Local geographical variations in levels of ill-health also suggest that deprivation is a significant factor (Senior et al., 2000). Yet these factors do not entirely explain urban–rural differences, and some effect must also be attributed to specifically rural issues in health and healthcare, notably problems in healthcare provision and the social and economic conditions of rural lifestyles.

The provision of rural healthcare

The delivery of healthcare to rural areas is faced by a number of difficulties, most of which are related to the relative isolation and sparse population density of rural regions. Health facilities are more expensive to provide in rural areas, levels of use and occupancy are frequently lower than in urban centres, and the viability of specialist units is more tenuous. Even in countries with comprehensive public health services, such as the UK, the accessibility of health services in rural regions can be variable. A third of rural residents in England live more than 2 kilometres away from the nearest doctor's surgery and over half live 4 or more kilometres from the nearest hospital (see Chapter 7). In larger continental states such as Australia, Canada and the United States these distances can be multiplied several-fold. The problem of rural health provision in the US has been further intensified by the closure of rural hospitals due to pressures of specialization and cost-effectiveness, and by the comparatively low proportion of the rural population covered by health insurance – only 53.7 per cent in 1996 (Vistnes and Monheil, 1997). Health facilities that do exist may be poorly equipped, with half of the rural hospitals in New South Wales lacking diagnostic equipment to treat respiratory and heart diseases, for example (Lawrence, 1990).

The recruitment and retention of trained health professionals is a further problem. A combination of personal and professional factors often militates against doctors and other health specialists choosing to locate in rural areas. These include the limited opportunities for specialization, the absence of back-up support facilities, issues of finding suitable accommodation and employment for families, and the expectations placed on sole health practitioners in small towns (Gordon et al., 1992). The consequence is that many peripheral rural regions have a shortage of health professionals. In New South Wales, Australia, for example, the ratio of physicians to population in the rural west of the state is 1:1500 compared with 1:30 in the metropolitan areas (Lawrence, 1990).

Health and rural lifestyles: stress and drugs

The countryside is popularly associated with peace and tranquillity, yet for many people rural life can be a stressful experience, provoked by isolation, pressures to conform, the inability to escape or hide in close-knit communities, the lack of diverting entertainment and the strain of economic restructuring, particularly in agriculture. One survey of British farmers in 2001 reported that 40 per cent found running their farm business to be 'continually stressful' (Lloyds TSB Agriculture, 2001). Gray and Lawrence (2001) summarize that stresses associated with agricultural restructuring in Australia could be inferred to lead to marriage breakdown, ill-health, insomnia and aggressive and violent behaviour. Moreover, initiatives established to deal with problems of rural stress in the foot and mouth epidemic in the UK in 2001 reported an increase in suicides by farmers. Many of these responses, including suicide, have also been observed in other sections of the rural population. The National Rural Health Alliance in Australia, for example, found that suicide rates among men aged between 15 and 24 in rural areas were more than twice those in urban areas (Gray and Lawrence, 2001).

Mental health in general can be a significant issue in rural areas, in part because of the challenges faced in providing appropriate support, and in part because of the heightened visibility of those suffering mental health problems in small rural communities (see Philo and Parr, 2003).

High stress levels are also associated with high levels of substance abuse. Whilst the problem of alcoholism in rural communities has long been recognized, concerns about drug abuse have gained publicity in the past ten years. A 1998 survey in Britain suggested that over a quarter of children aged 14 and 15 in rural schools had tried illegal drugs – a higher percentage than in urban areas (Schools Health Education Unit, 1998). Similarly, research in neighbouring rural and urban communities in Scotland found that four out of ten 14- and 15-year-olds had tried illegal drugs in both localities, with cannabis use by far the most common (Table 16.3) (Forsyth and Barnard, 1999). Yet, the study also found that levels of drug use varied more significantly between schools in the rural district, and that this variation could not be explained by deprivation measures. Rather, Forsyth and Barnard (1999) argue, experimentation with drugs in rural communities is influenced by the availability of particular drugs and local sub-cultures. Such findings correspond with the experiences of one young rural drug-user quoted in the *Guardian* newspaper:

> What else were we meant to do? You couldn't go to the pub because the chances are your parents would be there. You can't buy a drink from the shop because you've grown up around the people behind the counter and they know your age. No youth club, pub out of bounds, and the parents of the younger kids didn't like us hanging out in the kiddies' play area. So we would go down

Table 16.3 Percentage of children aged 14–15 in neighbouring rural and urban areas of Scotland reporting use of drugs

	Perth & Kinross (rural)	Dundee (urban)	All schools
Any drug use	43.0	44.7	43.9
Cannabis	42.4	43.0	42.7
Amphetamine	11.0	13.1	12.1
Psilocybin	9.8	7.3	8.5
LSD	5.4	9.6	7.5
Temazepam	5.9	5.9	5.9
Ecstasy	3.3	5.4	4.4
Cocaine	2.6	3.3	2.8
Heroin	2.6	1.4	2.0

Source: After Forsyth and Barnard, 1999

to a recreation field on the outskirts of the village and get stoned … Pot, magic mushrooms and LSD are the drugs we did mostly, and occasionally a bit of speed when it came into the village. (Rural youth quoted in the *Guardian*, 11 March 1998)

Rural drug abuse is not restricted to young people. In West Virginia, health authorities have expressed concern about levels of abuse of OxyContin, a prescription pain-killer that is also known as 'hillbilly heroin', among rural residents of all ages (Borger, 2001). Such are levels of addiction that 80 per cent of crime in some rural towns is believed to be OxyContin-related. In a different context, Cocklin et al. (1999) describe the widespread use of cannabis in Northland, New Zealand, quoting the observation of a local police officer that,

Cannabis use is a widespread problem right throughout the whole community. Generations are going through, we're locking up young and old people, whereas once it was always seen to be the young person's drug and everything like this. You do get it across the board; it would be misleading to say that there is significantly less problem for the Maori than it is for the Pakeha, it's a problem for the whole community. (Cocklin et al., 1999, p. 249)

Cannabis was described to Cocklin et al., as a 'coping mechanism', but cannabis use in Northland is also linked to availability and the significance of the region as a core cannabis producer. As Cocklin et al. report, Northland accounted for around a quarter of all cannabis plants seized by police in New Zealand. Yet, despite the social consequences, cannabis cultivation was credited by a number of residents as bringing in income, improving living conditions and 'helping to keep afloat one of the country's most economically depressed regions' (Cocklin et al., 1999, p. 248).

Crime and Rural Communities

The third element of the rural idyll myth relating to the quality of life is that the countryside is a safe and crime-free place to live. Here the statistical evidence is more positive than for health. Levels of victimization are significantly lower in rural areas than in urban areas, as figures from the UK and Canada show (Table 16.4). In general, rural residents are half

Table 16.4 Percentage of persons or households reporting that they had been a victim of crime

	England and Wales (1995)			Canada (1987)	
	Rural	**Urban**	**Inner City**	**Rural**	**Urban**
Vehicle-related theft	15.7	20.1	26.0	3.6	5.9
Vandalism	8.0	10.9	10.6	4.2	7.6
Burglary	3.9	6.3	10.3	3.2	6.4
Personal crime	3.9	6.3	10.3	11.4	15.8

Source: Cabinet Office, 2000; Norris and Johal, 1992

as likely to be victims of property crime, such as burglary or vehicle theft, than inner urban residents, although the difference for violent crime against the person is smaller. Overall, analysis of reported offences in Scotland suggests that for every crime committed in a rural area, more than four are committed in urban districts (Anderson, 1999). Although there is some evidence of increasing crime levels in rural areas, particularly for violent crime and burglary, which has narrowed the difference with urban areas, the base is such that the long-term trend in most regions is as McCullagh (1999) remarks for rural Ireland, 'from negligible to low' (p. 32).

Anderson (1999) observes that, despite some differences, the overall profile of crime in rural areas is not significantly different from that in urban areas. However, there are three types of offence that commentators commonly identify with rural crime. The first are what might be called 'activities of lawlessness' (perhaps reflecting a longer rural tradition, see Mingay, 1989). Okihoro (1997), for example, in describing crime in a small Canadian fishing community, places an emphasis on poaching, moonshining and petty corruption. Similarly, Yarwood (2001) notes that studies in the United States have identified rural crime with illegal drug production, militia group activity and gun offences (Weisheit and Wells, 1996), and larceny involving gasoline and vehicles (Meyer and Baker, 1982); and that the rustling of livestock has been reported as a problem in the UK.

The second type of offence associated with rural crime is inter-personal violence, which comprises a greater proportion of recorded crime in rural areas than in urban areas. This includes violent disorder, often alcohol-fuelled, between small town rivals (Gilling and Pierpoint, 1999), and an under-reported current of domestic violence (McCullagh, 1999; Williams, 1999).

The third commonly identified type of rural crime is social disorder. Vandalism is a major component of rural crime, whilst farmers in Britain have complained about trespass, damage to farm property and the worrying of livestock (Yarwood, 2001). Many perceived 'crimes' of social disorder, however, simply involve behaviour that is deemed to be 'out of place' (Cresswell, 1996) and reflect inter-community tensions between age groups or classes (Stenson and Watt, 1999). The conflation of actual experience of crime and perception of crime is illustrated in a case study by Yarwood and Gardner (2000) of a rural parish in Worcestershire, England. Reported crime in the parish is relatively low and not untypical for rural communities in the region (Table 16.5), but knowledge of crime is broader, if uneven. Over half of residents claimed to know someone in the parish who had been the victim of burglary and of vehicle theft, but knowledge of specific crimes affecting farmers, such as trespass, tended to remain in the agricultural community. This skewed perceptions of the risk from crime, but generally residents were fairly unconcerned about crime. Only

Table 16.5 Experience of crime in an English parish

	Respondents who had been victims of crime in the parish (%)	Respondents who knew somebody who had been a victim of crime in the parish (%)
Burglary	15	69
Attempted burglary	7	47
Damage to property	4	6
Damage to vehicle	7	9
Theft of vehicle	7	50
Theft of property from vehicle	5	31
Mugging	0.3	2
Pickpocketing/bag snatching	1	5
Interference with livestock	2	6
Trespass	13	13
Violence	0.3	8
Threats of violence	3	3
Verbal abuse	2	2
Racial harassment	0.3	1
Sexual assault	0.3	2

Source: Yarwood and Gardner, 2000

Table 16.6 Percentage of residents of an English parish perceiving various issues to be a problem

	'Problem' or 'big problem' (%)	'Not a problem' (%)
'Young people hanging around'	53	44
Rubbish/litter	39	55
Traffic	55	39
Dogs	65	31
Drug dealing	7	69
Drunks	12	74
Loud music/parties	11	73
Graffiti	14	75
'Travellers'	47	46

Source: Adapted from Yarwood and Gardner, 2000

burglary (about which 32 per cent of residents were concerned), vehicle theft (26 per cent) and theft of property from a vehicle (23 per cent) provoked significant levels of concern. In contrast, issues of 'anti-social behaviour', including 'young people hanging around', were identified as problems by much larger proportions of residents (Table 16.6). The two sets of concerns were merged by some residents who blamed crime on those groups who were most commonly perceived to be 'out of place':

I believe 90 per cent of burglaries are committed by travellers who distribute leaflets for jobs of work in the area.

We have a big problem with vandals, the 12–15-year-olds have nothing much to do so they find fun in other ways. (Residents quoted by Yarwood and Gardner, 2000, p. 407)

Perceptions of 'out-of-place' behaviour, which have in some places been intensified by the effects of restructuring, have contributed to a growing fear of crime in rural areas, along with the importation of urban concerns and expectations by in-migrants and high-profile cases such as that of Tony Martin, a British farmer who was imprisoned for shooting dead a burglar at his remote Norfolk farm in 2000. Expressions of this concern have included campaigns for increasing policing cover in rural areas and the development of alternative security strategies such as 'Neighbourhood Watch' schemes (Yarwood and Edwards, 1995), gated communities (Phillips, 2000), closed-circuit television surveillance (Williams et al., 2000), corporate sponsorship of police provision, mobile police stations and the employment of private security firms.

Summary

Life in the countryside is not a homogeneous experience. Whilst some residents – generally the wealthier ones – are able to enjoy a lifestyle that at least aspires to the model of the rural idyll, for many rural people the quality of life is blighted by poor housing, poor health or the fear of crime and social disorder. The availability of good quality, affordable housing is a major problem in many rural communities. As counterurbanization has fuelled demand for rural property, households on lower incomes have found themselves unable to compete and been forced into cheap but sub-standard accommodation. Ill-health is also pervasive in many rural areas, with above average rates of chronic illness worsened by problems of accessing health services and facilities. The stresses of isolation, insularity and rural restructuring have also contributed to illness, alcoholism and drug abuse. And although crime rates are lower in the countryside than in the city, a significant minority of rural residents are victims of crime and many more live in fear of crime or of the cultural threat of 'out-of-place' behaviour. Moreover, it is frequently the same people who are affected by these different problems. Residents of sub-standard housing are more likely to suffer ill-health; there is a link between drug abuse and crime, such that communities with drug problems are also likely to have higher rates of crime, as are settlements with poor housing. The rural residents who get in this trap tend not to be the 'Mr Averages' described by Philo (1992) as the traditional focus of rural research (see Chapter 15). They are rather the vulnerable in society, the 'neglected rural others': the elderly, individuals living in poverty, indigenous communities. The rural lifestyles of these groups will be examined in more detail in the next few chapters.

Further Reading

The topics discussed in this chapter have been written about in a wide range of focused books and papers. One of the few studies to draw together the themes is Janet Fitchen's *Endangered Spaces, Enduring Places: Change, Identity and Survival in Rural America* (Westview Press, 1991), which includes short sections on housing and health in rural New York State. Roy Jones and Matthew Tonts, in 'Transition and diversity in rural housing provision: the case of Narrogin, Western Australia', *Australian Geographer*, volume 34, pages 47–59 (2003), provide a good empirically based discussion of rural housing problems that has a relevance beyond Australia. For more on the restructuring of social housing provision in Britain, see Paul Milbourne, 'Local responses to central state restructuring of social housing provision in rural areas', *Journal of Rural Studies*, volume 14, pages 167–184 (1998). Rural health has an extensive literature of its own, including a number of specialist journals. The edited volume by Wilbert Gesler and Thomas Ricketts, *Health in Rural North America: The Geography of Health Care Services and Delivery* (Rutgers University Press, 1992) provides an overview of some of the key issues from a geographical perspective. For an overview of research on rural crime see Richard Yarwood, 'Crime and policing in the British countryside: some agendas for contemporary geographical research', *Sociologia Ruralis*, volume 41, pages 201–219 (2001). 'Fear of crime, cultural threat and the countryside', by Yarwood and Gardner, in *Area*, volume 32, pages 403–412 (2000), is a good empirical study of perceptions of rural crime and cultural threats.

Websites

The Housing Assistance Council's website (www.ruralhome.org) includes a range of information about rural housing in the United States, whilst the website of the National Center on Rural Justice and Crime Prevention (www.virtual.clemson.edu/groups/ncrj) similarly includes information on rural crime in the United States. Data on house prices in both rural and urban areas of the UK can be found on the website of the Land Registry (www.landreg.gov.uk). Also in the UK, the Sussex Crime and Disorder Partnership's website (www.caddie.gov.uk) includes interactive maps of reported crimes by ward, enabling viewers to compare crime patterns in rural and urban areas. For more on rural health issues, see the Center for Rural Health Policy Analysis (www.rupri.org/healthpolicy/) in the US, and the Institute of Rural Health (www.rural-health.ac.uk) in the UK.

17

Growing Up and Growing Old in the Countryside

Introduction

Mainstream rural studies have traditionally focused on those elements of rural activity that are predominantly experienced by the working age population: economic activity, employment, farm management, property ownership and migration decision-making, to cite a few examples. Studies of rural communities have similarly focused on social interaction between the active adult population. Comparatively little attention has been paid to those rural residents at either end of the age spectrum – the young and the elderly. Yet, arguably, it is these groups whose lifestyles are most significantly influenced and shaped by the rural context. This chapter examines the experience of rural life for groups at three points in life: children, young people entering adulthood and the elderly. It explores their perceptions of rurality and rural communities and the geographies of their rural existence.

Rural Childhoods

The countryside is a popular setting for children's literature. From *Winnie-the-Pooh* and *Wind in the Willows* to *Swallows and Amazons* and the *Famous Five* to contemporary stories such as *The Animals of Farthing Wood*, children's literature has not only portrayed a rural idyll, but also the countryside as a site of idyllic childhoods. In these stories the countryside is represented as a place of fun, adventure and freedom, but also of safety and security. As Jones (1997) observes, these literary associations form a powerful cultural discourse that continues to inform popular ideas about rural

childhood (although Horton, 2003, argues that children's literature has promoted a variety of representations of the rural). The notion that the countryside is a 'safe place' to bring up children is commonly cited as a reason for migration to rural areas, and 'safety' is a recurrent theme in adult narratives of rural childhood:

> I think that it was a conscious decision on our part to move, to have somewhere where they could have you know, more freedom to run about and you know, make friends and all the rest of it, without having to worry that they weren't actually

safe. (In-migrant mother, England, quoted by Valentine, 1997a, p. 140)

Well I think that it's a nice environment, it's relatively quiet, safe, in terms of traffic and things like that kind. Pleasant community. So I think it has idyllic prospects for children ... and you can watch things grow and play in the stream. (Father, England, quoted by Jones, 2000, p. 33)

Children's own narratives of rural life also tend to reflect these beliefs. In a case study in northern Scotland, Glendinning et al. (2003) found that over 80 per cent of children aged between 11 and 16 agreed that 'this is a good place for children to grow up in', and over 80 per cent of children aged 15 and 16 agreed that 'this is a safe place for young people to live' (p. 137). They quote two teenage girls commenting,

I think it is good for young children. It's much safer.

You don't have to lock your doors all the time.

It's quite safe. Your Mum and Dad can let you go down the street, or wherever, when you're young, or go across the road on your own sort of thing.

You can go to the park on your own, or with a friend, without them having to take you and take you back.

(15- and 16-year-old girls, Scotland, quoted by Glendinning et al., 2003, p. 138)

'Safety' in this context, clearly, has a number of different, parallel, meanings. These include safety from traffic and other environmental dangers associated with urban space, as well as safety from criminal threat. They also include safety from undesirable cultural influences, as one parent is quoted by Valentine

(1997a) remarking, 'there aren't the pressures on them in terms of you don't have to have Adidas tracksuits, you don't have to have the latest video games it seems' (p. 140). Furthermore, the perception of the rural as a 'safe place' for children has a geographical manifestation in the degree of autonomy that parents are prepared to give children to go where they like unsupervised – at least within the spatial bounds of the village (Jones, 2000).

Thus, the geographies of rural children are characterized by a dualism that contains a number of contradictory assertions. On one side, the rural is a space of freedom and independence for children, but only within a framework of adult regulation. On the other side, the rural is a place of dependency, in which children are often reliant on parents for transport, but where the spatial dynamics of, for example, travel to school, create independent sites for interaction and identity formation. These two aspects can be examined in turn.

Spaces of freedom and regulation

The idyllic country childhood of literature and the popular imagination involves considerable freedom to roam at will across open rural space. Autobiographical accounts of rural childhoods in the early to mid-twentieth century often present memories of walking or cycling over significant distances and using fields, woodland and rivers as a vast playground (Jones, 1997; Valentine, 1997a). Jones (2000) observes that children still use the natural and built features of the rural landscape for play, albeit within a smaller spatial range:

Some children have limited autonomy to exploit certain spaces. For example, at one end of the village the stream is followed by a footpath, and in one place is overhung by trees where the valley is quite steep and this makes what feels like a private, secretive space. Various cohorts of children have used this place, known as 'the den', for meeting up, and a base for activities. This was marked on a number of the maps drawn by the children, and two friends ... told me the den was 'somewhere you can go and sit and talk away from everyone else ... everyone's in the house, it's really cramped, so you go to the den'. Another space used by successive cohorts of village children is one of the remaining farm yards with its two barns, which was also depicted in the children's maps. (Jones, 2000, p. 35)

In the search for spaces beyond adult surveillance, rural children become adept at exploiting the permeability of physical and metaphorical boundaries, from garden fences to rules about not venturing onto private property. This often involves transgressing the spatial order placed on the countryside by adults. Ward (1990) laments the 'over-ordering' of rural space that restricts children's mobility, including the fencing off of fields and woods and increase in residential and industrial land uses, arguing that rural children are 'victims of the municipal urge to tidy up everywhere and cut each blade of grass ... [and] ... to turn every patch of ground to commercially viable use' (Ward, 1990, p. 94; see also Philo, 1992). As such, children's own narratives of their geographies frequently include reference to confrontations with adults, both protective landowners and residents who perceive children 'hanging out' to be a nuisance or cultural threat:

Oh, I was seven and I was walking around with Holly and she was about eight. We went up through the gardens in Moreton Pinkney and we went across this bridge that goes over Westly Hill and we were standing at the top and we were throwing apples down into the brook and this man came out of a house and ... he just sort of blew up and he went bright red like a tomato and shouted: 'GET OFF THIS BRIDGE NOW', and the ground started shaking. So we ran down and he said: 'If I see you up there once more I'm going to call the police.' (Girl aged 10, England, quoted by Matthews et al., 2000, pp. 144–145)

I was ... on the green. I had all my friends there, it was one of my birthdays. We were all riding about on the green ... and we got to the gate on the green and this lady said, 'You're not allowed here, stop riding on the green you will spoil the grass' and she hadn't even bought the land. (Girl aged 9, England, quoted by Matthews et al., 2000, p. 146)

Children's rural geographies are also regulated by parents who impose rules about where they are allowed to go. As Matthews et al. (2000) note, 'parental interpretations of the Good Life rarely exceeded the immediacy of the physical fabric of the village' (p. 145), such that the actual distance that rural children are permitted to go unaccompanied can be considerably shorter than that for urban children. Further regulation is in effect introduced by the time consumed in travelling to school in another settlement, and by childcare arrangements in evenings, weekends and holidays (see Box 17.1).

Box 17.1   Rural childcare

Traditional discourses of gender in the countryside emphasized the women's domestic role and the importance of the family as a framework for raising children. Increasingly, however, women are employed in the rural labour market (see Chapter 18) and the

---

Box 17.1 (Continued)

recomposition of the rural population has fragmented care networks based on extended families and neighbourhoods, creating a demand for more formal childcare arrangements. A study in Devon, south-west England, for example, found that 28 per cent of parents surveyed used a nursery or registered childminder (including 50 per cent of respondents in full-time employment), and 47 per cent used a playgroup or toddler group for childcare (Halliday and Little, 2001). Yet, formal childcare facilities in rural areas are less extensive than in urban areas.

Consequently, studies in both the UK and the United States have found that rural parents are still more likely than urban parents to rely on informal childcare arrangements, including family members and neighbours (Casper, 1996; Halliday and Little, 2001). Similarly, rural families can be prepared to travel significant distances to find appropriate childcare (Halliday and Little, 2001). The type of childcare used has an impact on children's landscapes of play. Smith and Barker (2001) report that out-of-school clubs are important sites to meet and play with other children and offer opportunities for play that may not be provided elsewhere in the rural community. In some cases this included explicit activities to engage with the rural environment, by for example, permitting access to fields and making farm visits. For some children, therefore, it is through these controlled and supervised activities that they are able to experience something of the 'idyllic country childhood' rather than in their own unstructured play time.

For more see Joyce Halliday and Jo Little (2001) Amongst women: exploring the reality of rural childcare. Sociologia Ruralis, 41, 423–437; Fiona Smith and John Barker (2001) Commodifying the countryside: the impact of out-of-school care on rural landscapes of children's play. Area, 33, 169–176.

Places of dependency

The myth of the idyllic country childhood also assumes that the life of rural children is strongly focused on the settlement in which they live. However, the closure of rural schools, decline of rural services and changing family patterns of shopping and leisure (see Chapter 7) have all expanded the social space of rural children. Friends made at school may well live in different towns and villages. When combined with parents' reluctance to allow their children to walk or cycle long distances on their own, rural children wanting to meet with friends – or to use facilities such as shops, youth clubs or cinemas – are increasingly dependent on transport provided by their parents. In a comparison of rural and urban districts in Sweden, Tillberg Mattson

(2002) analysed children's travel diaries to show that whereas urban children travelled an average of 2 km per day by bicycle or walking, for rural children the average was just 0.3 km, despite rural children on average travelling up to four times the distance of urban children to school each day, and twice as far for leisure activities. Over half of the trips to leisure activities by rural children, and a third of trips to visit friends, were made by their parent's car. As Glendinning et al. (2003) record, dependence on parental transport restricts children's activities:

> I very definitely felt left out of things when I was younger, because it was just so difficult. So much arranging had to go into everything. Every little thing you wanted

to do, somebody had to give you a lift there. You couldn't just pop out and come back anytime. And you had to be there at this time to come back, or else you were dead. (17-year-old girl, Scotland, quoted by Glendinning et al., 2003, p. 140)

The dependence on parents for transport reflects the limited service of public transport in rural areas. Where children and young people are able to use public transport independently, buses and trains can be important sites of social interaction. Ward (1990) notes the 'culture of the school bus' and the forms of sociability and micro-spatial organization that are developed, including seating arrangements on the bus. The desire of rural youth to travel into local towns is also in part about finding spaces to 'hang out' where they are beyond the visibility of their parents and neighbours. Indeed, neighbouring towns have become such a locus for both the education and leisure time activities of rural children, that Tillberg Mattson (2002) questions how far they do 'actually *grow up* in the rural area' (p. 446).

As children grow older, the pressure to be independently mobile means that there often is an expectation that young people will learn to drive as soon as they are legally able to and that they will obtain a car of their own – both of which can be expensive (Storey and Brannen, 2000). Transport costs in general are one factor that helps to make raising children in the countryside resource-heavy, thus contributing to social polarization in rural communities. Davis and Ridge (1997), for example, report that nearly half of children in lower income households studied in Somerset, England, did not have access to a car, further restricting their ability to participate in social activities. As one girl commented,

Sometimes you miss out on things badly … I mean if I wanted to get somewhere really badly and like it was only until the evening … and there was like only a bus in the day, I would go in the day and stay till the evening. But it's like there's not, some days like there's just not a bus. (Girl, aged 13, England, quoted by Davis and Ridge, 1997, p. 51)

The expense involved in joining clubs and societies and paying for after-school activities can also be prohibitive, such that children from lower income households have a different experience of rural childhood from their counterparts. Moreover, these differences may be more visible and pronounced in rural communities than in urban areas because there is less spatial segregation of social groups and therefore children are more likely to make friends from a range of social backgrounds.

Young Adults in Rural Communities

The experience of rural life for young adults is coloured by broader cultural reference points. Compared with the perceived lifestyle of urban youth, young people in rural areas commonly express a sense of amenity deprivation. Glendinning et al. (2003), for example, found that 87 per cent of girls aged 15 and 16 interviewed in rural northern Scotland, and 75 per cent of boys, agreed with the statement that 'there's nothing for young people like me to do', whilst 67 per cent of girls and 53 per cent of boys felt that there were too few shops selling things that they wanted. Accordingly, young people's satisfaction with rural life decreases with age (Figure 17.1).

As well as a paucity of shops and entertainment facilities, rural communities often lack specific amenities for young people. Only half of rural parishes in England, for example, have a youth club or other formal group for young people (Countryside Agency, 2001). The shortage of formal facilities for young people has been identified as a contributing factor in

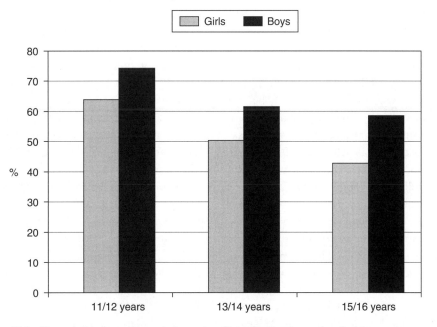

Figure 17.1 Percentage of young people in rural northern Scotland agreeing that the rural community is 'a good place for young people like me to live'
Source: Based on Glendinning et al., 2003

problems of under-age drinking, drug abuse and vandalism in rural areas (see Chapter 16). Additionally, tensions can develop within communities as the gathering of young people in public spaces is perceived by other residents to be threatening. More aggressive behaviour can also erupt as youths contest the dynamics of social restructuring, with conflict between 'locals' and 'newcomers' (Jones, 2002).

The lifestyles of rural young people, therefore, have their own spatial and political dynamics, as demonstrated by Panelli et al. (2002) in a case study of Alexandra, a small town of 4,600 population in New Zealand. Through interviews and focus groups with teenagers aged between 13 and 18, Panelli et al. identify a number of ways through which young people structure their community, including the skateboard park, high school, park, the Kentucky Fried Chicken outlet and the main street. However, these sites are described as spaces of marginalization, places where young people congregate not through choice but because they feel excluded from other spaces or do not have facilities such as shopping malls:

The town library is a bit dodgy. The library ladies peer at you funny, like we're gonna steal books. We were told we were not allowed jackets in the library in case we steal books. The school library is cool though. (15-year-old females, New Zealand, quoted by Panelli et al., 2002, p. 116)

Some shops don't like teenagers going in. They think they might steal stuff. (17-year-old males, New Zealand, quoted by Panelli et al., 2002, p. 116)

The experience of marginalization varied between young people depending on their interests and local connections. Some noted, for example, that most organized activities for youth in the town revolved around sport and there was little for non-sporting individuals to do. Moreover, whilst many young people felt that they were part of the community because they knew people, others reported experiencing exclusion on racial and other grounds.

In this context, Panelli et al. (2002) suggest that young people in the town actively assemble a 'politics of youth' through a series of negotiations, of which they illustrate three strategies. First, young people negotiate their place in the community through subtle forms of space occupation. This involves both implicit activities such as the sharing of the public space of main street with other users and more explicit struggles over, for example, skateboarding on the town's streets. Secondly, direct challenges are sometimes mounted to the regulation of space, with skateboards again being a focus of contestation. One 'skatie' described contesting the designation of skateboards as 'vehicles' (which excluded them from footpaths) by attempting to use the 'drive-through' section of a fast food outlet. Thirdly, young people constructed their own sense of 'creative participation' through activities which, whilst marginalized, give a sense of agency, such as smoking and lighting firecrackers in a disused shed. The marginal nature of these activities, however, means that they are often temporary, existing only until they attract adult attention.

Rural sexualities

The insularity of rural communities can be particularly unforgiving for young people as they discover and experiment with their sexuality. Traditional rural discourses have reproduced strict stereotyped representations of gender roles to which young people are expected to conform. These representations are both geographically and historically constructed and have their origins in concerns for the reproduction of labour for agriculture. Thus femininity is constructed in terms of domestic accomplishment, including motherhood, but also through the representation of rural young women as coy, demure and wholesome (Little, 2002). Masculinity, meanwhile, is also constructed through notions of the rural, particular through an association of the rural with a rugged masculinity as exemplified through images of the cowboy or the pioneer (Campbell and Bell, 2000). The performance of rural masculinity is therefore identified with agricultural work (Liepins, 2000c; Saugeres, 2002), but as this outlet has declined it has found alternative expressions, for example through a macho drinking culture (Campbell, 2000).

These rural constructs of gender, Little (2002) argues, are 'heavily inscribed with an assumption of heterosexuality' (p. 160). As Bell and Valentine (1995) contend, representations of gender and sexuality in rural society follow a strong moral code, such that 'rurality conflates with "simple life", with hegemonic sexualities (church weddings, monogamy, heterosexuality)' (p. 115). The historic origins of this moral code in the agricultural economy also structures the rituals of partnering within rural society, with a concern for perpetuating the core tenet of the family farm by enabling young farmers to find future farmers' wives. To facilitate this, and to overcome the isolation of farm life, some rural regions developed matchmaking events, such as the annual matchmaking festival at Lisdoonvarna in western Ireland (Figure 17.2) and a similar farmers' ball in Middlemarch, New Zealand (Little, 2003).

The hegemony of heterosexual values in rural society traditionally supported a strong prejudice against homosexuality. As Fellows (1996) describes for the rural Mid-West of the United States, homosexuality was seen as

Figure 17.2 Lisdoonvarna, Ireland, location of an annual match-making festival for the rural farming community
Source: Woods, private collection

'an unnatural phenomenon of the city that has no relevance to rural life' (p. 18) (see also Bell, 2000). Kramer (1995) similarly recounts the experiences of gays and lesbians in the remote rural town of Minot (34,000 population) in North Dakota. Kramer details the covert strategies employed by gay men and women to perform their gay identity and meet sexual partners, including the identification of clandestine 'sexual marketplaces' such as a highway rest area, a park and a railway viaduct. Use of these sites involved risk, and for many gay and lesbian residents the preferred strategy was to make periodic trips to large cities. In 1979 an attempt was made to give the gay community a more public presence in the town through the founding of a gay and lesbian organization and an adult bookstore. The organization folded in less than a

year after public opposition and intimidation. The bookstore survived longer and provided access to gay media and literature. However, the wider community continued to be ill-informed about homosexuality, such that, as Kramer observes,

Many of the men I came into contact with in Minot possessed such inaccurate imagery of the meanings of being gay, defining gay men as being effeminate, as being transvestites who live in large cities (an image propagated by local media coverage of gay pride events), as being pederasts or otherwise immoral or deviant. These men instead saw themselves as too 'normal' to be gay, or saw their own behaviour as a temporary phase, attributable to high libido or the effects of alcohol. (Kramer, 1995, p. 210)

The drawing of a distinction between homosexual activity and homosexuality identity is, Kramer notes, a prominent feature of rural homosexualities. Whilst militating against the mainstream acceptance of homosexuality, it does form part of a culture of experimentation that permits the conventional sexual mores of rurality to be contested by young people through both homosexual and heterosexual behaviour. As well as overt gay and lesbian performances, this includes experiments with fashion and promiscuous heterosexual activity.

Moving out and staying behind

In spite of criticisms of the lack of amenities and the insular nature of rural communities, many young people do actually want to remain in their home area – but the opportunities for them to do so may be limited. Pursuing higher education commonly means leaving rural communities to attend colleges and universities in urban areas and the shortage of graduate jobs in rural areas restricts the opportunities of those who leave for university to return. Whilst higher education provides an entry route into national labour markets, those young people who stay behind are often restricted to a local labour market that is characterized by poorly paid, insecure and unrewarding jobs with few prospects for career advancement (Rugg and Jones, 1999). Finding appropriate, affordable housing is also difficult for young people in rural areas, and forms an additional 'push' factor for youth out-migration.

Moreover, Ni Laoire (2001), in a study of rural Ireland, suggests that there are positive and negative connotations associated with decisions to migrate and to stay respectively, especially for young men. Whilst out-migration is associated with heroism and freedom, a stigma can be attached to those who 'stay behind'. Even though men are less likely to move away

from rural areas than women, the devaluation of remaining in a rural area contributes to a reworking of rural masculinities and potentially to problems of depression, mental illness and the risk of suicide (Ni Laoire, 2001). Hajesz and Dawe (1997) similarly record a perception in Newfoundland and Labrador that leavers are considered smart and stayers are 'losers', which informs young people's decisions to migrate. Yet although they find that two-thirds or more of young people believe that they will leave their rural home area, Hajesz and Dawe also find that most young people would prefer to stay if there were no constraints upon doing so, and that many who leave retain permanent residence status at home and often return to live in their rural home area later in life.

Elderly Rural Residents

The rural population is an ageing population. In the United States 13 per cent of the population of non-metropolitan counties in 1980 were aged over 65; by 2001, the proportion had increased to 20 per cent (Laws and Harper, 1992; ERS, 2002). Similar trends are evident elsewhere in the developed world (see also Chapter 6). The ageing of the rural population is a product of a number of separate but parallel processes. First, the family structure of the rural population has changed. Whereas farming was once associated with large families, the decline in the farm population and the modernization of agriculture such that it is less dependent on family labour have contributed to a decrease in the average family size and hence a residual elderly bias in the farming community (Laws and Harper, 1992). Secondly, the out-migration of young people means that older people are over-represented in the remnant rural population compared with the national average. Thirdly, the older rural population has in some areas been boosted by retirement in-migration.

These processes have also produced a differentiated geography of the rural elderly, with particular concentrations of older people in certain rural areas. These include very small settlements with a strong agricultural presence; remote and economically deprived areas from which there is significant out-migration; and areas that are attractive as retirement destinations, notably coastal locations (Laws and Harper, 1992).

Laws and Harper (1992) comment that 'on almost any indicator of health, income, access to services etc., the rural elderly have been shown to be disadvantaged' (p. 102). This generalization, however, understates the degree of stratification within the rural elderly population. Poverty is a major problem for many older people in rural areas, particularly those who are dependent on small pensions. Elderly people are more likely to live in poverty than rural residents as a whole, and elderly people in rural areas are more likely to live in poverty than those in urban areas. However, there are also many relatively wealthy retirees in rural communities who have contributed to the processes of gentrification and class recomposition (see Chapter 6). Moreover, the resources available to elderly rural residents do significantly shape their lifestyles. Living in a rural area as an elderly resident is considerably easier if one can afford to run a car, to pay for private healthcare, for domestic and gardening help and for purchases to be delivered. Those living on restricted funds find that their ability to access services and participate in social activities is severely curtailed. Yet, one experience shared by the majority of elderly rural residents is the transition with age from a state of independence to a state of dependency.

The very physical environment of rural areas makes the loss of independence problematic, but as Chalmers and Joseph (1998) discuss in the context of New Zealand, the difficulties that result have been increased by the effects of rural restructuring. This is most prominently manifest in two ways. First, there is a tension between the restricted mobility of many elderly residents and the rationalization of rural services. Gant and Smith (1991), in a study of the Cotswolds in England, found considerable differences in the monthly geographical mobility of 'independent' elderly residents, who were in good health, and those in 'dependent' households where one or more members suffered from some serious disability. Whilst 'independent' residents travelled quite extensively and quite frequently to nearby towns to visit grocers, chemists, post offices and hospitals, 'dependent' households were largely restricted to their own village, making only occasional visits to amenities in towns. The declining provision of shops, post offices, banks and other basic services in villages and small towns (Chapter 7) therefore further disadvantages those elderly rural residents with least independence. Elderly rural residents in New Zealand interviewed by Chalmers and Joseph (1998) expressed anger towards the companies and government agencies whose rationalization strategies had closed local branches:

> I'm very cross with the Bank of New Zealand. They lost five accounts from our family when they moved. We weren't happy when the post office closed, or when the Matamata County Council disappeared from the face of the earth.

> By taking services away from Tirau, it's no longer attractive for older people to live here.

> (Elderly residents, New Zealand, quoted by Chalmers and Joseph, 1998, p. 162)

Secondly, older residents have found themselves dislocated from the community as social recomposition has taken place. Rowles (1983, 1988) argued that older residents he studied in rural Appalachia were able to draw on 'social credit' that they had accumulated over the course of their life in the community to receive help from neighbours as they grew less independent. However, for many elderly residents the 'communities' in which they built up social credit have been substantially eroded. Population restructuring, the closure of village services, the fragmentation of families, increased daily mobility and the inevitable deaths of friends and associates, mean that the community networks in which they participated are now sometimes non-existent. They may find that they now know few people in the community and may find it difficult to comprehend the new patterns of rural life. This sense of dislocation is conveyed in the narratives of elderly rural residents in mid-Wales recounted by Jones (1993):

> It's nice to see people coming to live in the valley again after the houses lay empty all those years. A lot of them went nearly derelict as a result of people having to go away and look for jobs. I like to see a light in the window and smoke coming out of the chimney again. Transport is no problem now that everybody has a car. But if you haven't got a car, there's only the Friday bus. There used to be a bus every day till everybody got cars. The buses were cut to three times a week and then once a week. You have no choice anymore when to go into town. There used to be the cheap Market Day tickets on the train every Monday, so a lot of people would go in together to sell and buy. Now there's only me and Mrs Daniel on the Friday bus, because everybody else has their own car. (Elderly female, Wales, quoted by Jones, 1993, p. 24)

The experience of older people living in rural areas hence has a temporal as well as a spatial dimension. Their lifestyle and the constraints that are placed upon it are in part described and understood in terms of rurality and comparisons with other larger settlements. But for residents who have lived in the community for some time, perhaps all their life, the present-day experience of rural life is also understood in terms of long-term rural change and a memory of how life used to be.

Summary

The experiences of children, young people and the elderly living in the countryside are all shaped by the conditions of rurality. Problems of access to services, poor public transport and dependency on others are shared by young and old alike. In contrast, the rural community values that are often held to by elderly residents against the direction of the tide, are also the same values that are frequently seen by young residents as oppressive and stifling. Both young and old adopt strategies to cope with the pressures of rural life, and particularly the pressures of rural restructuring. In doing so they create their own rural geographies and experiences that are strikingly different from those of the working age population which has so often been the predominant focus of rural studies.

Further Reading

In recent years there have been a significant number of articles and books published on children and young people in rural areas. These include a series of papers in a special issue of the *Journal of Rural Studies* on 'young rural lives' in 2002 (volume 18, number 2). Other key readings include Glendinning and colleagues 'Rural communities and well-being: a good place to grow up?', in *The Sociological Review*, volume 51, pages 129–156 (2003); Owain Jones (2000) 'Melting geography: purity, disorder, childhood and space', in Sarah Holloway and Gill Valentine (eds), *Children's Geographies: Playing, Living, Learning* (Routledge, 2000); Hugh Matthews et al., 'Growing up in the countryside: children and the rural idyll', in the *Journal of Rural Studies*, volume 16, pages 141–153 (2000); Ruth Panelli et al., '"We make our own fun": reading the politics of youth with(in) community', in *Sociologia Ruralis*, volume 42, pages 106–130 (2002); and Gill Valentine, 'A safe place to grow up? Parenting, perceptions of children's safety and the rural idyll', in the *Journal of Rural Studies*, volume 13, pages 137–148 (1997). Issues relating to gender identities and sexualities in the countryside are discussed at greater length in Jo Little's book, *Gender and Rural Geography* (Prentice Hall, 2002), and in a themed issue of *Rural Sociology* on rural masculinities in 2000 (volume 65, number 4). Less has been published on the elderly in rural areas, but a good empirically based discussion is the study of elderly residents in Tirau, New Zealand, in Lex Chalmers and Alan Joseph, 'Rural change and the elderly in rural places: commentaries from New Zealand', *Journal of Rural Studies*, volume 14, pages 155–166 (1998).

Websites

There are a number of websites that give young people in rural areas an opportunity to comment about their experiences, or which are linked to organizations providing support for rural youth. These include the Young Australians Rural Network (www.yarn.gov.au), a group for young people working in rural industries; the website of Heywire (www.abc.net.au/heywire/default.htm), a long-running Australian radio programme aimed at young people in rural areas; Rural Youth Voice (www.ruralyouthvoice.org.uk), an initiative working with rural young people aged between 13 and 19 in western England; and the National Rural Youth Network Council (realm.net/rural) in Canada. First-hand accounts of gay, lesbian and bisexual young people in rural areas of the United States can be found on the Youth Resource website (www.youthresource.com/our_lives/rural.index.cfm).

18

Working in the Countryside

Introduction

The restructuring of the rural economy not only changed the economic sectors in which rural people worked, but also the nature of work in the countryside itself. As employment in agriculture, forestry, mining and other traditional rural industries has declined, and that in the service sector has increased, so the requirements for the rural labour force have been redefined. This chapter examines the transformation of rural employment over the past century and the changing experience of working in the countryside. It first charts the way in which the rural workforce has been restructured before moving on to discuss contemporary experiences of finding work in rural areas, the changing gender dynamics of participation in the rural labour market, the experiences of migrant workers and finally the significance of commuting and experiences of commuters.

The traditional model of rural employment is epitomized by Howard Newby's study of farm labourers in eastern England, *The Deferential Worker* (1977). Newby describes a form of employment that was framed by paternalistic and particularistic relations between the employer and the worker. Most farmworkers tended to hold one job for life, many lived in tied housing and a significant number worked in the same community in which they had been born, or at least in a neighbouring community. The work itself was hard, manual labour, mainly done outside and in all weathers. It did not require any formal qualifications or training, but it did involve particular specialist knowledge that was either picked up on the job or passed down between generations in the rural community. It was not, by modern standards, highly skilled work, but it was secure and stable and, in a society that largely revolved around farming, it was work that was valued.

The modernization of agriculture during the twentieth century changed the role of farm labour (see also Chapter 4). In Britain the number of hired farmworkers fell sharply from over 800,000 in the 1940s to under 300,000 in the 1990s (Clark, 1991). Whereas there had been nearly three hired farmworkers for every farmer in 1931, by 1987 the ratio was 1.1 : 1. For those farmworkers who remained, the nature of their work changed with mechanization, becoming an increasingly skilled job. Formal training schemes were established and the higher skill level was recognized through the payment of above-average wages compared with equivalent positions (Clark, 1991). Additionally, more farm workers were employed on a casual basis, working for a number of different farms as required, or as a part-time or seasonal activity, whilst the expansion of corporate farming contributed to a 61 per cent increase in the number of salaried farm managers in England and Wales between 1972 and 1977 (Clark, 1991).

Outside agriculture, the growth of the service sector as the countryside's largest employer contributed to a polarization in the nature of rural work. At one level, the relocation of service sector headquarters and hi-tech industries and the expansion of the public sector in rural areas have increased the amount of managerial, professional and skilled technical employment. However, such jobs frequently recruit from a regional or national labour pool and can be associated with middle class migration rather than providing employment for the existing local population. At another level, employment in tourism, catering, retailing, call centres and the delivery end of public services – all of which have been growth areas – is identified with low skilled, low paid, insecure and often temporary or seasonal jobs. By 2000, some 42 per cent of workers in the rural United States were employed in jobs with low skill requirements, compared with 36 per cent of workers in the US as a whole, but significantly the gap between levels of low skill and high skill jobs in rural areas had narrowed noticeably in the previous decade. Analysts suggest that this trend reflects an ongoing transition in rural employment from manufacturing to the service sector, but note that it was slightly offset by a shift towards low skill jobs within the service sector (Gibbs and Kusmin, 2003).

A further feature of the contemporary rural labour force is the concentration of employment in small enterprises. Although there are national variations, generally around half of employees in predominantly rural regions work in establishments with fewer than 20 employees, compared with around a third of employees in predominantly urban regions (Table 18.1). Figures from Scandinavia also suggest that the degree of concentration of employment in smaller enterprises increased between the mid-1980s and mid-1990s (Foss, 1997). Similarly, rural regions tend to have higher levels of self-employment than urban areas, and there is also a strong element of people in rural areas holding multiple part-time or low paid jobs, including mixing self-employment and paid employment.

Finding Work in Rural Areas

The opportunities for employment in rural areas are shaped as much by regional and local factors, including the structure of the economy, history of industrialization and demographic profile, as by their rural situation. Thus, there is no consistent pattern of levels of employment and unemployment in rural areas compared with urban areas. In a number of countries, including the UK, Belgium and Japan, unemployment rates in rural regions are significantly lower than those in urban

Table 18.1 Percentage of employees working in establishments of differing size

	Number of employees per establishment				
	1–9	10–19	20–49	50–99	100 +
Norway					
Predominantly rural regions	40	13	15	11	21
Predominantly urban regions	27	12	15	12	34
Finland					
Predominantly rural regions	39	14	17	8	22
Predominantly urban regions	18	11	16	12	43
Switzerland					
Predominantly rural regions	34	16	19	12	18
Predominantly urban regions	22	12	16	12	38
UK					
Predominantly rural regions	25	20	17	12	27
Predominantly urban regions	17	14	13	13	44

Source: Adapted from Foss, 1997

regions, yet in many other countries, including Canada, Italy and New Zealand, rural unemployment rates are higher than those for urban centres (von Meyer, 1997). In both contexts, however, the experience of finding appropriate employment in rural communities can be beset by a number of common difficulties, many of which result from the structural characteristics of rural environments and societies.

Research in three areas in England and Wales – an accessible rural locality in Suffolk, a less accessible locality in Lincolnshire (Hodge et al., 2002; Monk et al., 1999), and four remote rural localities in mid-Wales (Cloke et al., 1997) – has identified a number of barriers to participation in paid employment. First, the traditionally close-knit nature of rural communities and the small size of enterprises mean that many employers recruit largely through informal networks. Around a fifth of employees surveyed in mid-Wales had found work through the recommendation of friends or direct personal enquiries, and similar experiences were recorded in Lincolnshire:

Lincolnshire's very, very much word of mouth. A lot of firms don't even advertise for jobs. [One firm] I can't ever remember them advertising, but I got a job by going round there putting an application form in and chasing it up, word of mouth. Got the job where I am now originally by word of mouth. (Male, England, quoted by Monk et al., 1999, p. 25)

Well, I mean, you know, things just turn up, don't they, down the pub or somebody says something. Yeah, that is the only way I find work. (Male, England, quoted by Monk et al., 1999, p. 25)

Such informal recruitment practices discriminate in favour of 'known' local residents and can make it difficult for in-migrants especially to find out about vacancies and to apply for jobs.

Secondly, access to transport can be a major barrier. The limited employment opportunities in most rural communities require people to travel into neighbouring towns, or elsewhere in a wider rural region, to find work. For individuals who are dependent on public transport, in particular, this can severely restrict their options. The problem of transport can become self-reproducing as individuals may be unable to afford to run a car without getting job, but cannot get a job without having a car:

> I did have a couple of interviews, one was at [a food factory] and they said, because of my transport, buses were not good enough, because of the timetable, that was the only thing that stood in my way. The same with [another factory], they said I could have had a permanent job if I had reliable transport and not the bus service. (Female, England, quoted by Monk et al., 1999, pp. 26–27)

> So I was in a bit of a catch-22 situation. I couldn't get a job, I couldn't get a car until I got a job, and I couldn't get a job until I got a car. (Male, England, quoted by Monk et al., 1999, p. 27)

Thus, thirdly, the costs of participating in the labour market can be prohibitive given the low rate of wages paid by many rural jobs. In the three research areas of Lincolnshire, Suffolk and mid-Wales, average male earnings were between 77 per cent and 84 per cent of the national mean (Cloke et al., 1997; Monk et al., 1999). After tax, a quarter of full-time workers surveyed in Lincolnshire took home less than GB£150 per week, out of which they needed to pay transport and housing costs that were generally higher than they would have been in an urban area, and in many cases, childcare costs as well. As such, Monk et al. (1999) report examples of individuals who had restricted themselves to low skill, low paid jobs in or near their place of residence because the additional income from higher skilled, better paid jobs further away would have been offset by the transport costs.

Fourthly, when these various constraints are taken on board, many rural workers find that the skills and qualifications that they have are mismatched to the jobs that are available.

This can be a problem of over-qualification, for example as a result of lack of graduate opportunities in rural areas or because clerical skills developed in an urban-based job have no obvious application in a rural area. It can also be a problem of skills developed in traditional industries such as farming not being transferable as the opportunities for work in those sectors decline:

> I suppose I am limited by what I can do really, because I've had previous experience on the land, and in the garden and that, you know, I suppose without the relevant qualifications, you're restricted as to what you can go for or apply for really. I did look for other jobs, but only as a gardener. (Male, England, quoted by Monk et al., 1999, p. 24)

Therefore, even in rural regions where unemployment is not considered to be a significant problem many people are restricted to jobs that do not fully use their skills or qualifications. This gives rise to two models of employment history in two areas. The first is of frequent short-term, often temporary or seasonal employment, interspersed with periods of unemployment. Looker (1997), for example, reports that more young people surveyed in rural localities in Canada had drawn unemployment insurance than those in urban localities (Table 18.2). The second model is of longer-term employment in the same job, but less through choice than because of the absence of alternatives. In traditional occupations, such as farm work, the ability to look for alternative employment may be further restricted by a dependence on tied housing (Monk et al., 1999).

Gender and Rural Employment

One of the most prominent changes in rural employment has been the shifting gender

Table 18.2 Work experiences of young people in rural and urban localities in Canada

	Rural respondents (%)	Urban respondents (%)
Drawn unemployment insurance	50	23
Prefer seasonal work plus unemployment insurance	18	5
Accessed a government programme	20	15
Held a full-time job	68	74
Quit a job	32	46
Started own business	3	8

Source: After Looker, 1997

balance of the workforce. The number of rural women in paid work increased significantly in the second half of the twentieth century, although the rate of female participation in the labour force remains lower in rural areas than in urban areas (Little, 1997; von Meyer, 1997). In part, this reflects changes in attitudes to gender and employment within the agricultural community. Historically, women were fully involved in farm work, but as Hunter and Riney-Kehrberg (2002) describe, during the late nineteenth and early twentieth centuries new constructs of gender roles emerged in which agricultural work was identified with masculinity, and farm women were associated with a domestic role. This gendering of roles understated and undervalued the contribution of women to the functioning of the farm economy. As Little observes,

> farmer's wives were nearly always responsible for the majority – if not all – of the domestic work of the farm household. They were found to be in charge of the cooking, cleaning, shopping and child-care duties associated with the household, and expected to perform these duties regardless of what other work they did on the farm – on either a routine or emergency basis. (Little, 2002, p. 105)

The 'other' work on the farm included both administrative and manual tasks. A study in southern England found that 85 per cent of farm women dealt with enquiries and ran errands, 70 per cent were involved in manual work on the farm and 65 per cent were responsible for book-keeping and other clerical duties (Whatmore, 1991). Up to a third of the women in the study regularly undertook manual work on the farm – in keeping with evidence from Scandinavia that women are increasingly challenging gender stereotypes by becoming involved in all aspects of farming and establishing themselves as independent farmers in their own right (Silvasti, 2003).

Women have also been at the forefront of adaptation strategies to agricultural restructuring through involvement in on-farm and off-farm diversification. On-farm, women have frequently been responsible for developing new initiatives such as farm shops, bed and breakfast accommodation, craft enterprises and educational activities (Gasson and Winter, 1992; Little, 2002). Off-farm, income earned by women from full-time or part-time employment in a diverse range of occupations has provided an important addition to farm finances at a time when income from agricultural production has been under pressure (Kelly and Shortall, 2002). Participation in paid employment by women in the agricultural population in Canada surpassed the national average in the early 1970s and is now over 60 per cent (Dion and Welsh, 1992). A separate study in Manitoba found that 55 per cent of farmers' wives had off-farm employment in 1992 (Stabler and Rounds, 1997). As well as the financial benefits, off-farm employment also gives women an identity and role that is independent of their association with the farm and which challenges conventional gender relations in the farm household (Kelly and Shortall, 2002). However, if farm women are still expected to take a lead in domestic responsibilities, employment either on-farm or off-farm may simply increase the amount of work that farm women are compelled to do.

At the same time, many of the new jobs created in expanding sectors of the rural economy have been filled by women. Female employment in tourism, for example, is significantly higher in relative terms in rural areas than in urban areas, and over half of rural tourism jobs in the UK, Canada and Germany were held by women in 1990 (Bontron and Lasnier, 1997). Overall, the jobs commonly taken by rural women reflect the full range of occupations, from professional positions, notably in teaching and healthcare, to clerical work, manufacturing production lines, cleaning and childminding (Little, 1997). In some areas, rural development agencies have implemented specific strategies to increase employment opportunities for women (Little, 1991), but as Little (2002) notes, most rural development strategies pay little attention to the particular issue of female participation in the labour market. As such, the growth in women's employment in rural areas has been produced more by demand than by supply. There has been a desire on the part of rural women to break stereotypes and to establish themselves as independent earners. Professional women in-migrants, in particular, have sought to maintain a career presence. Women's employment, can, however, also be a

response to a need for households to have a dual income in order to afford high rural property prices.

The incentives for employment are balanced by costs and constraints, especially from family responsibilities. Little (1997), for example, cites women interviewed in two English villages who felt restricted in their employment options by the expectation placed on them to be full-time mothers, or by difficulties in finding appropriate and affordable childcare (see also Chapter 17):

> The strain with younger children is always on the mother to provide the childcare. Before my present job [as a secretary] I worked as a cleaner and drove a fish van to fit in with school hours. (Mother, England, quoted by Little, 1997, p. 150)

Jobs within rural communities that can be fitted within the school day are at a high premium. Accordingly, studies in England and Canada have identified very high rates of part-time employment for rural women (Little and Austin, 1996; Leach, 1999) – and as such, women in rural areas often experience relatively poor employment conditions. The time constraints on women's employment also produce significant underemployment. Over half of the employed women surveyed by Little and Austin (1996) in an English rural community in 1993 were in jobs that did not use their qualifications or training.

The working lives of rural women therefore are highly complex and frequently involve a mixture of formal and informal, paid and unpaid activity. In a study in rural Vermont, Nelson (1999) found that a majority of both men and women were engaged in income-generating activity beyond their main work, and that most households undertook some form of self-provisioning, such as car maintenance, growing vegetables, keeping animals for meat or eggs or cutting firewood. However, Nelson notes that there were significant gender differences in the way in which men and women approached these additional activities. Men were more likely to have a formal second job, more likely to be formally self-employed and more involved in self-provisioning activity. Significantly, whilst for men these activities often took place away from the home and were devoted dedicated time, women's supplemental economic activity tended to be more casual and more home-based, including sewing, knitting, craft-making, baby-sitting and caregiving, detailed work in home decoration and growing vegetables. These tasks were often fused with their other activities, especially domestic work. Whereas men would regard childcare as time when they were not engaged in more economically productive activity, for women childcare often was performed alongside other chores:

> When he baby-sits, he can't do anything else ... I can cook and do laundry and clean the house and work at my desk and take care of the baby, but *he* can't. (Mother, Vermont, quoted by Nelson, 1999, p. 528)

This differential approach to supplemental activity, Nelson argues, continues to reinforce male privilege in rural households and to undervalue the work done by women in the rural milieu.

Migrant Workers in the Rural Economy

The modernization of agriculture may have reduced the farming workforce, but there are types of agriculture, notably forms of vegetable and fruit cultivation, that are still very labour-intensive, albeit on a seasonal basis. However, the labour requirements of these farms are increasingly filled by migrant workers. As noted in Chapter 3, the presence of migrant workers from the developing world in the rural labour force of the developed world can be seen as a dimension of the globalization of mobility, with employers recruiting for low skilled, marginal and usually temporary jobs through transnational networks. An estimated 69 per cent of all seasonal farmworkers in the United States are foreign-born, including more than 90 per cent of the seasonal workforce in California (Bruinsma, 2003). In Europe, the dependency on migrant workers is less extensive but significant none the less. Hoggart and Mendoza (1999) report that migrant African workers comprised more than 5 per cent of the agricultural workforce in the three Spanish provinces of Murcia, Almeria and Cáceres in 1995, and that 32 per cent of African migrant workers in Spain were employed in agriculture. Similarly, in the UK there are estimated to be some 20,000 foreign workers on farms in East Anglia, including Lithuanians, Russians, Portuguese, Macedonians, Latvians, Poles, Ukrainians, Bulgarians and Chinese.

Californian agriculture has been heavily dependent on migrant Mexican labour since the mid-twentieth century. The rapid development of intensive, capitalist agriculture in California at the start of the century (see Chapter 4), first attracted mass migration from other parts of the United States. The harsh work, exploitation and poverty of the migrants was documented in the writing of John Steinbeck, notably *The Grapes of Wrath* (1939). The radicalization and unionization of the farm workforce in a struggle for better conditions, however, generated conflict with employers and it was thus in search of 'docile' workers, with 'no political ambitions', that the industry started to recruit foreign migrant labour (Mitchell, 1996). Between 1924 and 1930, an estimated 58,000 Mexican and Hispanic workers arrived in the San Joaquin Valley each year, with many more employed in the Los Angeles Basin (eastern Asia, including

China, Japan and the Philippines, was also for a short period a key recruiting ground). As Mitchell (1996) describes, the employment of migrant workers was framed by racist attitudes and practices from the beginning. Brutal working conditions and poor wages were a matter of course and migrants lived in poverty in racially stratified labour camps. The exploitation was disguised by a representation of the rural idyll that was used to attract migrants, promising that families 'could find healthful, fulfilling living in the countryside should they choose to spend their summers helping bring in the crop' (Mitchell, 1996, p. 83). Yet, as Mitchell later concludes, the promised 'rural idyll' 'was built on the constant, consistent objectification and racialization of labor' (p. 107).

The farmworkers of California won union recognition from growers in 1975, but by 2002 only 27,000 of the 600,000 agricultural workers in the state were unionized and relations remained exploitative. Three-quarters of migrant farmworkers earned less than $10,000 a year and 90 per cent had no health insurance (Campbell, 2002). Housing availability has remained limited, with many migrants forced to sleep in overcrowded camps and hostel accommodation or outside. As one worker told the *Los Angeles Times*: 'When it comes to providing us with a place to sleep, many owners look the other way. They tell you "It isn't my problem". They don't care what happens at night as long as you show up for work at dawn the next day' (Glionna, 2002, p. B1). The problem was partially addressed in 2002 when vineyard owners and local residents in the Napa Valley voted for an ordinance to treble the amount of housing for migrant workers in the county. Also in 2002, farm workers began a political mobilization that spread across the United States in support of unionization and better working conditions.

The introduction of migrant workers in European agriculture has a more recent heritage and is connected to the broadening of employment opportunities for rural residents. Hoggart and Mendoza (1999) explain the increase in African migrant workers in Spain in terms of Spanish workers rejecting seasonal agricultural labour in favour of better opportunities, for instance in tourism. Agriculture hence provides an entry point into the Spanish labour market for African migrants who themselves aspire to move on to other jobs. Although conditions are rarely exploitative, the employment undertaken by migrant workers in Spain is characterized as 'unskilled work, on poor pay, in occupations associated with inferior social status, with short periods of employment, in jobs that are rarely part of a promotional ladder' (Hoggart and Mendoza, 1999, p. 554).

The employment of migrant workers has begun to spread from agriculture to other sectors of the rural economy. However, whereas in agriculture migrant workers could be described as resolving a labour shortage, in other industries they can displace an existing workforce. Selby et al. (2001), for example, discuss the employment of Mexican women in the crab houses of North Carolina. Mexican workers were recruited in response to pressure from foreign competition and a perceived inability to cut labour costs in the local, predominantly black, workforce. The women studied by Selby et al. (2001) were employed as 'crab pickers' in a small crab house, extracting meat from crabs. Reflecting the gendering of the crab industry, all the crab pickers in the company were women, including 12 Mexicans and three, elderly, white women with strong connections to the proprietors who had not been displaced by the shift to migrant workers. Visa regulations meant that the Mexican workers were essentially tied to the job, but working conditions were reasonable and with pay linked to productivity, they could theoretically earn significantly above the minimum wage. Yet, the spatial and social organization of the working environment clearly indicated the secondary status of the migrant workers. Although Selby et al. (2001) note that the Mexican and white women had much in common, they observe that there was little interaction between them:

> In the main room, three white women sit at tables while picking the meat out of the crabs. The women always sit together, working and sometimes singing hymns, talking and laughing. On the other side of the room, 12 Hispanic women stand around a table and work in silence. ... There is no discernible contact between the two groups of pickers as the day wears on. (Selby et al., 2001, p. 239)

Thus, whilst the Mexican women saw themselves as being empowered by the leverage their earnings gave them to send money home for the education of their children or improvements to their homes, their working existence was marked out very clearly as being temporary and separate from that of the local, white, community.

Commuting

Rural restructuring has produced a dislocation of places of work and places of residence for most rural people. Whereas in an agriculturally dominated economy work and residence were closely bound together within discrete, coherent communities, today the limited number of employment opportunities in rural communities means commuting to work outside their home settlement has become the norm for most rural workers. Three-quarters of non-metropolitan counties in the United States have out-commuting rates from their main settlements of more

than 35 per cent. In Austria, nearly 30 per cent of residents in predominantly rural regions are commuters, as are over 15 per cent in Canada, and around 10 per cent of residents in significantly rural regions in Britain and Germany (Schindegger and Krajasits, 1997). Rates of out-commuting predictably increase with decreasing settlement size (see Figure 18.1), as well as with proximity to larger urban centres. Overall, commuting is a growing practice. The number of commuters in significantly rural regions in Canada increased by over 50 per cent between 1980 and 1990, and in the UK by around 25 per cent in the same period (Schindegger and Krajasits, 1997).

These aggregate figures disguise some of the dynamics of commuting within rural areas. Analysis of trends in Canada, for example, reveals that over 20 per cent of all commuting is within rural areas or between different rural areas. Commuting from rural areas into urban centres constitutes 11 per cent of all commuting, nearly three times as much as travel from urban areas to employment in the countryside (Green and Meyer, 1997a). A separate study of Wilmot Township, within the commuting zone of Kitchener/Waterloo, Ontario, demonstrates this complexity further (Thomson and Mitchell, 1998). Around half of household members in the township commuted to employment in the urban centres of Waterloo, Kitchener or Cambridge. Of the remaining households, most residents worked at home with a minority working elsewhere in the township. Whilst commuting rates were higher for in-migrants than for longer-term residents, Thomson and Mitchell (1998) note that in nearly a quarter of newcomer households both partners worked at home. As such, they conclude that although commuting is dominant, 'one cannot ignore the fact that new residents are finding, or creating, gainful employment in the countryside' (pp. 196–197).

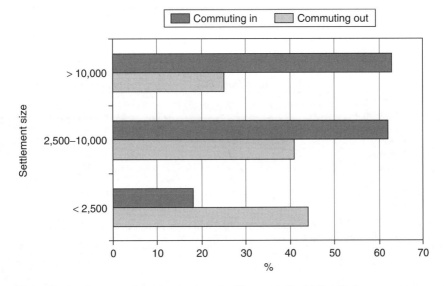

Figure 18.1 Aggregate commuting rates by size of settlement in the United States
Source: Based on Fuguitt, 1991

Canadian research also challenges the perception that commuting is associated with middle class in-migrants. Although slightly more workers employed in professional and managerial occupations in total are commuters than for other occupations, commuting by rural residents is no more significant for these occupation types than for manual or lower skilled occupations (Green and Meyer, 1997b). Indeed, Monk et al. (1999) in the UK note that some lower skilled workers are prepared to travel long distances, involving long hours, for employment. As they quote one man describing, long-distance commuting on a daily basis can have a detrimental effect on health and family relations:

You've got to think how far Peterborough is from [place of residence]. It's 50 miles, right. We got picked up ... but having to get up at 4 o'clock in the morning, get picked up at 5, got there for 6 ... It was 12 hours a day ... not getting home again till 8 o'clock ... spend two hours with your wife ... [the children] were in bed by then, I never got to see them at all ... And it was, it was making me rather ill ... And that took me three months to recover. (Male, England, quoted by Monk et al., 1999, pp. 27–28)

Commuting can also have a detrimental effect on communities. Research by Errington (1997) on a village in Berkshire, in the London commuting belt, found that residents working outside the village were significantly less likely than those working in the village to use shops and facilities in the village and more likely to use facilities in other places (Table 18.3). For communities where a majority of residents work in other settlements, this can mean that levels of patronage fall below the level at which services and amenities can be maintained, forcing their closure. This in turn reduces employment opportunities in the communities concerned and contributes to a process in which some villages effectively become specialist 'dormitory' settlements.

Table 18.3 Proportion of total visits per year to selected facilities in a Berkshire village made by members of survey panel (*n* = 55) working in the village and working outside the village

	Residents working in village (%)	Residents working outside village (%)
Bank	88	12
Post office	64	36
Newsagent	58	42
Baker	63	37
Chemist	67	33
Grocer	64	36
Clothes shop	66	34
Pub	53	47
Church	46	54
Doctor's surgery	46	54

Source: After Errington, 1997

Summary

Economic restructuring has reshaped the rural labour market. As the dependency on farming and other resource-exploitation-based employment has declined, the range of occupational opportunities in rural areas has expanded vastly. The nature of the rural labour force has also changed, with greater female participation and the recruitment of foreign migrant workers to fill labour shortages in basic, unskilled work in agriculture and elsewhere. However, despite these broad trends, the experience of finding work for individuals living in rural areas can still be difficult, obstructed by barriers including problems of access to transport, the availability of childcare and the shortage of appropriate, skilled, occupations. As such, many people are employed in jobs that do not fully use their skills, qualifications or training. This also means that many rural residents are earning less than their theoretical income potential, but are instead 'trapped' in low wage employment, contributing to problems of poverty and deprivation in rural society that are discussed further in the next chapter.

Further Reading

Statistical analysis of a wide range of issues concerning rural employment patterns, including commuting, with examples from Europe and North America, is contained in an edited volume by Ray Bollman and John Bryden, *Rural Employment: An International Perspective* (CAB International, 1997). The strong economic focus of this work, however, reveals little of the actual experience of people working in the countryside, more of which is conveyed by Ian Hodge and colleagues in 'Barriers to participation in residual rural labour markets', *Work, Employment and Society*, volume 16, pages 457–476 (2002). For more on gender and rural employment, see two chapters by Jo Little: 'Employment marginality and women's self-identity', in P. Cloke and J. Little (eds),

Contested Countryside Cultures (Routledge, 1997) and Chapter 5 in her *Gender and Rural Geography* (Prentice Hall, 2002).

Don Mitchell's book *The Lie of the Land: Migrant Workers and the California Landscape* (University of Minnesota Press, 1996) discusses historical dimensions of migrant workers in Californian agriculture. For discussion of more contemporary experiences, see the work of Keith Hoggart and Cristobal Mendoza, 'African immigrant workers in Spanish agriculture', *Sociologia Ruralis*, volume 39, pages 538–562 (1999); and Emily Selby, Deborah Dixon and Holly Hapke, 'A woman's place in the crab processing industry of Eastern Carolina', *Gender, Place and Culture*, volume 8, pages 229–253 (2001).

Websites

More information on migrant agricultural workers in the United States, and campaigns to improve their conditions, can be found on the Farmworkers website (www.farmworkers. org) and the website of the Rural Coalition (www.ruralco.org).

19

Hidden Rural Lifestyles: Poverty and Social Exclusion

Introduction

The previous three chapters have highlighted a number of processes and experiences that have contributed to deprivation and poverty in rural areas: problems of access to good quality, affordable housing and the burden of debt placed on many rural households in seeking to pay for property (Chapter 16); problems of the dependency of elderly residents on local services that are being rationalized (Chapter 17); and problems in finding appropriate work that leads to underemployment and a prevalence of low wage employment (Chapter 18). However, this landscape of rural poverty is often hidden. As Furuseth (1998) observes, 'for most residents of the industrialized world, who live in urban and suburban communities, the term *rural* conveys a comfortable image of picturesque small towns and open countryside populated by prosperous farmers and other middle-class or similar residents' (p. 233).

The marginalization of rural poverty (or its near synonyms, 'deprivation' and 'social exclusion' – see Box 19.1) has three facets. First, the experience of poverty in rural areas is fragmented. Deprived households tend not to be clustered together in identifiable territorial units, but rural communities often incorporate wide discrepancies in income and wealth. Milbourne, for example, notes that,

> households living in poverty in small and scattered rural settlements tend to remain physically hidden, in contrast to the visual concentration of poverty in the urban, and more specifically inner city arena. Indeed, in many areas of the countryside, the marked physical segregation of 'rich' and 'poor' households – suburbia and inner city – tends to be absent, with the 'rural poor' often living cheek-by-jowl with more affluent residents. (Milbourne, 1997b, pp. 94–95).

> ### Box 19.1   Key term
>
> **Poverty, deprivation and social exclusion:** These terms are often used inter-changeably but in fact have subtly different meanings. *Poverty* is an absolute condition that relates to the economic position and power of a household or individual. Households can be identified as living 'in poverty' or 'below the poverty line' against official definitions (as in the United States) or academic definitions (as in the UK). *Deprivation* is a relative term that suggests that some communities, households or indi-viduals have less of a resource than others. Deprivation is commonly used to refer to economic circumstances but it can also be related to health, education, transport, access to services. The use of the term 'deprivation' in a rural context has, however, been criticized because many rural residents do not accept that rural households can be deprived (see Woodward, 1996). *Social exclusion* has gained popularity in recent years among policy-makers and academics alike. It is again a broader term than poverty, focusing on the ways in which households and individuals are marginalized from mainstream society. However, social exclusion has been criticized for not enga-ging with the root causes of poverty by implying that remedies lie in education, train-ing and programmes for social integration, rather than in the redistribution of wealth. The debate about the appropriate use of these terms in rural studies continues. In this chapter, the term 'poverty' is usually used; 'deprivation' is occasionally used when referring to relative disadvantage, and 'social exclusion' when referring to marginal-ization within rural society.

Secondly, the discourse of the 'rural idyll' masks the existence of rural poverty. The idealistic images of the rural idyll do not appear to allow for the possibility of poverty within the countryside and thus opinion both inside and outside the countryside that is informed by this discourse presumes that deprivation cannot be present (Cloke, 1997b; Woodward, 1996). Moreover, the discourse of the rural idyll can exacerbate rural deprivation because it celebrates aspects of country life, including isolation, the shortage of housing and absence of industry, that contribute to social exclusion. Similarly, the shared situation of rural residents of all social positions in an aesthetically valued natural landscape that is associated with peace and tranquillity, is also portrayed by some as 'compensation' for material deprivation, such that whilst rural households may have the same level of material deprivation as urban households the experience of poverty is judged to be less severe:

> the poor can thus be disregarded as being 'content' with their (undemanding) rural life, and the not-so-poor will not be able to reconcile the idea of poverty with the idyll-ised imagined geographies of the village, so any material evidence of poverty will be screened out culturally.
> (Cloke, Goodwin et al., 1995, p. 354)

Thirdly, the discourse of the rural idyll also informs a set of moral values that discriminate against the recognition of poverty in rural areas. In these, rural life is associated with resilience, perseverance and self-help. Those individuals and households who slip into poverty, therefore, may be perceived not only as an affront to the 'rural idyll', but also as 'failures' and 'undeserving poor'. As such, even households experiencing material deprivation may be unwilling to admit their circumstances and become complicit in the reproduction of the notion that rural poverty does not exist:

> In this respect, the poor unwittingly conspire with the more affluent to hide their own poverty by denying its existence. Those values which are at the heart of the rural idyll result in the poor tolerating their material deprivation because of the priority given to those symbols of the rural idyll: the family, the work ethic, and good health. And when material deprivation becomes so chronic by the standard of the area that it has to be recognised by the poor themselves, shame forces secrecy and the management of that poverty within the smallest possible framework. (Fabes et al., 1983, pp. 55–56)

In seeking to move beyond stalled debates about the identification of deprivation in rural areas and the appropriate use of terminology, Woodward (1996) suggests that the gap between academic and lay understandings needs to be bridged and calls for the attitudes and beliefs of different groups of people living in rural areas to be taken account of in research on rural lifestyles. This chapter seeks to honour this call by discussing both the evidence for rural poverty and the experiences of rural people living in or with deprivation. It first engages with evidence for rural poverty in the United States, Canada and the UK before discussing particular case studies and narratives of rural poverty. It then focuses on the particular circumstance of the homeless in rural areas and concludes by considering responses to the problem of rural poverty.

Evidence for Rural Poverty

The measurement of poverty in rural areas is notoriously problematic. In addition to the cultural perceptions of rurality and poverty discussed above, indicators of deprivation developed in urban contexts do not translate smoothly to rural situations. Although there are interconnections between rural and urban poverty, there are also a number of differences in emphasis in terms of the key problems. Limited accessibility, high per capita service costs, weak service provision and problems of housing provision are all, for example, major components in rural deprivation that are less significant in urban areas (Furuseth, 1998). Conversely, problems of overcrowding, high crime and a blighted physical environment are generally less important in rural deprivation than urban deprivation. Similarly, as noted in Chapter 18, the underemployment of skills is more of a problem in rural areas than actual unemployment, the latter being a key indicator of urban deprivation. Milbourne (1997b) also observes that rural poverty 'tends to be characterized by higher proportions of households in employment, married couple families and the elderly, and a lower incidence of single parent households than in metropolitan areas' (p. 98).

Further problems arise in the spatial scale at which indicators of deprivation are constructed. Local government units in rural areas tend to cover more extensive and more diverse territories than in urban areas and consequently there is a moderating effect on the statistics collected. Moreover, as noted above, there is greater interspersion of households of differing income levels in rural communities than in urban neighbourhoods such that even at a small area level the presence of a minority of deprived households may be disguised by the prosperity of the majority of households.

In response, attempts have been made to develop more spatially sensitive indicators but much of this work is still in its initial stages.

These qualifications notwithstanding, existing indicators of deprivation do provide evidence that rural poverty is more pervasive than popular perceptions envisage. In the United States, where there is an official poverty line defined in terms of income and necessary budgets for food and essential household items, 15.9 per cent of the population in non-metropolitan counties were calculated to be living in poverty in 1997, compared with 13.2 per cent of the metropolitan population (Nord, 1999). Similarly, in Canada a 'low income cut-off' is fixed at 62 per cent or more of household income spent on food, clothing and shelter, and 16 per cent of rural households passed this threshold in 1986 (Reimer et al., 1992). This was a lower percentage than in urban centres, but partly because of difference within rural regions. The proportion of households in rural settlements of less than 5,000 residents, including farms, falling below the low income cut-off was roughly equivalent to those in cities of over 50,000 population and Reimer et al. (1992) also note that differences in the cost of living meant that rural households reached the threshold at lower levels of income than urban households.

There is no equivalent official definition of poverty in the UK, but research in the 1990s applied the Townsend indicator that defines households in or on the margins of poverty as those with an income that is less than 140 per cent of their state income supplement entitlement. Using this measure the research found that 23.4 per cent of households in 12 rural study areas were in or on the margins of poverty, with rates in the individual study areas ranging from 12.8 per cent in Cheshire to 39.2 per cent in Northumberland (Cloke, 1997b; Cloke et al., 1994; Milbourne, 1997b). Two alternative indicators, comparing household income to mean and median income respectively, suggested even higher levels of poverty (Cloke, 1997b).

The figures from both North America and the UK further suggest that there are significant variations in the degree of poverty between different social groups in the countryside and between different rural localities. In Canada, for example, 28 per cent of rural families were calculated to live below the low income cut-off in 1986 compared with 13 per cent of unattached people in rural areas (Reimer et al., 1992). This bias is replicated in the United States, where 61 per cent of the rural poor have been reported to be in two-adult households (Porter, 1989), and where 24 per cent of rural children were calculated to be living in poverty in 1996, compared with 22 per cent of children in metropolitan areas (Dagata, 1999). Porter (1989) also suggests that compared with the urban poor, the rural poor in the United States are disproportionately white and disproportionately elderly. Similarly, Cloke et al.'s (1994) research in England indicates that levels of poverty are more prevalent for particular social groups, notably single elderly households, long-term resident households and those with close relatives nearby, households in social housing and those without access to a car (Table 19.1). These findings appear to support the notion of class recomposition taking place with counterurbanization (see Chapter 6), with poverty being more extensive in the 'local' population than among in-migrants, but it should also be noted those areas with the most extensive poverty also attracted significant numbers of in-migrants on low incomes.

The geographical pattern of rural poverty reflects the spatial distribution of these at-risk groups combined with factors concerning the structure of the local economy and labour market. In 1990, there were 765 non-metropolitan counties in the United States

Table 19.1 Social groups with 20 per cent or more of households
in or on the margins of poverty in 12 rural localities in England

	% in or on margins of poverty
One-person elderly households	41.8
Two-person elderly households	27.4
Two-person non-elderly households	20.5
Household resident for less than 5 years	24.1
Household resident for 5–15 years	31.8
Household resident for more than 15 years	42.4
Close relatives living nearby	60.0
Housing fully owned	34.1
Households in socially rented property	47.1
Households with no access to a private car	42.4

Source: After Milbourne, 1997b

where more than 20 per cent of the population lived below the poverty line – considerably less than the 2,083 counties that were in the same position in 1960. However, in 535 counties, the poverty rate had exceeded 20 per cent of the population in each of the years 1960, 1970, 1980 and 1990. As Figure 19.1 shows, most of these 'persistent poverty counties' are located in the southern states and in Appalachia, contributing to a regional geography of rural poverty in the US. Nearly a third of the rural poor were concentrated in these counties in 1990, with an average of 29 per cent of the population in the persistent poverty counties living below the poverty line. Unemployment rates were significantly higher than the rural average in these counties, and average income levels markedly lower. The persistence of poverty in these areas results from a combination of physical, social and economic factors, with the counties generally characterized by sparse settlement patterns, the decline of staple industries, long-standing low wage economies and high levels of disabilities that affect labour market participation (Lapping et al., 1989).

One of the most prominent features of rural poverty is the presence of the 'working poor', individuals who are in employment, often in the service sector, but who have low incomes and limited employee benefits such as health insurance (Lapping et al., 1989).

Porter (1989) found that nearly two-thirds of rural households in poverty in the United States had at least one paid worker in the household, and nearly a quarter had two, whilst the comparative figures for urban poor households were 51 per cent and 16 per cent respectively. As such, a key factor in explaining rural poverty is the prevalence of a low wage economy, especially when combined with higher mean expenditure by rural households on basics such as fuel and transport. In 2002, one in four rural workers aged 25 and over in the United States earned less than the weighted poverty threshold of $18,390 per year, compared with one in six urban workers (ERS, 2003a). Low wage earners are particularly concentrated in agriculture, manufacturing, retailing and services – sectors that collectively comprise 71 per cent of rural employment in the United States. Indeed, across the board average weekly earnings in non-metropolitan counties in 2002 were 20 per cent below those in metropolitan counties (ERS, 2003b).

A similar picture has been found in the UK, with average earnings in peripheral rural localities as much as 25 per cent lower than the national mean (Cabinet Office, 2000). In over a quarter of households in rural Wales the gross annual salaries of the first two adult earners in the mid-1990s were less than £5,000 – with localities within the region where the proportion

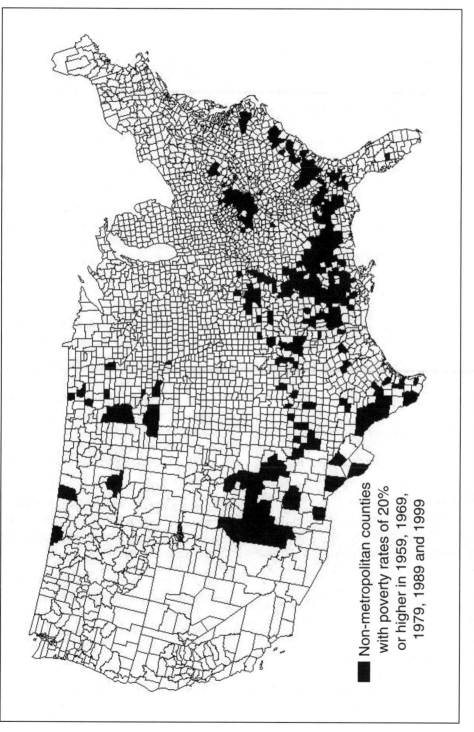

Figure 19.1 Persistent poverty counties in the rural United States, 1990
Source: Based on information from the Rural Policy Research Institute

Non-metropolitan counties with poverty rates of 20% or higher in 1959, 1969, 1979, 1989 and 1999

Table 19.2 Households where the gross annual salaries of the first two adults was less than £8,000 in eight case study areas in England and Wales, mid-1990s

England	%	Wales	%
Northumberland	53.4	Betws-y-Coed	43.6
North Yorkshire	50.5	Devil's Bridge	41.1
Devon	46.9	Tanat Valley	37.0
Shropshire	33.0	Teifi Valley	36.3

Source: After Cloke et al., 1997

was considerably higher (Cloke et al., 1997). Furthermore, case study research in England and Wales suggested that between a third and a half of households in rural areas received less than £8,000 from the gross annual salaries of the first two adults (Table 19.2).

Experiences of Rural Poverty

Individuals' experiences of rural poverty are strongly shaped by their geographical context. Households in poverty in generally affluent rural communities, for example, can experience 'double deprivation' in that they are not only in a deprived situation against the national mean, but local 'peer-pressure' about the type of expected lifestyle is based on an above-average income level. Yet, publicly at least, individuals in this position will commonly not acknowledge their poverty. As discussed at the beginning of this chapter, the denial of rural deprivation is a core element in narratives of poverty in the UK countryside. Woodward (1996) observes that rural poverty is often perceived as an historical anachronism, such that people charged with administering village charities remark that their function is becoming defunct. Behind this perception is an association of poverty with squalor and an assumption that it must have a clear, material, expression. This has a two-fold implication. Poverty that is hidden behind fairly respectable, well-maintained, property façades gets ignored, whilst in searching for examples of rural poverty, people may point to rundown cottages and the like whose residents may not actually consider themselves to be deprived.

The expectation that poverty has a material expression is also evident in descriptions by rural residents in England and Wales of the deprivation that they perceive to be present in their locality, as reported by Cloke et al. (1997) and Milbourne (1997b). These frequently refer to low incomes, limited employment opportunities, environmental conditions and the lack of public transport and services:

[There is a] shortage of local employment and low wages [and] deprivation associated with agricultural policy. (Retired male, Wales, quoted by Milbourne, 1997b, p. 110)

Employment [is] not available, no real jobs. [We need] higher income and better employment. [There are] too many odd jobs – not enough real jobs. (36-year-old man, Wales, quoted by Milbourne, 1997b, p. 110)

[There is a] lack of affordable accommodation, lack of money, lack of choice of jobs. (50-year-old man, Wales, quoted by Milbourne, 1997b, p. 111)

[We need] improved water supply, improved telephone lines, improvement in the pollution of rivers in the area ... improvement in police services. (Resident, Wales, quoted by Cloke et al., 1997, p. 131)

These statements, however, convey a rather disjointed, detached engagement with rural poverty. They reveal nothing about how material deprivation actually impacts on the lifestyle choices and decisions of those affected, or how individuals may experience multiple forms of deprivation. More evidence of this comes from rural regions where poverty is both more pervasive and more widely recognized, such as Appalachia in the eastern United States and the area north into New York State. The region is the location of a cluster of persistent poverty counties, although Cloke (1997b) notes many people in Appalachia are not disposed to accept the 'stigmatic tag of "poor" or "deprived"' (p. 265). Fitchen (1991) contends that poverty in the region is typically manifest in three types of residential landscape, and her descriptions of these sites provide some indication of the dynamics of rural poverty.

First, Fitchen argues that there are pockets of long-term, inter-generational poverty in the open countryside. These are generally locations with poor conditions for agriculture where the resident families did not adapt to the employment opportunities provided in the towns and cities of the region. As Fitchen describes, the social exclusion of such households is based on economic status but reinforced by spatial isolation and cultural stereotyping:

> Their social separation from the larger urban-based community is revealed verbally: They refer to it as 'the outside world.' Reciprocally, the people of the rural pockets of poverty are referred to by the larger community in such derogatory terms as 'poor white trash,' 'the shack people,' or 'people who live like animals.' Their social life is almost entirely confined to the immediate neighborhood or within a cluster of similar depressed neighborhoods that are linked by geographic proximity, kinship, marriage, car trading, and shared poverty and stigma. (Fitchen, 1991, p. 119)

Secondly, poverty is associated with people on low incomes living in rented housing in small towns and villages. The towns and villages concerned tend to be in more peripheral rural areas away from major sources of employment. The lack of employment opportunities has encouraged out-migration, leaving surplus, and consequently cheap, housing that has been filled by tenants unable to afford accommodation in larger towns. Much of the housing is sub-standard yet the rents charged are still demanding on the budgets of tenants, many of whom are dependent on welfare payments. Many of the villages have no public transportation service, such that unemployment is exacerbated not only by a shortage of local jobs but also by the difficulties of travelling to work elsewhere.

Thirdly, Fitchen locates poverty in the growing number of trailer parks and informal clusters of trailers in the region. Trailer accommodation has increased as a relatively affordable housing option (see Chapter 16), although the costs of purchasing trailers have also increased with demand and the maintenance of trailers can involve numerous hidden costs that are significant for low income residents. Trailers can have inadequate space for families and many are expensive to heat, with poor insulation. Increasingly trailer parks are also stigmatized as sites of poverty. The cumulative deprivation experienced by many rural people in poverty is illustrated by Fitchen's pen portrait of one trailer park resident:

> Sandy is twenty years old. She lives in a trailer park in a small village and is sole provider for herself and her one child. The welfare department wanted her to get a job, and she herself desperately wanted to get off welfare. The only work she could get was a thirty-hour a week job, at $4.05 an hour, in a supermarket in the town. This leaves her below the poverty level; even with continued food-stamp and medicaid benefits, she has insufficient income. Sandy interviewed at a fast-food restaurant

for a second part-time job but she found it impossible to combine the shifts of two jobs, as the restaurant would not inform her until each Friday what her next week's schedule would be. Besides, she still would not have had health benefits. She decided not to pursue the job. Eventually, when she got too far behind in the rent, Sandy moved in with a friend. (Fitchen, 1991, p. 132)

The cumulative experience of rural poverty does not apply just to individuals but can also be inter-generational. Cloke (1997b), for example, recounts the story of one small town resident in Kansas, as told to *Newsweek* magazine:

Poverty is passed from one generation to another: it is the only legacy of the poor. Ida Swalley married at 15 to escape a hard-drinking stepfather. She has no marketable skills. Now 43, she is separated from her fourth husband and is living in a squalid $200-a-month apartment that could be owned by an urban slumlord. Swalley shares the hovel with her 17-year-old son and a menagerie of bugs and mice. An old fly swatter is the only decoration on one wall. The Kansas heat pushes the fetid air towards 100 degrees and aggravates Swalley's heart problems.

She says things may improve once her new boyfriend gets out of jail. Her fondest hope is that life will somehow be better for her daughter, Carol Sue, 26, and her two-year-old granddaughter, Jacqueline Ruth. But that dream may be illusory. Carol Sue Stevens earns just $3.85 an hour as a nursing-home aide. Her life, like her mother's, has been a succession of small-town romances with men prone to drunkenness and violence. Little Jacqueline Ruth was fathered by Carol Sue's current boyfriend, but the toddler doesn't carry either parent's surname ... 'If we end up in some custody fight, I don't want her in court already using her daddy's last name.' (McCormick, 1988, p. 22)

Experiences of rural poverty are shaped by over-arching processes of restructuring, but in each individual's case initial situations of unemployment or low income have the potential to multiply into spiralling problems of ill-health, crime, drug abuse, alcoholism, family breakup and homelessness (see Box 19.2). Many of these experiences are shared with urban households in poverty, but the particularities of rural localities influence the ability of individuals to escape from poverty and inform the attitudes of wider society towards the disadvantaged.

Box 19.2   Rural homelessness

The relative invisibility of rural poverty is particularly marked with respect to the problem of homelessness. Not only is homelessness disguised by the discourse of the rural idyll (Cloke et al., 2001a, 2002), but also because rural homelessness is often literally less visible than urban homelessness. Rural homelessness is identified less with rough street sleeping than with transitory residence in temporary accommodation, hostels, disused buildings or involuntary residence with friends and family. The rural homeless population also tends to be more diffuse than in urban areas and is systematically undercounted in official surveys (Cloke et al., 2001b; Lawrence, 1995). As such, the scale of the problem of rural homelessness can go unacknowledged by responsible local government agencies as well as by the public.

> **Box 19.2 (Continued)**
>
> The shortcomings of official counts notwithstanding, statistical evidence nevertheless points to a significant and growing problem of rural homelessness. In 1996, for example, there were nearly 16,000 registered homeless households in rural England, or 14.4 per cent of the national total (Cloke et al., 2002). Whilst this figure was both below the national average and marginally down on 1992, homelessness in 'deep rural' districts had increased by over 12 per cent since 1992 and over a quarter of rural local authorities reported an increase in homelessness of more than 25 per cent. In the United States, Lawrence (1995) reports estimated rates of homelessness in rural counties in Iowa as high as 70 persons per 1,000 population, with an average rate of around 20 persons per 1,000 population that was higher than rates in New York, Los Angeles or Washington.
>
> Homelessness in rural areas may also have different causes from that in urban areas. Housing factors including the termination of shorthold tenancies, mortgage arrears and the loss of rented or tied accommodation through other reasons are all more significant in rural than in urban homelessness (Cloke et al., 2002). Individuals' accounts of rural homelessness collected by Cloke et al. (2002) emphasize the multiple events that are often involved in the process of becoming homeless, including loss of employment, relationship breakdown, family disputes and illness, but the shortage of affordable housing, including difficulties in accessing social housing, is a common factor that tips otherwise vulnerable people into homelessness. They also note the interconnection of rural and urban homelessness, with economic migrants from rural areas experiencing homelessness and homeless individuals from cities and towns moving to rural areas that are perceived to be safer, cheaper and a more pleasant environment. Indeed, the accounts quoted by Cloke et al. (2002) indicate that the periodic mobility of homeless individuals between urban and rural locations is not uncommon.
>
> *For more see Paul Cloke, Paul Milbourne and Rebekah Widdowfield (2002) Rural Homelessness (Policy Press); Paul Cloke, Paul Milbourne and Rebekah Widdowfield (2001) Homelessness and rurality: exploring connections in local spaces of rural England. Sociologia Ruralis, 41, 438–453; and Mark Lawrence (1995) Rural homelessness: a geography without a geography. Journal of Rural Studies, 11, 297–307.*

Summary

Poverty is prevalent and persistent in rural areas, yet its presence is often disguised by diffusion and by the powerful discourse of the rural idyll. The hidden nature of rural poverty can frustrate the development of policies and initiatives to tackle the problem. In general, attempts to address rural poverty take one of two forms. First, the alleviation of poverty can be an objective of rural development strategies (see Chapter 10). However, this approach has been argued to be only partially successful. Economic development initiatives may create more jobs, but there is usually no guarantee that the new jobs will go to local residents experiencing poverty, that obstacles to employment such as transport will be overcome, that wages will be sufficient to raise income levels or that alteration will result in non-economic factors in deprivation. Secondly, individuals and households experiencing poverty or near-poverty are supported by welfare payments from the state. Yet, again, welfare payments are often insufficient to lift recipients out of poverty, and programmes implemented as part of national welfare systems may not be attuned to the particular circumstances of rural poverty. Furthermore, Cloke suggests that welfare reforms in line with New Right ideologies in the 1980s and 1990s contributed to the problem of rural poverty by removing the safety net for many households. Indeed, newer 'workfare' and 'welfare to work' programmes have been criticized as ineffective in rural areas because of the different nature of rural poverty and of the rural economy and labour market.

As such, increasing emphasis is placed on self-help and voluntary action as strategies for dealing with rural poverty. These include forms of community-based mutuality such as soup kitchens, food banks and credit unions, as well as informal networks and coping mechanisms developed by deprived households themselves. Additionally, the historic response to rural poverty is still an option – migration. Public officials in California have sought to address problems of unemployment among former agricultural workers by encouraging them to migrate eastward to states such as Kansas, Iowa and Nebraska where low skilled jobs exist in industries like meat-packing. Migration itself, however, is an expensive process than can be beyond the means of many rural households trapped in conditions of multiple deprivation.

Further Reading

The characteristics, dynamics and relative neglect of rural poverty are discussed, with evidence from England and Wales, by Paul Cloke in 'Poor country: marginalization, poverty and rurality', in P. Cloke and J. Little (eds), *Contested Countryside Cultures* (Routledge, 1997); and by Paul Milbourne in 'Hidden from view: poverty and marginalization in rural Britain', in P. Milbourne (ed.), *Revealing Rural 'Others': Representation, Power and Identity in the British Countryside* (Pinter, 1997). The perceived incompatibility between deprivation and the discourse of the rural idyll is discussed in more detail by Rachel Woodward in her article ' "Deprivation" and "the rural": an investigation into contradictory discourses', in *Journal of Rural Studies*, volume 12, pages 55–67 (1996). Janet Fitchen's study of rural communities in New York State includes a detailed discussion of problems of rural poverty, highlighting the way in which multiple factors contribute to deprivation; see *Endangered Spaces, Enduring Places: Change, Identity and Survival in Rural America* (Westview Press, 1991).

Websites

More information on rural poverty in the United States can be found at the Rural Poverty Research Center (www.rprconline.org). The Countryside Agency in England has a concern with tackling social exclusion in rural areas, as detailed on its website (www.countryside.gov.uk).

20

Rurality, National Identity and Ethnicity

Introduction

The countryside has long played an important role in the constitution of national identities. Cities might be celebrated as symbols of civilization and may provide the stages for monumental landscapes of power that celebrate national prowess, but they also attract suspicion as 'melting pots' of different peoples and ideas, where national values and principles might be compromised by association with foreign peoples and influences. Short (1991) notes that this moral geography was expressed as early as the first century BCE by the Roman writer Cicero and still finds articulation today. The countryside, in contrast, was represented as an innocent, purer, space in which national values and national identities were held true.

It is, however, a dangerously short step from representing rural areas as places of *national* purity, to representing them as places of *ethnic or racial* purity. In the developed world, that means representing the countryside as a 'white space' from which people of non-white ethnic backgrounds are implicitly or explicitly excluded. This prejudice is indeed reinforced by historical social and economic factors that have tended to concentrate non-white populations in urban centres, such that the non-white population of many rural regions is very small, compounding experiences of isolation and discrimination. At the same time, established non-white rural populations, notably the black population of the southern United States and first nation peoples in North America, Australia and New Zealand have been marginalized and discriminated against by a white national elite, with their rural situation frequently taken advantage of in their exploitation.

In order to explore these themes further, this chapter starts by discussing in more detail discursive connection between rurality and national identity, and the construction of the rural as a white space. The remainder of this chapter then investigates non-white experiences of rurality, examining the exclusion and racism experienced by people of different ethnic backgrounds living in rural areas or using rural space for recreation. The chapter then describes an exception to the model of the white countryside in the predominantly black rural counties of the southern United States, but notes that such areas have experienced systematic collective exclusion and marginalization. It finally addresses a second exception to the model in the first nation ruralities of indigenous peoples of North America, Australia and New Zealand, observing again how such communities have been systematically excluded and marginalized from mainstream rural society.

Rurality and National Identity

The association of national identity with rurality has involved ideas both of landscape and of rural life. Landscapes, as Daniels (1993) notes, picture the nation, proving visual shape to constructs of identity: 'As exemplars of moral order and aesthetic harmony, particular landscapes achieve the status of national icons' (p. 5). Distinctive rural landscapes are hence venerated as iconic symbols of national identity – the American prairies, the Australian outback, the highlands of Scotland and the rolling hills and vales of England. Such landscapes can be inspirational and comforting, Daniels (1993) again remarking that, 'protective images of landscape have played a role in cultural resistance to outside aggression' (p. 7).

Rural life, meanwhile, is constructed in nationalist discourses as being purer and more honourable than life in the city. The eighteenth-century French philosopher Jean-Jacques Rousseau, for example, claimed that 'it is the rural people who make the nation' (quoted by Lehning, 1995, p. 12), and the peasant class has frequently been held up as the prime example of national character. Not only were rural people celebrated for feeding the nation, but they were also constructed as being less 'contaminated' with alien ideas and affections than city-dwellers, and as being closer to a traditional way of life that somehow recalled the origins of the nation. As Ramet (1996) describes in an article tracing the connections between the rural population and Serbian nationalism in the 1990s, the 'rural claims to be more pure than the city; to preserve the old values which the city has sullied' (p. 71). This representation of rural life is clearly premised on the persistence of an agrarian society, yet the emphasis that it places on tradition and stability chimes with the contemporary discourse of the rural idyll.

Rural space can be both the *heartland* and the *frontier* of the nation. The latter identification is an important element in the national identities of the United States, Canada and Australia:

> For states in the New World, nation-building has been intimately related to conquering the wilderness. Throughout America and in Australia the national histories have consisted of creating a country from the forest and the grasslands. The transformation of the wilderness has a special place in their national identity. (Short, 1991, p. 19)

Thus, the expansion of the United States into the wilderness of the west not only represented a symbolic progress away from Europe, but also provided a space in which the young nation could prove itself through its conquest of nature. In this 'frontier thesis', the key figure was not the peasant but the *pioneer* – the adventurer whose bravery, determination and resourcefulness supposedly epitomized the national character, and whose spiritual descendants are claimed to be the family farmers and ranchers of contemporary rural America. Furthermore, as discussed in Chapter 13, the wilderness also provided the United States with sites of cultural and natural importance that gave the young country an instant heritage to rival European nations, establishing the rationale for the creation of national parks.

The representation of the countryside as the national heartland is exemplified by the case of England. Although England has been a predominantly industrial and urban country since 1861, the ideology of Englishness is, as Howkins (1986) observes, 'to a remarkable degree rural. Most importantly, a large part of the English *ideal* is rural' (p. 62; original emphasis). Howkins traces this identification of Englishness with rurality to the late nineteenth and early twentieth century and the era of imperial expansion. The process of colonization was led by military officers and administrators drawn from the minor gentry, many of whom had been raised on country estates. Thus,

> The very global reach of English imperialism, into alien lands, was accompanied by a countervailing sentiment for cosy home scenery, for thatched cottages and gardens in pastoral countryside. Inside Great Britain lurked Little England. (Daniels, 1993, p. 6)

This discourse was cemented in the popular imagination by the First World War and the reproduction of images of England, such as John Constable's iconic painting *The Haywain*, as the depiction of the country that the armed forces were fighting to defend (Daniels, 1993; Howkins, 1986). At the same time, however, the war and its immediate aftermath witnessed rapid urbanization which

threatened the very rural landscape reproduced as the epitome of Englishness. Spurred by this perceived threat, the vision of rural England became introspective, paradoxically represented both as timeless and enduring and as fragile and endangered. These interpretations were famously appealed to by the interwar prime minister Stanley Baldwin, in a speech that celebrated the agrarian countryside as the seat of continuity in English national identity:

> To me, England is the country, and the country is England ... The sounds of England, the tinkle of the hammer on the anvil in the country smithy, the corncrake on a dewy morning, the sound of the scythe against the whetstone, and the sight of a plough team coming over the brow of a hill, the sight that has been seen in England since England was a land, and may be seen in England long after the Empire has perished and every works in England has ceased to function, for centuries the one eternal sight of England. (Speech by Stanley Baldwin, 1924, quoted by Paxman, 1998, p. 143)

Yet, as Paxman observes, the rural scene described by Baldwin was a historical anachronism even by the time he was speaking in 1924. As in other countries, the rural idyll that was reproduced at the heart of English national identity was always more of a historical fiction than a tangible reality. It was also based on a particular regional landscape, that of the 'south country' of central southern England, distinguished by 'a uniform landscape type of smooth, bare, rolling hills dotted with woodlands' (Brace, 1999, p. 92). The more peripheral rural landscapes of the western moors, northern uplands and eastern fens, as well as the more industrialized countryside of the Midlands, were all excluded from vision of the ideal England.

In constructing an association between the countryside and national identity in a manner that celebrated the purity of rural people and the lack of contact with foreign influences, and positioned rural space as a repository of historic national values, representations of this type explicitly or implicitly identified the rural with a homogeneous ethnic group. Consequently, people of ethnic groups other than the dominant national ethnicity have often been excluded from discursive representations of the rural idyll, and have experienced racist discrimination in their occupancy or use of rural space. The remainder of this chapter investigates non-white experiences of rurality. It first explores the construction of the rural as a white space and the exclusion and racism experienced by people of different ethnic backgrounds living in rural areas or using rural space for recreation.

Contesting the Rural as a White Space

The identification of the rural idyll with national identity has provided spurious legitimation to racist constructs of the countryside as a 'white' space that have been further reinforced by the spatial dynamics of immigration. Proximity to airports and major seaports, the existence of established ethnic communities and greater levels of institutional support for new arrivals all favour cities as the initial destination. For instance, only 5 per cent of immigration into the United States between 1990 and 1999 was directly into rural counties (Isserman, 2000). However, even settled and domestic-born ethnic minority populations tend to be spatially concentrated in urban areas. In 1991, ethnic minorities formed 6.2 per cent of the UK population, but only 1.6 per cent of the population in rural districts; similar patterns are repeated elsewhere, such that white ethnic groups form the significant majority of the rural population in most of Europe, Australia, New Zealand, Canada and the northern parts of the United States. The main exceptions to this pattern – the primarily black and Hispanic counties of the southern United States and first nation communities – are discussed later in this chapter.

The combination of demographic trends and cultural prejudices means that the identification of the countryside as a 'white' space becomes self-reproducing. Racist attitudes are a small but notable factor in counterurbanization (see Chapter 6), reinforcing white perceptions in which,

'ethnicity' is seen as being 'out of place' in the countryside, reflecting the Otherness of people of colour. In the white imagination people of colour are confined to towns and cities, representing an urban, 'alien' environment, and the white landscape of rurality is aligned with 'nativeness' and the absence of evil or danger. The ethnic associations of the countryside are naturalised as an absence intruded upon by people of colour. (Agyeman and Spooner, 1997, p. 199)

Exclusionary discourses of this type mean that for many people of colour rural areas are perceived to be threatening places where they are unwelcome. This geography of fear and exclusion has been articulated by the black British photographer Ingrid Pollard, whose 'Pastoral Interludes' collection involved self-portraits in the rural landscape. In the caption to one image Pollard writes, 'I thought I liked the Lake District, where I wandered lonely as a Black face in a sea of white. A visit to the countryside is always accompanied by a feeling of unease, dread'; whilst another simply reads: 'feeling I don't belong. Walks

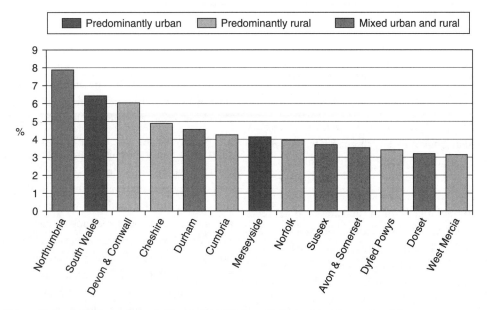

| | Predominantly urban | | Predominantly rural | | Mixed urban and rural |

Figure 20.1 Police force areas in England and Wales with highest incidence of racial crime, by percentage of ethnic minority population affected, 1999–2000
Source: Based on information in the *Observer*, 18 February 2001

through leafy glades with a baseball bat by my side' (quoted in Kinsman, 1995, p. 310 and p. 302).

These representations of race and rurality impact on the everyday lived experiences of people of colour living in rural areas. Although there is no single, standard experience and although many people of colour are welcomed and integrated into rural communities, Agyeman and Spooner (1997) highlight a number of reports in rural England that have found 'an extensive amount of racial violence, harassment, condescension and bigotry, provoked by a mixture of ignorance, the uncritical acceptance of stereotypes and a resistance to the arrival of incomers' (p. 203). They note that institutional racism is frequently evident in rural areas and that public service providers and employers often pay only lip-service to equal opportunities.

Incidents of racial crime, including violent attacks and racial abuse, have also been reported to be higher in many rural parts of England and Wales than in urban areas, relative to the size of the ethnic minority population (Figure 20.1).

The perception of the countryside as a threatening and unwelcoming environment also deters people of colour from visiting rural areas for recreational activities. Studies in Britain have shown that members of ethnic minorities are among the social groups that are least likely to participate in rural recreation. As Agyeman and Spooner (1997) note, this is in part due to economic and time factors, but it also reflects a sense of fear, as recorded by Malik:

it appeared to be more the anticipation of abuse or unacceptance in the countryside, rather than any direct experiences of racism on previous visits, that deterred many people from going at all, and prevented a complete feeling of safety and relaxation for others while there. (Malik, 1992, p. 32)

Yet, rural racism in the UK is a largely hidden phenomenon, disguised by the relatively small number of people of colour living in rural areas and by devices that are used to explain away racist sentiments. Thus, racially motivated migration to rural areas is disguised as 'quality of life' migration and racial prejudice is dismissed as local antipathy to incomers of all backgrounds (Agyeman and Spooner, 1997). Initiatives have been launched to challenge these attitudes, and to encourage more people of colour to visit rural areas, including the work of the Black Environmental Network and schemes operated by national parks, and there is evidence that these attempts have met a positive response, particularly among younger black and Asian Britons who see the enjoyment of the countryside as their right as much as it is for other citizens.

The Rural Black Experience in America

Prominent exceptions to the construct of rural areas as a 'white space' include the 77 non-metropolitan counties in the southern United States where African Americans form the majority of the population. These counties, which are predominantly located in the Mississippi valley and the former cotton- and tobacco-growing belt of Alabama, Georgia and South Carolina, are both highly visible and strongly rooted in a particular social, economic and political history. The concentration of African Americans in these areas reflects the history of slavery and its abolition, but the continued social, economic

and geographical marginalization of the communities is a legacy of entrenched racial discrimination and repression in the region. So-called 'Jim Crow' laws introduced in the late nineteenth century legislated for segregation of black and white populations and the provision of separate schools, hospitals, public parks, transport, housing, restaurants and theatres in many southern states. As the standard of the facilities provided for the black community was invariably inferior to that of the facilities for the white community, segregation enforced the social and geographical isolation of rural African Americans in the South (Snipp, 1996). Additionally, the rural black communities were economically disadvantaged not only by poor education and limited opportunities but also by employment discrimination and the practice of sharecropping, whereby freed slaves who had no land of their own worked a portion of land owned by whites in return for a limited share of the profit. The sharecropping system, however, was brutally exploited by white landowners such that the black farmer was often 'shown' to owe money to the landowner and could rarely earn enough to enable an escape from the hard, physically and mentally damaging work (Harris, 1995).

Between 1920 and 1950 out-migration reduced the rural African American population in the southern states by 20 per cent. However, in recent decades the population has stabilized and in 1990 around 15 per cent of the total US African American population lived in the rural South (Snipp, 1996). In the long term rural–urban black migration has increased social polarization within the African American population as the emergence of an urban black middle class has not been matched in the rural counties of the South, which remain economically marginalized:

Despite their persistence, rural black communities have become 'places left behind' in many respects. Although much has been written about the return of African-Americans to the South, and the southern economic boom of the 1970s and 1980s, these developments have not rejuvenated rural black communities. There is ample evidence that economic development in the South is highly uneven, concentrated in urban areas, bypassing African-Americans in rural places. (Snipp, 1996, p. 131)

Most rural black counties are classified as persistent poverty counties (see Chapter 19) and in 1989 nearly half (47.8 per cent) of black households in rural counties where blacks formed the majority were classified as living in poverty (Cromartie, 1999). As well as being economically marginalized, rural African Americans have also historically been excluded from positions of political power in much of the rural South by the persistence of a minority white elite whose position was originally founded on segregation. It was only at the end of the twentieth century that this political marginalization began to be seriously challenged as blacks started to be elected to local political office in significant numbers.

Some of the most brutal discrimination has been that faced by black farmers in the United States. In 1920, there were over 925,000 black farmers in the US, or one in seven of all American farmers. By 1982, only one in sixty-seven farmers was black and by 1992 there were fewer than 19,000 black farmers still working. The virtual eradication of the black agricultural community has resulted from a combination of pressure from economic restructuring and institutional racism. Black-operated farms have always been small units and therefore ill-equipped to compete in the increasingly commercialized and globalized agricultural industry that developed during the twentieth century (see Chapter 4). However, whereas white farmers were heavily supported and subsidized by the government to adapt, assistance for black farmers was more restricted and more conditional. Black farmers experienced difficulty in obtaining loans from commercial banks and hence were especially dependent on loans from the USDA, more generally perceived to be the 'lender of last resort'. Yet the USDA processed applications from black farmers more slowly than those from white farmers and charged a higher interest rate (Sheppard, 1999). Black farmers were also confronted by considerable racism, discrimination and abuse from local USDA officials in many areas of the South. It was not until 1999 that the settlement of a lawsuit substantiated many of the black farmers' allegations of institutional racism and the USDA agreed to pay compensation of up to $300 million in total to black farmers whose civil rights had been violated.

First Nation Ruralities

Exceptions to the rule of the rural as a 'white space' are also presented by the ruralities of first nation indigenous peoples in North America, Australia and New Zealand. Prior to European colonization the first nations were essentially rural societies; however, in the process of colonization indigenous people were dispossessed of their lands and forced into reservations that were again predominantly located in rural situations. At the same time, the reinvention of the countryside of the new nations – Australia, New Zealand, Canada and the United States – denuded rural space of its first nation references and meanings and imposed new meanings that ignored the continuing presence of indigenous communities (see Box 20.1). The rural geographies of first nation groups are hence geographies of oppression and subordination. As Snipp (1996) argues with reference to Native American reservations in the United States,

The original motive for creating reservations in the 19th century was to isolate and contain American Indians in areas distant from the mainstream of American society. It was expected that eventually, with education, Christian conversion, and other measures designed to 'civilize' American Indians, reservations would no longer be needed. (Snipp, 1996, p. 127)

Box 20.1   The myth of the Canadian rural north

The association of rurality and national identity and the discursive exclusion of indigenous peoples from such imagined geographies are both illustrated in representations of the rural north of Canada. Shields (1991) describes the reproduction of the myth of the 'True North Strong and Free' as a core component in the discourse of Canadian national identity, which positioned the northern regions of the country as a spiritual national heartland and a counterbalance to the cities of southern Canada where the vast majority of the population reside. As Shields observes, 'for most English-speaking Canadians the "North" is not just a factual geographical region but also an imaginary zone: a frontier, a wilderness, an empty "space" which, seen from southern Canada is white, blank' (p. 165). It is, Shields continues, 'an empty page onto which can be projected images of the essence of "Canadian-ness" and also images to define one's urban existence against' (p. 165). The representation of the North as a 'blank space', however, denies the presence and heritage of the Inuit communities in the region. To the extent that the Inuit presence is acknowledged, the forms of representation employed reinforced a paradoxical perception of the Inuit lifestyle both as being harsh and difficult, thus symbolizing the resilience of the Canadian national character, and as inferior to the civilization of the urban south. For example, Shields notes that in films of the North, 'the removal of the roofs of igloos to permit the filming of life inside necessitated the inhabitants' being fully clothed in the sub-zero temperatures, introducing the idea that igloos are uncomfortable and cold habitations' (p. 176).

Moreover, the tension between representations of the North as a resource-rich hinterland and as a cultural heartland in need of protection, 'furnished the basis for paternalistic policies on Northern development and the "civilising" of the Inuit with little power exercised by Northern inhabitants' (p. 165). The constitutional status of the Northwest Territories and the Yukon Territory denied them the autonomy of the southern provinces and meant that they were largely governed from the south in 'the national interest'. From the 1980s onwards, demands for self-government formed a central objective of the indigenous rights campaign in Canada, eventually leading to the establishment in 1999 of the new territory of Nunavut, which means 'Our Land' in the Inuktitut language. Inuit comprise 85 per cent of the population of just 29,000 residents who occupy the two million square kilometre territory carved from the Northwest Territories to the north and west of Hudson Bay.

For more see Chapter 4 in Rob Shields (1991) Places on the Margin (Routledge).

Concentration on the reservations was, in the nineteenth century at least, associated with the deprivation of the Native Americans' self-sufficiency, as they were prohibited from hunting outside the reservation or from possessing firearms, and thus became increasingly dependent on rations supplied by the military. During the twentieth century the rural focus of first nation populations was diluted by out-migration to urban centres. In 1990, just under half of the Native American population in the United States still lived in rural areas, mostly on reservations, but with some regional variation (Snipp, 1996). Whilst the Native American population in California is predominantly urban, in the mountain states and Alaska it is predominantly rural, forming a majority in 27 non-metropolitan counties (Brewer and Suchan, 2001; Snipp and Sandefur, 1988).

Economic factors have been influential in Native American migration from the reservations to cities. Labour force participation is significantly higher for Native Americans in urban areas than rural areas and average annual earnings are also around 20–25 per cent higher (Snipp and Sandefur, 1988). The economic capacity of reservations has been limited by the marginal quality of land, cultural attitudes towards the exploitation of the environment, the absence of industrialization and a lack of capital. Although many reservation lands are rich in minerals and other natural resources, they have been dependent on external capital to fund exploitation such that much of the wealth generated does not reach the first nation community. Native American reservations are therefore frequently associated with high poverty levels. In 1989, half of Native Americans in rural counties where they are predominant lived below the poverty line, whilst in 1997 a majority of the labour force were recorded as unemployed on a number of reservations, including Cheyenne River Sioux Reservation (South Dakota,

80 per cent unemployment), Rocky Boy's Reservation (Montana, 77 per cent), and Red Lake Chippewa Reservation (Minnesota, 62 per cent) (Cornell, 2000).

Yet, Cornell (2000) also notes that there are substantially lower unemployment rates on many other reservations, suggesting a polarization in the circumstances of different Native American groups in different rural regions that is also indicated by figures for changes in poverty levels on reservations in the 1980s (Figure 20.2). The relative affluence or deprivation of reservations can reflect the degree of their geographical isolation and hence their ability to tap into local urban labour markets, but they also reflect the extent to which individual reservations have taken advantage of their sovereign status to develop niche market activities, such as the sale of tax-free tobacco products and, most significantly, gambling-related tourism (Snipp, 1996).

Snipp (1996) observes that as the last remaining land base, Native American reservations have become the central sites of indigenous culture and social life. A series of treaties and legal agreements have permitted tribal groups a degree of self-government on reservations that allows them to manage the land according to their own traditions. However, this reassertion of first nation ruralities is strictly limited to the territories of the reservations themselves. A more radical process is taking place in Australia in the wake of the Native Title Act 1993 (with the issue of land rights also high on the political agenda in New Zealand). This enables 'native title communities' of indigenous (Aboriginal) people to make title claims to rural territories to which they can demonstrate a connection under the body of traditional law and custom. Recognition of a title claim gives the native title community a right to involvement in the governance of the territory and hence requires changes in practice from existing

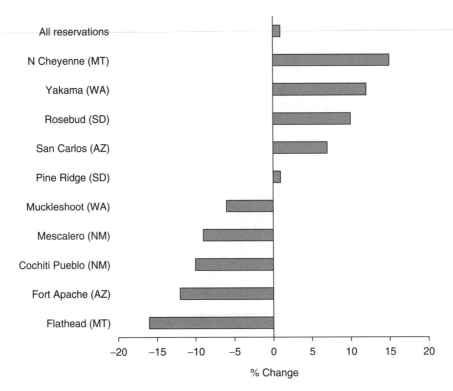

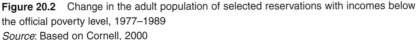

Figure 20.2 Change in the adult population of selected reservations with incomes below the official poverty level, 1977–1989
Source: Based on Cornell, 2000

owners of rural land and governing authorities. It can also potentially bring economic benefits through agreements negotiated over mining and exploration activity and from compensation for certain past government actions (Davies, 2003). By June 2001, more than one thousand native claim applications had been made covering much of rural Australia, including most land in Western Australia, although the slow recognition process meant that only a handful had actually been determined. As Davies comments, the programme has wide-reaching implications for rural Australia and the place of indigenous communities within it:

Recognition of native title has opened the door for indigenous people to be involved as collaborators in management

of their traditional country. The claim process has made traditional ownership of country, the basis of indigenous law and governance mechanisms, increasingly more visible. The legal limitations on recognition of native title are making the conservative perception that native title claims present a threat to non-indigenous property rights and interests increasingly difficult to sustain. (Davies, 2003, p. 41)

Yet, the recognition of native title stops short of returning land to indigenous peoples or of fully empowering indigenous groups. Little of the land held by or on behalf of Aboriginal and Torres Strait Islanders organizations under the Indigenous Land Corporation has been submitted for native claim because there is perceived to be no significant additional benefit. Furthermore, as Davies acknowledges, there is a disjuncture between the geography of native title claims and the geography of where Australia's indigenous people actually live. Although indigenous people comprise around a fifth of the 'outback' population, nearly three-quarters of indigenous people now live in urban areas. 'Native rights communities' making claims to territory need not be resident in the lands they claim, such that the native rights process in itself does not present a framework for reducing the social and economic deprivation of indigenous communities in rural Australia.

Summary

The discursive association of rurality with national identity has contributed to the marginalization and exclusion of ethnic populations who do not fit with dominant constructions of national ethnicity. In much of rural Europe and many rural parts of North America, Australia and New Zealand where the population is predominantly white, the reproduction of such discourses reinforces the sense of threat and exclusion experienced by many non-white residents and visitors. In regions with a greater historical presence of non-white populations, including the majority black rural communities of the southern United States and first nation indigenous communities in North America, Australia and New Zealand, the relative social and economic deprivation of these groups has been intensified by histories of oppression and economic, cultural and political marginalization. Although efforts have been made in recent years to acknowledge and correct or compensate for historic injustices, including strategies to support economic development, to recognize native title claims and to introduce greater degrees of self-government, the situation of many non-white people in rural areas continues to be marked by poverty and isolation. Moreover, the association of rurality and national identity still has purchase, making rural areas sites for nationalistic and racist activity. In their most moderate form these include the appropriation of national symbols by campaigns that have sought to defend rural interests. For example, the series of marches in London organized by the Countryside Alliance in defence of hunting with hounds have employed nationalistic iconography such as flags, and have identified the perceived threat to their rural way of life with a threat to 'British values'. More extreme examples include the rural basis of electoral support for the short-lived anti-immigration One Nation party in Australia at the turn of the twenty-first century, and the establishment in the rural United States of extreme right-wing, racist 'militia' groups. The militia groups form one of the 'alternative rural lifestyles' considered in the next chapter.

Further Reading

The connections between rurality and national identity have been discussed by a wide range of writers. Stephen Daniels's book *Fields of Vision: Landscape Imagery and National Identity in England and the United States* (Polity Press, 1993) is a comprehensive study of the importance of landscape to national identity in England and the United States, whilst Chapter 8 of Jeremy Paxman's book *The English: A Portrait of a People* (Michael Joseph, 1998) is an accessible discussion of the role of rurality in English national identity. Julian Agyeman and Rachel Spooner discuss the experience of ethnic minorities in rural areas at length in their chapter in P. Cloke and J. Little (eds), *Contested Countryside Cultures* (Routledge, 1997), with a particular focus on the UK. Matthew Snipp (1996) 'Understanding race and ethnicity in rural America', *Rural Sociology*, volume 61, pages 125–142, meanwhile, provides a comprehensive overview of the study of race and rural societies in the United States. The process of native rights claims to land in Australia is examined in detail by Jocelyn Davies in 'Contemporary geographies of indigenous rights and interests in rural Australia', *Australian Geographer*, volume 34, pages 19–45 (2003).

Websites

Further information relating to a number of the topics discussed in this chapter can be found on a range of websites, including those of pressure groups and government agencies involved with some of the issues described. These include the Black Environmental Network (www.ben-network.org.uk), a UK organization that encourages greater use of the countryside for recreation by ethnic minorities; the National Black Farmers Association (www.blackfarmers.org) and the Black Farmers and Agriculturalist Association (www.coax.net/people/lwf/bfaa.htm), both lobby groups representing the interests of black farmers in the United States; and the National Native Title Tribunal (www.nntt.gov.au), which is the body charged with administering the process of native title claims in Australia.

21

Alternative Rural Lifestyles

Introduction

The idea of 'escaping to the countryside' is an important rhetorical notion in discourses of rurality. The desire to escape the city is a significant component in middle class counterurbanization (see Chapter 6), where it is commonly constructed in terms of fleeing from the stresses and intensity of urban life to a slower, more peaceful environment. However, as Valentine (1997b) argues, 'the focus within rural geography on white middle-class visions of a "rural idyll" obscures the fact that "other" groups have also idealized "the rural" as a peaceful, safe place and sought to establish their own versions of "community" life away from the city' (p. 119). Frequently, the 'rural idylls' pursued by these other groups involve not a lifestyle change, but an escape to the countryside in order to find space (both physically and psychologically) to pursue lifestyles which they feel are inhibited by prejudice and social and economic pressure in urban areas. However, the aspirations of such 'alternative rural lifestyles' are not always complementary to those of the middle class rural idyll, particularly if they challenge conventional understandings of property rights or promote non-conventional sexualities, such that conflict can be generated. This chapter explores three examples of alternative rural lifestyles – new age travellers in the UK; experimental utopian communities based on ecological principles or sexuality; and the militia movement in the United States.

Travellers and Rurality

Despite the popular representation of rural communities as places of stability and even insularity, there is a long history of mobile populations in the countryside. In Europe, gypsies have pursued a distinctive lifestyle and culture that has been part of the rural experience for centuries, intersecting with mainstream rural culture through events such as horse fairs. Similarly, Cresswell (2001) describes the historical experience of the tramp in America, whose existence was both urban and rural, in some cases relying on agricultural work and following harvest cycles through the Mid-West or down the west coast, and in others moving through rural space as they journeyed between cities. The life of both the gypsy and the tramp were romanticized as embodying the freedom of the open countryside – think for instance of Toad's adventure with the gypsy cart in Kenneth Grahame's *Wind in the Willows*. But both groups were also subjected to considerable discrimination and oppression and were often portrayed as a threat to more sedentary rural lives with prejudice stoked by both their mobility and their 'outsider' status (MacLaughlin, 1999).

The emergence of new nomadic cultures as alternative modern lifestyles has been inspired by the romantic vision of the itinerant rural traveller, but also confronted by suspicion and hostility from other rural inhabitants. Among the most prominent of the new nomads are the 'new age travellers' in the UK, a counterculture that developed from the festival circuit in the 1970s (McKay, 1996). In rejecting modern, consumerist society, the traveller culture developed an identification with a romantic ruralism and adopted a semi-nomadic lifestyle that is mainly performed in rural space. By the late 1980s, the traveller community throughout the UK was estimated to number 8,000 people, travelling and living in some 2,000 vehicles (McKay, 1996). A number of new age traveller groups became involved in protests against the construction of new roads in the late 1980s and 1990s, with the traveller community gaining new members from the ranks of eco-protesters.

The discourses of rurality articulated by travellers draw heavily on the idea of the rural idyll, frequently employing language and imagery that would resonate with that of middle class in-migrants:

the appeal is quite romantic. It's the English dream really isn't it? – the fantasy most English people have: trees, fields, all those images from [Thomas Hardy's] *Tess of the d'Urbervilles*.

('Jeremy', traveller, England, quoted by McKay, 1996, pp. 47–48)

One traveller quoted by Lowe and Shaw (1993) even identified the lifestyle with that portrayed in *The Archers*, a long-running BBC radio soap opera centred on a rural village which once branded itself as 'an everyday story of country folk':

The Archers ... is so popular. It's a bit of a cult with travellers. It's my soap habit. It features people like us. ('Jay', traveller, England, quoted by Lowe and Shaw, 1993, p. 59)

The identification of the rural with freedom is fundamental to this discourse, but travellers frequently find that their freedom to pursue their idyllized rural lifestyle is compromised by hostility from landowners and local residents. As such, they challenge the principles of private property and traditional biases of rural society:

I don't think it should be a crime to want to live like that, to want to live in a rural area rather than in a big city. And it's impossible to go out and rent a cottage or a farmhouse or buy one. Apart from people who are born to it, rural Britain is for the rich. It's for people who can afford to buy themselves a weekend place or go and retire in the country. For me, if I want to live with space around me and trees and hills and woods, the only possible way apart from sleeping out is to buy a vehicle and live like that. ('Shannon', traveller, quoted by Lowe and Shaw, 1993, p. 240)

It's blatant. The sheep have it all. We have nothing. Look out of the window. Every field you look in has got sheep in it ... They're not actually earning their keep, they just sit in the field and say, 'I am a sheep. This is a sheep field, therefore nothing else is allowed in here'. You can't walk your dog there, you can't even go in a sheep field. It's just so obvious to me how much land the sheep have, so many acres per square sheep [*sic*], and how little land we have. ('Decker John', traveller, quoted by Lowe and Shaw, 1993, p. 104)

Tensions between travellers and landed rural interests developed most explicitly into open conflict over access to the ancient stone circle at Stonehenge. From the early 1970s, travellers had converged on Stonehenge, a national icon that is steeped in mysticism, for an annual festival at midsummer. The increasing numbers attending the festival in the 1980s, however, provoked opposition from local landowners and from the heritage agency responsible for managing the site who took legal action to prevent the travellers from reaching Stonehenge. As an exclusion zone was created around Stonehenge at midsummer, stand-offs developed in several years between police and travellers attempting to reach the site, notoriously erupting into a skirmish in 1985 that was later known as 'the battle of the beanfield', which Sibley (1997) suggests was the turning point in attitudes towards the travellers' place in the countryside.

The conflict at Stonehenge was hence a significant factor in the introduction of legislation to control and regulate mobility in the countryside, particularly in the Criminal Justice and Public Order Act 1994 (Sibley, 1997). As Halfacree (1996) describes, the parliamentary debates on the Act articulated the anxieties of middle class rural Britain about the threat that they perceived to be posed to their rural idyll by travellers. These included a number of elements. Travellers were represented as disrupting the tranquillity of rural life, with one Member of Parliament claiming that,

The new age travellers displayed some dreadful antics: they invaded peaceful countryside, decimated peaceful villages, went on the rampage and had raves lasting two or three days, showing a total disregard for the area. (Quoted by Halfacree, 1996, p. 62)

They were also portrayed as a 'visual menace' or 'an untidy aspect of urban life' (Halfacree, 1996, p. 63), that disrupted the rural landscape, and they were accused of disrupting the spatial and social order of rural space, for example, by refusing to conform to property conventions:

New age travellers appear to have no wish to establish themselves or reside on authorised sites, but simply want to roam through the countryside unchecked. (Member of Parliament, quoted by Halfacree, 1996, p. 58)

Significantly, new age travellers were explicitly represented as *not the same* as gypsies, positioning them outside the rural community, in part because of a perceived non-compliance with an imagined rural work ethic:

True gypsies have been with us for centuries. They have been tolerated – indeed welcomed – in the rural community, where they regularly assisted with the harvest and did other casual jobs around the farms and houses. But today they have acquired their parasites, the hippies or drop-outs – generically referred to as New Age travellers – who do not work, who do not want to work, but who believe that because the gypsies have the apparent right to roam the countryside at will, they can do the same at the expense of the local taxpayer. (Member of the House of Lords, quoted by Halfacree, 1996, p. 59)

Alternative Rural Communities

Whilst new age travellers embody a strategy of semi-nomadism in the countryside, alternative rural lifestyles are also performed through the development of new forms of settled communities, albeit based on very different principles from the traditional notion of the rural community. Some of these communities have been established as an off-shoot of the traveller counterculture described above. Most famous is Tipi Valley in west Wales, a community of around two hundred residents accommodated in some fifty to sixty tents in a remote rural valley, established in 1976. With an obvious reference to Native American culture, Tipi Valley represents itself as an experiment in a form of eco-friendly rural living. Yet, as McKay (1996) observes, by 'deliberately placing itself at the margins, away from the centre of majority culture – almost from any culture – it has developed into a central space inhabited by *authentic* veterans and idealists, surrounded by its own marginal types, problem cases' (p. 57; original emphasize). Tipi Valley has also been criticized for its lack of self-sufficiency and dependence on state benefits and outside resources; and has faced attempts at eviction by the local government authority which claimed that the community was 'an unauthorised shift from agricultural to residential land use' (McKay, 1996, p. 52).

The pioneers at Tipi Valley were part of a long tradition of utopian groups who have appropriated the seclusion, space and isolation afforded by rural locations to establish communities that have sought to promote new ways of living. These have included, for example, various religious groups and communities identified with particular forms of farming or environmental management. They also include initiatives designed to establish communities that can liberate members from oppressive structures of racism, ableism or homophobia. Valentine (1997b) discusses the creation of separatist lesbian communities in the rural United States, which developed into a significant movement in the 1970s. Despite the common association of rural society with homophobia (see Chapter 17), rural locations

were selected by the lesbian separatists because of the potential to control an extensive territorial space and thus to develop radical forms of social and economic organization:

> We view our maintaining lesbian space and protecting these acres from the rape of man and his chemicals as a political act of active resistance. Struggling with each other to work through our patriarchal conditioning, and attempting to work and live together in harmony with each other and nature. (Resident of Wisconsin Womyn's Land Co-operative, quoted by Cheney, 1985, p. 132)

Within the separatist communities, attempts were made to construct a lesbian feminist society that was non-hierarchical and self-sufficient. Old skills such as fire-making and producing herbal medicine were rediscovered and the communities actively fostered a distinctive women's culture expressed through language, music, literature and histories. As such, Valentine (1997b) notes, 'they constructed very politicised visions of a "rural idyll"' (p. 112).

Yet, Valentine also documents the tensions and differences that emerged within the communities, with conflicts developing around issues including the management of the land, the practice of monogamy versus non-monogamy, and the presence in the communities of male children. Thus, Valentine (1997b) concludes, 'lesbian separatist attempts to establish "idyllic" ways of living in the countryside appear to have unravelled because, in common with traditional white middle-class visions of "rural community", attempts to create unity and common ways of living produced boundaries and exclusions' (pp. 118–119).

Militant Reactionary Ruralities

The examples discussed above all represent attempts to develop what might be labelled broadly progressive alternative lifestyles in rural areas and the conflicts that have been noted have often arisen when the progressive values of the new communities have collided with the conservatism of traditional rural society. However, alternative rural communities have also been established by more reactionary, right wing groups, for whom the countryside is attractive in part because of its association as a 'pure' rural space and relatively monocultural society (see Chapter 20), and in part because isolated locations offer the potential of operating beyond the surveillance of the state. The most prominent such group is the militia movement in the United States. As Kimmel and Ferber (2000) note, the militia movement is a loose collection of paramilitary groups who share a distrust of government and a paranoia that global politics are controlled by an elite conspiracy, and who have armed themselves to fight back. Its world-view is racist and anti-Semite, constructed on particular notions of masculinity, and informed by a fundamentalist interpretation of Christianity. Many militia members also believe that they become sovereign individuals by taking actions to remove themselves from the authority of the federal government, for example by refusing to pay taxes. They assert the right to do this by considering themselves to be 'natural citizens', born and bred in the United States, as opposed to 'Fourteenth Amendment citizens', who include immigrants who swear their allegiance to the constitutions and Americans who accede to the authority of the federal government by paying taxes, receiving social security cards and obtaining drivers' licences, birth certificates and the like (Dyer, 1998).

The, mostly male, members of the militia movement are drawn from across the United States, often from the lower-middle classes, such as small farmers, shopkeepers, craftsmen and skilled workers – occupational groups who feel aggrieved at the tax burden and lack of economic assistance from government, or threatened by competition from non-white

ethnic groups in the labour market (Kimmel and Ferber, 2000). Yet, the militia community has gravitated towards remote rural areas, with a particular presence in Montana and Idaho. Kimmel and Ferber argue that militia members move to rural areas because

> they seek companionship with like-minded persons, in relatively remote areas far from large numbers of non-whites and Jews, where they can organize, train, and build protective fortresses. Many groups seek to establish a refuge in rural communities, where they can practice military tactics, stockpile food and weapons, hone their survivalist skills, and become self-sufficient in preparation for Armageddon, Y2K, the final race war, or whatever cataclysm they envisage. (Kimmel and Ferber, 2000, p. 590)

In advance of the year 2000 (Y2K), some militia groups established 'covenant communities' in preparation for the arrival of non-white refugees whom they believed would leave cities for the American countryside as computer systems collapsed and welfare payments ceased. In the anticipated fight for food supplies, the 'covenant communities' were armed and trained to 'defend' the resources of their all-white members (Kimmel and Ferber, 2000).

Additionally, the extremist groups have identified rural areas as recruiting grounds. The mythic history of the militia movement makes connections to a strong tradition of rural producer radicalism in the United States, based on ideologies of localism and vigilantism (Stock, 1996), such that it positions itself as defending 'rural America'.

This message has an alluring appeal in rural communities that have suffered from the farm crisis, depopulation, the loss of services and infrastructure, environmental degradation, poverty and ill-health (Dyer, 1998; Kimmel and Ferber, 2000; Stock, 1996). As Dyer (1998) remarks, sovereigntist ideas of withholding taxes have found favour with rural people struggling to make ends meet and who perceive government expenditure to be biased towards (non-white) urban communities. Thus, Dyer locates in the militia discourse an explicit threat about the relationship between rural America and the US government: 'This idea is that if the federal government won't help rural America, then rural America will simply govern itself by ignoring federal authority' (p. 174).

Summary

No one has exclusive ownership of the 'rural idyll'. For a diverse range of groups from very different backgrounds and with very different ideological, cultural and philosophical influences, the countryside offers an escape from the pressures and demands of city living and provides the space in which to construct a new, idealistic, way of life. As such, rural areas increasingly play host to a diversity of alternative lifestyles and communities that do not conform to conventional understandings of rural life and rural communities. Many of the participants in such communities – whether they be fixed place settlements such as Tipi Valley and the separatist lesbian communities; looser networks of mobile populations such as the new age travellers; or regionally focused groupings based on particular ideological positions and ways of life, such as the militia movement – believe that the open spaces of the countryside will permit them sufficient autonomy and seclusion to allow them to pursue the lifestyle of their choice without interference. However, the practices of such alternative rural lifestyles are frequently incompatible with many of the values, principles and prejudices of established rural communities, creating tensions between the two groups. As such, rural areas can be less of a place of escape than a place of conflict in which power struggles are played out between different lifestyle groups, each in pursuit of their own 'rural idyll'.

Further Reading

The contested rural lifestyles of the new age travellers and the experimental alternative community at Tipi Valley are both discussed in more detail by George McKay in his book *Senseless Acts of Beauty* (Verso, 1996). The depiction of travellers as a threat to the middle class rural idyll by British parliamentarians, meanwhile, is examined by Keith Halfacree in 'Out of place in the countryside: travellers and the "rural idyll"', *Antipode*, volume 29, pages 42–71 (1996). For more on lesbian separatist communities in the rural United States, see Gill Valentine's chapter in P. Cloke and J. Little (eds), *Contested Countryside Cultures* (Routledge, 1997). The right-wing extremist militia movement in the rural United States is discussed at length by a number of studies, but see particularly Joel Dyer's book, *Harvest of Rage* (Westview, 1998) and Michael Kimmel and Abby Ferber's article '"White men are this nation": right-wing militias and the restoration of rural American masculinity', in *Rural Sociology*, volume 65, pages 582–604 (2000). Carol McNichol Stock's book *Rural Radicals: Righteous Rage in the American Grain* (Cornell University Press, 1996) provides a detailed historical context.

Part 5

CONCLUSIONS

22

Thinking Again About the Rural

The Differentiated Countryside

There are many different countrysides. They are distinguished by different landscapes and natural environments; by different histories, settlement patterns and densities of population; by relative isolation or proximity to metropolitan centres; by different economic structures, types of farming, industrial developments and experiences of economic change; and by different patterns of migration and population recomposition. Even within a single rural territory people will have very different ideas about what it means to be rural. Some will claim that rural people need to be born into the countryside, that they need to be steeped in traditional rural folk knowledge and to practise traditional rural pursuits such as hunting. Others will assert their right to buy into the countryside, purchasing property or enjoying rural recreation in pursuit of an idea of rural life that is often strongly influenced by media representations. Experiences and expectations of rural life are informed by an individual's socio-economic status, gender, age, ethnicity and sexual orientation, and by other personal characteristics.

This diversity means that it is difficult to produce a single, objective, definition of 'rural' that can be employed to delimit rural and urban spaces, or to differentiate rural societies and economies from urban societies and economies. Yet, it does not mean that the concept of the 'rural' is devalued for geographers and social scientists. 'Rural' is still a tremendously powerful concept in modern society and the different ways in which people and institutions socially construct 'rurality' have a very real effect on the restructuring of those areas that are claimed to be rural, on the responses that are adopted, and on individuals' experiences of rural change.

The diversity of the countryside is not new. The differences indicated above have a long historical presence, in spite of attempts by geographers and social scientists to impose strict definitions of 'rurality', or those of policy-makers to reduce the rural to manageable unifying concepts such as agriculture. Yet, it can be argued that economic and social restructuring in recent decades has increased the differentiation of the countryside, whilst reducing the differentiation between rural and urban space.

Processes, Responses and Experiences

This book has sought to analyse contemporary rural restructuring by examining in turn the main processes of restructuring, the responses of communities and governments, and the experiences of people living in, working in or consuming rural space. The contemporary era of rural restructuring is set apart from earlier changes in the rural world by the *pace and persistence* of change and by the *totality and interconnectivity* of change as experienced during the twentieth and early twenty-first centuries. These characteristics in turn are an expression of the way in which recent rural change has been driven by two over-arching processes of transformation that have impacted on both rural and urban areas: modernization and globalization.

Many processes of rural restructuring involve a notion of modernization. Changes to farming practices, for example, were advanced under the banner of 'agricultural modernization', which meant mechanization, specialization, larger farm units and the use of agri-chemicals and other technologies to maximize production (Chapter 4). As a consequence, the number of people employed in agriculture plummeted, such that farming lost its place as the major source of rural employment, and problems of over-production and environmental degradation were created. Elsewhere in the rural economy, new employment opportunities were created by 'modernization' in other industries and sectors which enabled the relocation of plants and offices to rural areas (Chapter 5). 'Modernization' in the form of technological innovation changed the patterns of everyday life in the countryside, particularly through developments in transport, food preservation and communications. On the one hand, such developments raised the quality of life in rural areas, making them more desirable places to live and encouraging counterurbanization (Chapter 6); on the other hand, technological innovations that enabled people to become more mobile and to be less dependent on shopping locally for fresh food removed trade from village retailers, contributing to the rationalization and closure of many rural shops and services (Chapter 7).

Globalization is closely related to modernization, one of the effects of modernization being the compression of time and space. As noted in Chapter 3, there are many different strands of globalization and three in particular have impacted on rural areas. First, economic globalization means that food and other agricultural products, as well as other traditional rural products such as timber, are increasingly traded in a global market and that the economic circumstances of farmers and rural producers are strongly influenced by the conditions of the global market (Chapters 3 and 4). Additionally, these markets tend to be dominated by a small number of transnational corporations and 'commodity chain clusters'. Secondly, the globalization of mobility has increased the flow of people in, out and through rural space not just on regional and national scales, but also internationally. A number of rural areas have become heavily dependent on foreign tourism (see Chapter 12); whilst in several rural areas, significant amounts of property have been purchased by non-nationals as permanent or second homes. At the other end of the economic scale, foreign migrant workers comprise a crucial part of the labour force for seasonally intensive forms of agriculture, including fruit-farming and viticulture (Chapter 18). Thirdly, a globalization of values has eroded historic rural cultures. The conservatism and conformity of traditional rural society has been challenged by the assertion of values of plurality and tolerance, expressed particularly through a reworking of rural gender relations and attitudes towards sexuality and race (see Chapters 15, 17 and 20). Similarly, the ascendancy of globalized values of conservation and animal rights over rural lay discourses of nature has both supported the adoption of more stringent measures to protect the rural environment and generated conflicts over farming practices, resource management and hunting (Chapters 13 and 14).

The effects of restructuring have demanded responses both from the state and from rural communities and residents. The state – including national, regional and local government and other public agencies charged with management of elements of the rural economy, society and environment – has an interest in responding to rural restructuring not just from a welfare perspective, but also in terms of supporting capitalist accumulation, and for practical governmental reasons of maintaining order by addressing regional inequalities and avoiding unmanageable population movements. Thus responses adopted by the state include the reform of agricultural policy to encourage a transition away from productivism (Chapter 4); actions to regulate world trade in agricultural and other commodities (Chapter 9); investment in rural development programmes (Chapter 10); the regulation of land use and development and initiatives to protect the rural environment (Chapter 13). The state response to restructuring also involved changing the way in which rural space is governed (Chapter 11) and incorporating new actors into the policy-making process (Chapter 9). Yet the response to rural restructuring has not been left to the state alone. At a grassroots level, rural communities and people themselves have responded through a variety of locally based self-help and entrepreneurial activity, through ventures to exploit the commodification of the countryside (Chapter 12) and through political mobilization to defend aspects of rural life, culture and environment that they perceived to be under threat (Chapter 14).

The processes of rural restructuring and the response to them have fundamentally changed the experience of living in the countryside. This is significant as it is by exploring the experiences of rural change that we can respond to Hoggart and Paniagua's (2001) challenge that *restructuring* (as opposed to mere

change) requires 'fundamental readjustments in a variety of spheres of life, where processes of change are causally linked' (p. 42), and which have both a quantitative and a qualitative expression. As the chapters in the final part of this book demonstrated, there is substantial evidence of a qualitative shift in the experience of life in the countryside, with change often experienced as a consequence of a number of inter-locking processes. Moreover, there is again no one common experience of rural change, rather many different, situated, stories.

The myth of the rural idyll might suggest that rural life is safe, peaceful and prosperous, but there are many rural residents for whom life is restricted or constrained by poverty, poor housing, ill health, prejudice and lack of opportunity (Chapters 16–20). In some cases, the experience of rural restructuring has been empowering – there is generally greater tolerance now towards racial and sexual diversity in rural communities; and there are more employment opportunities for rural women. In other cases, however, rural restructuring has compounded problems of unemployment and the absence of appropriately skilled work, poor accessibility to services and workplaces, and a shortage of affordable housing (Chapters 16 and 18).

Re-thinking Rurality

Tracing through the processes of rural restructuring and their consequences might lead us to re-think the way in which we approach the 'rural' as students and researchers. The concept of the rural as a social construction, as described in Chapter 1 and followed through the book, is very useful in revealing difference and conflict in the countryside, but more recently rural researchers have started to move beyond *perceptions* and *representations* of the rural to try to grasp the ways in which the rural is *performed* and *constituted*.

Examining the performance of rurality enables the researcher to move away from thinking of the rural as a spatially fixed entity and to focus on the way in which rurality (or particularly ways of being rural) is embedded in social practices. Thus, for example, traditional family-based farming may be regarded as a particular way of performing a rural lifestyle, as may participation in 'traditional rural sports' such as hunting or shooting. Seeing these activities as performances of rurality may help us to understand why perceived threats to their practice can generate deep anger and political mobilization (Woods, 2003a). At a more collective level, Liepins's (2000a) model of community, as discussed in Chapter 7, allows us entry points to thinking about the performance of rural community, both through everyday practices and through literal performances of fairs, shows and community events. Finally, the consumption of the commodified countryside increasingly involves not just the focusing of the tourist gaze on representations of an imagined 'rural idyll', but also active engagement with embodied rural performances through forms of adventure tourism (Cater and Smith, 2003).

Meanwhile, the re-examination of the constitution of the rural has begun to think about the rural as a 'hybrid space', a set of complex inter-relations of social and natural entities. The countryside, it is argued, is co-constructed by the agency of both human and non-human actors (think, for example, about the place of animals in the countryside, or about the unpredictable impact of diseases such as BSE and foot and mouth, or of extreme weather events). Once again, exploring the rural in terms of its hybridity serves to highlight the diversity and dynamism of the countryside. As Jonathan Murdoch suggests,

> The countryside is hybrid. To say this is to emphasise that it is defined by networks in which heterogeneous entities are aligned in a variety of ways. It is also to propose that these networks give rise to slightly different countrysides: there is no single vantage point from which the panoply of rural or countryside relations can be seen. Thus, a 'regionalised' perspective can be adopted only in the knowledge that network and fluid spaces will escape its purview; a focus on networks and fluid spaces will disrupt the notions of easily demarcated and fixed rural spaces but will generate contrasting and sometimes contradictory understandings of rural processes. (Murdoch and Lowe, 2003, p. 274)

Focusing on hybridity, networks and 'fluid space' resonates with other developments that are already occurring in rural research. First, there has been a renewed concern with the inter-relation of nature and rurality, which involves both the analysis of cultures of nature (see Milbourne, 2003c), and collaboration between human and physical geographers (and social and natural scientists more broadly) – the latter being the objective of a major research programme launched in the UK in 2003.

Second, there are new attempts to examine the interplay of the city and the countryside and instances of urban–rural hybridity. This might involve, for example, investigating how the processes shaping rural spaces also work through urban space and how finding solutions to policy problems might involve both rural and urban action. Counterurbanization evidently involves migration from urban districts as well as migration to rural districts. Similarly, the relocation of some types of manufacturing industry and service sector employment to rural locations is intrinsically connected to processes of urban economic restructuring. Furthermore, the use of rural space by urban residents, particularly for recreation, creates sets of overlapping urban and rural interests. Some rural research has begun to explore these connections, for instance by tracing the various components of the food chain from production to consumption, but there is scope for more studies of this nature.

Third, there is a growing concern with the interaction of global and local scales in rural restructuring, and thus with questions about rural change beyond the developed world. This book has deliberately focused on the rural areas of the developed world – defined as North America, Europe, Australia, New Zealand and Japan – and excluded any consideration of the rural developing world. Despite the differences discussed above, there are a number of structural characteristics that are common to rural areas in most of the developed world. First, agricultural production and other economic activity in rural space is almost entirely for commercial purposes rather than self-sufficiency. Secondly, commercial exploitation of rural resources is performed in a capitalist free market economy. Thirdly, all but the most remote rural areas are provided with basic infrastructure including electricity and water supply. Fourthly, the state accepts the principle of a universal entitlement to public services across its territory. Fifthly, the population as a whole is sufficiently affluent to pay for the commodification of rural landscapes, lifestyles, artefacts and experiences. Finally, there is a shared consumption of film, television, literature and music that informs perceptions of the rural.

None of these characteristics is found universally in the developing world. However, there are places in the developing world where some of the above characteristics do apply, and thus many of the features of rural

restructuring discussed in this book can also be found in particular contexts within the developing world. Furthermore, a number of the problems faced by rural areas in the contemporary developed world have developing world dimensions. For example, the resolution of negotiations about global trade in agriculture will have a significant impact on farming communities in both the developed and the developing world (Chapter 9). Similarly, migrant workers often form a bridge between the developed world rural economies in which they work and rural areas in developing world countries from which they have come (Chapter 18). These connections have been recognized by campaigning organizations that have formed linkages between groups in the developed and the developing worlds and one of the key challenges for rural geography is to engage more substantially with the interconnections of the rural experience in the developed and developing world.

As a student of rural geography, or a related rural social science, you need not be a passive observer of these developments, but can make your own contributions. The chapters and sections in this book have provided an inevitably brief introduction to a range of themes and topics in contemporary rural geography. By following the suggestions for further reading you will be able to find out more about subjects that interest you, but you will not find definitive statements that leave nothing more to be said. On-going processes of restructuring may have changed the situation, or new policies may have been adopted since the articles were written. Observations made in one rural context may not apply in another. And there is always the potential to find new perspectives on a topic, new ways of looking at or thinking about an issue. Hopefully, then, this book will have inspired you to ask your own questions about the contemporary countryside and the ways in which it is changing. The websites indicated at the end of chapters will enable you to look for yourself at up-to-date statistics and data, to read first-hand accounts and press reports, to find out about policies and to discover the positions of various rural campaign groups and agencies. With these resources you can carry out your own rural research for undergraduate and postgraduate projects and contribute to our broader understanding of the diverse, dynamic and complex place that is the twenty-first century countryside.

Bibliography

Agyeman, J. and Spooner, R. (1997) Ethnicity and the rural environment, in P. Cloke and J. Little (eds), *Contested Countryside Cultures*. London and New York: Routledge. pp. 197–217.

Aigner, S.M., Flora, C.B. and Herandez, J.M. (2001) The premise and promise of citizenship and civil society for renewing democracies and empowering sustainable communities, *Sociological Inquiry*, 71, 493–507.

Albrow, M. (1990) Introduction, in M. Albrow and E. King (eds), *Globalisation, Knowledge and Society*. London: Sage.

Anderson, S. (1999) Crime and social change in rural Scotland, in G. Dingwall and S.R. Moody (eds), *Crime and Conflict in the Countryside*. Cardiff, UK: University of Wales Press. pp. 45–59.

Arensberg, C.M. (1937) *The Irish Countryman*. New York, NY: Macmillan.

Arensberg, C.M. and Kimball, S.T. (1948) *Family and Community in Ireland*. London: Peter Smith.

Argent, N. (2002) From pillar to post? In search of the post-productivist countryside in Australia, *Australian Geographer*, 33, 97–114.

Banks, J. and Marsden, T. (2000) Integrating agri-environment policy, farming systems and rural development: Tir Cymen in Wales, *Sociologia Ruralis*, 40, 466–481.

Barnes, T. and Hayter, R. (1992) The little town that did: flexible accumulation and community response in Chemainus, British Columbia, *Regional Studies*, 26, 617–663.

Beesley, K.B. (1999) Agricultural land preservation in North America: a review and survey of expert opinion, in O.J. Furuseth and M.B. Lapping (eds), *Contested Countryside: The Rural Urban Fringe in North America*. Aldershot, UK and Brookfield, VT: Ashgate. pp. 57–92.

Beeson, E. and Strange, M. (2003) *Why Rural Matters 2003: The Continuing Need for Every State to Take Action on Rural Education*. Washington, DC: Rural Schools and Community Trust.

Bell, D. (2000) Farm boys and wild men: rurality, masculinity and homosexuality, *Rural Sociology*, 65, 547–561.

Bell, D. and Valentine, G. (1995) Queer country: rural lesbian and gay lives, *Journal of Rural Studies*, 11, 113–122.

Bell, M.M. (1994) *Childerley: Nature and Morality in a Country Village*. Chicago: University of Chicago Press.

Berry, B. (ed.) (1976) *Urbanisation and Counter-urbanisation*. Beverly Hills, CA: Sage.

Bessière, J. (1998) Local development and heritage: traditional food and cuisine as tourist attractions in rural areas, *Sociologia Ruralis*, 38, 21–34.

Biers, J.M. (2003) Bittersweet future, *The Times-Picayune*, 9 March, pp. F1–2.

Bollman, R.D. and Briggs, B. (1992) Rural and small town Canada: an overview, in R.D. Bollman (ed.), *Rural and Small Town Canada*. Toronto: Thompson Educational Publishing.

Bollman, R.D. and Bryden, J.M. (eds) (1997) *Rural Employment: An International Perspective*. Wallingford, UK: CAB International.

Bonnen, J.T. (1992) Why is there no coherent US rural policy?, *Policy Studies Journal*, 20, 190–201.

Bontron, J-C. and Lasnier, N. (1997) Tourism: a potential source of rural employment, in R.D. Bollman and J.M. Bryden (eds), *Rural Employment: An International Perspective*. Wallingford, UK: CAB International. pp. 427–446.

Borger, J. (2001) Hillbilly heroin: the painkiller abuse wrecking lives in West Virginia, *Guardian*, 25 June, p. 3.

Bourne, L. and Logan, M. (1976) Changing urbanization patterns at the margin: the examples of Australia and Canada, in B. Berry (ed.), *Urbanisation and Counterurbanisation*. Beverly Hills, CA: Sage. pp. 111–143.

Bové, J. and Dufour, F. (2001) *The World Is Not For Sale: Farmers against Junk Food.* London and New York: Verso.

Bowler, I. (1985) Some consequences of the industrialization of agriculture in the European Community, in M.J. Healey and B.W. Ilbery (eds), *The Industrialisation of the Countryside.* Norwich, UK: GeoBooks. pp. 75–98.

Boyle, P. and Halfacree, K. (1998) *Migration Into Rural Areas.* Chichester: Wiley.

Brace, C. (1999) Finding England everywhere: regional identity and the construction of national identity, 1890–1940, *Ecumene*, 6, 90–109.

Brewer, C.A. and Suchan, T.A. (2001) *Mapping Census 2000: The Geography of US Diversity.* Redlands, CA: ESRI Press.

Brittan, G.G. (2001) Wind, energy, landscape: reconciling nature and technology, *Philosophy and Geography*, 4, 169–184.

Browne, W.P. (2001a) *The Failure of National Rural Policy: Institutions and Interests.* Washington, DC: Georgetown University Press.

Browne, W.P. (2001b) Rural failure: the linkage between policy and lobbies, *Policy Studies Journal*, 29, 108–117.

Brownlow, A. (2000) A wolf in the garden: ideology and change in the Adirondack landscape, in C. Philo and C. Wilbert (eds), *Animal Spaces, Beastly Places.* London and New York: Routledge. pp. 141–158.

Bruckmeier, K. (2000) LEADER in Germany and the discourse of autonomous regional development, *Sociologia Ruralis*, 40, 219–227.

Bruinsma, J. (ed.) (2003) *World Agriculture: towards 2015/2030 – an FAO Perspective.* London: Earthscan.

Buller, H. and Morris, C. (2003) Farm animal welfare: a new repertoire of nature–society relations or modernism re-embedded?, *Sociologia Ruralis*, 43, 216–237.

Bunce, M. (1994) *The Countryside Ideal.* London: Routledge.

Bunce, M. (2003) Reproducing rural idylls, in P. Cloke (ed.), *Country Visions.* Harlow, UK: Pearson. pp. 14–30.

Butler, R. (1998) Rural recreation and tourism, in B. Ilbery (ed.), *The Geography of Rural Change.* Harlow, UK: Addison Wesley Longman. pp. 211–232.

Butler, R. and Clark, G. (1992) Tourism in rural areas: Canada and the United Kingdom, in I.R. Bowler, C.R. Bryant and M.D. Nellis (eds), *Contemporary Rural Systems in Transition, volume 2: Economy and Society.* Wallingford, UK: CAB International. pp. 166–183.

Buttel, F. and Newby, H. (eds) (1980) *The Rural Sociology of Advanced Societies: Critical Perspectives.* Montclair, NJ: Allanheld and London: Croom Held.

Cabinet Office (2000) *Sharing the Nation's Prosperity: Economic, Social and Environmental Conditions in the Countryside. A Report to the Prime Minister by the Cabinet Office.* London: Cabinet Office.

CACI (2000) *Who's Buying Online?* London: CACI.

Campagne, P., Carrère, G. and Valceschini, E. (1990) Three agricultural regions of France: three types of pluriactivity, *Journal of Rural Studies*, 4, 415–422.

Campbell, D. (2001) Greenhouse melts Alaska's tribal ways, *Guardian*, 16 July, p. 11.

Campbell, D. (2002) Farmworkers set out to harvest rights, *Guardian*, 17 August, p. 17.

Campbell, H. (2000) The glass phallus: pub(lic) masculinity and drinking in rural New Zealand, *Rural Sociology*, 65, pp. 562–581.

Campbell, H. and Bell, M.M. (2000) The question of rural masculinities, *Rural Sociology*, 65, 532–546.

Campbell, H. and Liepins, R. (2001) Naming organics: understanding organic standards in New Zealand as a discursive field, *Sociologia Ruralis*, 41, 21–39.

Carson, R. (1962) *Silent Spring.* Cambridge, MA: Riverside Press; (1963) London: Hamilton.

Casper, L.M. (1996) Who's Minding Our Preschoolers?, *Current Population Reports, Household Economic Studies P70–53.* Washington, DC: US Bureau of the Census.

Cater, C. and Smith, L. (2003) New country visions: adventurous bodies in rural tourism, in P. Cloke (ed.), *Country Visions.* Harlow, UK: Pearson. pp. 195–217.

Chalmers, A.I. and Joseph, A.E. (1998) Rural change and the elderly in rural places: commentaries from New Zealand, *Journal of Rural Studies*, 14, 155–166.

Champion, A. (ed.) (1989) *Counterurbanization.* London: Edward Arnold.

Cheney, J. (1985) *Lesbian Land.* Minneapolis, MN: Word Weavers.

Clark, G. (1979) Current research in rural geography, *Area*, 11, 51–52.

Clark, G. (1991) People working in farming: the changing nature of farmwork, in T. Champion and C. Watkins (eds), *People in the Countryside.* London: Paul Chapman. pp. 67–83.

Clark, M.A. (2000) *Teleworking in the Countryside.* Aldershot, UK: Ashgate.

Clemenson, H. (1992) Are single industry towns diversifying? An examination of fishing, forestry and mining towns, in R.D. Bollman (ed.), *Rural and Small Town Canada.* Toronto: Thompson Educational Publishing. pp. 151–166.

Cloke, P. (1977) An index of rurality for England and Wales, *Regional Studies*, 11, 31–46.

Cloke, P. (1983) *An Introduction to Rural Settlement Planning*. London and New York: Methuen.

Cloke, P. (ed.) (1988) *Policies and Plans for Rural People: An International Perspective*. London: Unwin Hyman.

Cloke, P. (1989a) Rural geography and political economy, in R. Peet and N. Thrift (eds), *New Models in Geography: The Political Economy Perspective, Volume 1*. London: Unwin Hyman. pp. 164–197.

Cloke, P. (1989b) State deregulation and New Zealand's agricultural sector, *Sociologia Ruralis*, 29, 34–48.

Cloke, P. (1992) The countryside: development, conservation and an increasingly marketable commodity, in P. Cloke (ed.), *Policy and Change in Thatcher's Britain*. Oxford, UK: Pergamon Press.

Cloke, P. (1993) The countryside as commodity: new rural spaces for leisure, in S. Glyptis (ed.), *Leisure and the Environment: Essays in Honour of Professor J.A. Patmore*. London: Belhaven Press. pp. 53–67.

Cloke, P. (1994) (En)culturing political economy: a life in the day of a 'rural geographer', in P. Cloke, M. Doel, D. Matless, M. Phillips and N. Thrift, *Writing the Rural*. London: Paul Chapman. pp. 149–190.

Cloke, P. (1997a) Country backwater to virtual village? Rural studies and 'the cultural turn', *Journal of Rural Studies*, 13, 367–375.

Cloke, P. (1997b) Poor country: marginalization, poverty and rurality, in P. Cloke and J. Little (eds), *Contested Countryside Cultures*. London and New York: Routledge. pp. 252–271.

Cloke, P. and Edwards, G. (1986) Rurality in England and Wales 1981: a replication of the 1971 index, *Journal of Rural Studies*, 20, 289–306.

Cloke, P. and Goodwin, M. (1992) Conceptualizing countryside change: from post-Fordism to rural structured coherence, *Transactions of the Institute of British Geographers*, 17, 321–336.

Cloke, P. and Le Heron, R. (1994) Agricultural deregulation: the case of New Zealand, in P. Lowe, T. Marsden and S. Whatmore (eds), *Regulating Agriculture*. London: David Fulton. pp. 104–126.

Cloke, P. and Little, J. (1990) *The Rural State?* Oxford, UK: Oxford University Press.

Cloke, P. and Little, J. (eds) (1997) *Contested Countryside Cultures*. London and New York: Routledge.

Cloke, P. and Milbourne, P. (1992) Deprivation and lifestyles in rural Wales: II Rurality and the cultural dimension, *Journal of Rural Studies*, 8, 359–371.

Cloke, P. and Perkins, H.C. (1998) 'Cracking the canyon with the awesome foursome': representations of adventure tourism in New Zealand, *Environment and Planning D: Society and Space*, 16, 185–218.

Cloke, P. and Thrift, N. (1987) Intra-class conflict in rural areas, *Journal of Rural Studies*, 3, 321–333.

Cloke, P., Goodwin, M. and Milbourne, P. (1997) *Rural Wales: Community and Marginalization*. Cardiff, UK: University of Wales Press.

Cloke, P., Goodwin, M., Milbourne, P. and Thomas, C. (1995) Deprivation, poverty and marginalisation in rural lifestyles in England and Wales, *Journal of Rural Studies*, 11, 351–366.

Cloke, P., Milbourne, P. and Thomas, C. (1994) *Lifestyles in Rural England*. London: Rural Development Commission.

Cloke, P., Milbourne, P. and Thomas, C. (1996) The English National Forest: local reactions to plans for renegotiated nature–society relations in the countryside, *Transactions of the Institute of British Geographers*, 21, 552–571.

Cloke, P., Milbourne, P. and Widdowfield, R. (2000) Partnership and policy networks in rural local governance: homelessness in Taunton, *Public Administration*, 78, 111–133.

Cloke, P., Milbourne, P. and Widdowfield, R. (2001a) Homelessness and rurality: exploring connections in local spaces of rural England, *Sociologia Ruralis*, 41, 438–453.

Cloke, P., Milbourne, P. and Widdowfield, R. (2001b) Making the homeless count? Enumerating rough sleepers and the distortion of homelessness, *Policy and Politics*, 29, 259–279.

Cloke, P., Milbourne, P. and Widdowfield, R. (2002) *Rural Homelessness: Issues, Experiences and Policy Responses*. Bristol, UK: Policy Press.

Cloke, P., Phillips, M. and Thrift, N. (1995) The new middle classes and the social constructs of rural living, in T. Butler and M. Savage (eds), *Social Change and the Middle Classes*. London: UCL Press. pp. 220–238.

Cloke, P., Phillips, M. and Thrift, N. (1998) Class, colonization and lifestyle strategies in Gower, in P. Boyle and K. Halfacree (eds), *Migration Into Rural Areas*. Chichester, UK: Wiley. pp. 166–185.

Clout, H.D. (1972) *Rural Geography: An Introductory Survey*. Oxford: Pergamon Press.

Cocklin, C., Walker, L. and Blunden, G. (1999) Cannabis highs and lows: sustaining and dislocating rural communities in Northland, New Zealand, *Journal of Rural Studies*, 15, 241–255.

Coppock, T. (1984) *Agriculture in Developed Countries*. London: Macmillan.

Cornell, S. (2000) Enhancing rural leadership and institutions, in Center for the Study of Rural America (eds), *Beyond Agriculture: New Policies for Rural America*. Kansas City: The Federal Reserve Bank of Kansas City. pp. 103–120.

Countryside Agency (2001) *Rural Services in 2000*. London: Countryside Agency.

Countryside Agency (2003) *State of the Countryside 2003*. London: Countryside Agency.

Cox, G. and Winter, M. (1997) The beleaguered 'other': hunt followers in the countryside, in P. Milbourne (ed.), *Revealing Rural Others: Representation, Power and Identity in the British Countryside*. London: Pinter. pp. 75–88.

Cox, G., Hallett, J. and Winter, M. (1994) Hunting the wild red deer: the social organisation and ritual of a 'rural' institution, *Sociologia Ruralis*, 34, 190–205.

Crang, M. (1999) Nation, region and homeland: history and tradition in Darlana, Sweden, *Ecumene*, 6, 447–470.

Cresswell, T. (1996) *In Place/Out of Place: Geography, Ideology and Transgression*. Minneapolis, MN: University of Minnesota Press.

Cresswell, T. (2001) *The Tramp in America*. London: Reaktion Books.

Cromartie, J.B. (1999) Minority counties are geographically clustered, *Rural Conditions and Trends*, 9, 14–19.

Cross, M. and Nutley, S. (1999) Insularity and accessibility: the small island communities of Western Ireland, *Journal of Rural Studies*, 15, 317–330.

Crump, J. (2003) Finding a place in the country: exurban and suburban development in Sonoma County, California, *Environment and Behavior*, 35, 187–202.

Dagata, E. (1999) The socioeconomic well-being of rural children lags behind that of urban children, *Rural Conditions and Trends*, 9, 85–90.

Daniels, S. (1993) *Fields of Vision: Landscape Imagery and National Identity in England and the United States*. Cambridge, UK: Polity Press.

Davies, J. (2003) Contemporary geographies of indigenous rights and interests in rural Australia, *Australian Geographer*, 34, 19–45.

Davis, J. and Ridge, T. (1997) *Same Scenery, Different Lifestyle: Rural Children on a Low Income*. London: The Children's Society.

Decker, P.R. (1998) *Old Fences, New Neigbors*. Tucson, AZ: University of Arizona Press.

DEFRA (Department for the Environment, Food and Rural Affairs) (2002) *England Rural Development Plan*. London: The Stationery Office.

DEFRA (2003) *Agriculture in the United Kingdom 2002*. London: The Stationery Office.

Dennis, N., Henriques, F.M. and Slaughter, C. (1957) *Coal is our Life*. London: Eyre and Spottiswoode.

Dion, M. and Welsh, S. (1992) Participation of women in the labour force: a comparison of farm women and all women in Canada, in R.D. Bollman (ed.), *Rural and Small Town Canada*. Toronto: Thompson Educational Publishing. pp. 225–244.

Diry, J-P. (2000) *Campagnes d'Europe: des espaces en mutation*. Documentation photographique no. 8018. Paris: La Documentation Française.

Dixon, D.P. and Hapke, H.M. (2003) Cultivating discourse: the social construction of agricultural legislation, *Annals of the Association of American Geographers*, 93, 142–164.

DoE/MAFF (Department of the Environment and the Ministry for Agriculture, Fisheries and Food) (1995) *Rural England: The Rural White Paper*. London: The Stationery Office.

Doremus, H. and Tarlock, A.D. (2003) Fish, farms, and the clash of cultures in the Klamath basin, *Ecology Law Quarterly*, 30, 279–350.

Dudley, K.M. (2000) *Debt and Dispossession: Farm Loss in America's Heartland*. Chicago: University of Chicago Press.

Duncan, J. and Ley, D. (eds) (1993) *Writing Worlds*. London: Routledge.

Dyer, J. (1998) *Harvest of Rage*. Boulder, CO: Westview Press.

Edwards, B. (1998) Charting the discourse of community action: perspectives from practice in rural Wales, *Journal of Rural Studies*, 14, 63–78.

Edwards, B., Goodwin, M. and Woods, M. (2003) Citizenship, community and participation in small towns: a case study of regeneration partnerships, in R. Imrie and M. Raco (eds), *Urban Renaissance: New Labour, Community and Urban Policy*. Bristol, UK: Policy Press. pp. 181–204.

Edwards, B., Goodwin, M., Pemberton, S. and Woods, M. (2000) *Partnership Working in Rural Regeneration*. Bristol, UK: Policy Press.

Edwards, B., Goodwin, M., Pemberton, S. and Woods, M. (2001) Partnership, power and scale in rural governance, *Environment and Planning C: Government and Policy*, 19, 289–310.

Errington, A. (1997) Rural employment issues in the periurban fringe, in R.D. Bollman and J.M. Bryden, *Rural Employment: An International Perspective*. Wallingford, UK: CAB International. pp. 205–224.

ERS (2002) Rural population and migration: rural elderly. USDA Economic Research Service, Briefing Room [Online]. Available at www.ers.usda.gov/Briefing/Population/elderly/

ERS (2003a) Rural labour and education: rural low-wage employment. USDA Economic Research Service, Briefing Room [Online]. Available at www.ers.usda.gov/Briefing/laborandeducation/lwemployment/

ERS (2003b) Rural labour and education: rural earnings. USDA Economic Research Service, Briefing Room [Online]. Available at www.ers.usda.gov/Briefing/laborandeducation/earnings/

Estall, R.C. (1983) The decentralization of manufacturing industry: recent American experience in perspective, *Geoforum*, 14, 133–147.

European Union (2003) *Europa: European Union Information On-line*, available at europa.eu.int

Evans, N. and Yarwood, R. (2000) The politicization of livestock: rare breeds and countryside conservation, *Sociologia Ruralis*, 40, 228–248.

Evans, N., Morris, C. and Winter, M. (2002) Conceptualizing agriculture: a critique of post-productivism as the new orthodoxy, *Progress in Human Geography*, 26, 313–332.

Fabes, R., Worsley, L. and Howard, M. (1983) *The Myth of the Rural Idyll.* Leicester, UK: Child Poverty Action Group.

Farley, G. (2003) The Wal-Martization of rural America and other things, *OzarksWatch, The Magazine of the Ozarks*, 2 (2), 12–13.

Fellows, W. (1996) *Farm Boys: Lives of Gay Men in the Rural Midwest.* Madison, WI: University of Wisconsin Press.

Fitchen, J.M. (1991) *Endangered Spaces, Enduring Places: Change, Identity and Survival in Rural America.* Boulder, CO: Westview Press.

Forsyth, A.J.M. and Barnard, M. (1999) Contrasting levels of adolescent drug use between adjacent urban and rural communities in Scotland, *Addiction*, 94, 1707–1718.

Foss, O. (1997) Establishment structure, job flows and rural employment, in R.D. Bollman and J.M. Bryden (eds), *Rural Employment: An International Perspective.* Wallingford, UK: CAB International. pp. 239–254.

Fothergill, S. and Gudgin, G. (1982) *Unequal Growth: Urban and Regional Employment Change in the UK.* London: Heinemann.

Fox, W.F. and Porca, S. (2000) Investing in rural infrastructure, in Center for the Study of Rural America (eds), *Beyond Agriculture: New Policies for Rural America.* Kansas City: The Federal Reserve Bank of Kansas City. pp. 63–90.

Frankenberg, R. (1957) *Village on the Border.* London: Cohen and West.

Frankenberg, R. (1966) *Communities in Britain.* Harmondsworth, UK: Penguin.

Friedland, W. (1991) Women and agriculture in the United States: a state of the art assessment, in W. Friedland, L. Busch, F. Buttel and A. Rudy (eds), *Towards a New Political Economy of Agriculture.* Boulder, CO: Westview. pp. 315–338.

Frouws, J. (1998) The contested redefinition of the countryside: an analysis of rural discourses in the Netherlands, *Sociologia Ruralis*, 38, 54–68.

Fuguitt, G.V. (1991) Commuting and the rural–urban hierarchy, *Journal of Rural Studies*, 7, 459–466.

Fuller, A.J. (1990) From part-time farming to pluri-activity: a decade of change in rural Europe, *Journal of Rural Studies*, 6, 361–373.

Fulton, J.A., Fuguitt, G. and Gibson, R.M. (1997) Recent changes in metropolitan to non-metropolitan migration streams, *Rural Sociology*, 62, 363–384.

Furuseth, O. (1998) Service provision and social deprivation, in B. Ilbery (ed.), *The Geography of Rural Change.* Harlow, UK: Longman. pp. 233–256.

Furuseth, O. and Lapping, M. (eds) (1999) *Contested Countryside: The Rural Urban Fringe in North America.* Aldershot, UK: Ashgate.

Gallent, N. and Tewdwr-Jones, M. (2000) *Rural Second Homes in Europe.* Aldershot, UK: Ashgate.

Gallent, N., Mace, A. and Tewdwr-Jones, M. (2003) Dispelling a myth? Second homes in rural Wales, *Area*, 35, 271–284.

Gant, R. and Smith, J. (1991) The elderly and disabled in rural areas: travel patterns in the north Cotswolds, in T. Champion and C. Watkins (eds), *People in the Countryside.* London: Paul Chapman. pp. 108–124.

Gasson, R. (1980) Roles of farm women in England, *Sociologia Ruralis*, 20, 165–180.

Gasson, R. (1992) Farmers' wives and their contribution to farm business, *Journal of Agricultural Economics*, 43, 74–87.

Gasson, R. and Winter, M. (1992) Gender relations and farm household pluriactivity, *Journal of Rural Studies*, 8, 573–584.

Gearing, A. and Beh, M. (2000) Let tiny towns die says expert, *Brisbane Courier Mail*, 5 July, p. 3.

Gesler, W.M. and Ricketts, T.C. (eds) (1992) *Health in Rural North America: The Geography of Health Care Services and Delivery.* New Brunswick, NJ: Rutgers University Press.

Gibbs, R. and Kusmin, L. (2003) Low-skill workers are a declining share of all rural workers, *Amber Waves*, June 2003 available online at www.ers.usda.gov/AmberWaves/June03/findings/LowskillWork.htm

Gilg, A. (1985) *An Introduction to Rural Geography.* London: Edward Arnold.

Gillette, J.M. (1913) *Constructive Rural Sociology.* New York, NY: Sturgis and Walton.

Gilling, D. and Pierpoint, H. (1999) Crime prevention in rural areas, in G. Dingwall and S.R. Moody (eds), *Crime and Conflict in the Countryside.* Cardiff, UK: University of Wales Press. pp. 114–129.

Gipe, P. (1995) *Wind Energy Comes of Age.* New York: Wiley.

Glendinning, A., Nuttall, M., Hendry, L., Kloep, M. and Wood, S. (2003) Rural communities and well-being: a good place to grow up?, *The Sociological Review*, 51, 129–156.

Glionna, J.M. (2002) Napa growers to build housing for harvesters, *Los Angeles Times*, 19 March, pp. B1 & B4.

Goffman, E. (1959) *The Presentation of Self in Everyday Life*. New York: Doubleday.

Goodman, D., Sorj, B. and Wilkinson, J. (1987) *From Farming to Biotechnology*. Oxford, UK and New York: Basil Blackwell.

Goodman, D. (2001) Ontology matters: the relational materiality of nature and agro-food studies, *Sociologia Ruralis*, 41, 182–200.

Goodwin, M. (1998) The governance of rural areas: some emerging research issues and agendas, *Journal of Rural Studies*, 14, 5–12.

Gordon, R.J., Meister, J.S. and Hughes, R.G. (1992) Accounting for shortages of rural physicians: push and pull factors, in W.M. Gesler and T.C. Ricketts (eds), *Health in Rural North America: The Geography of Health Care Services and Delivery*. New Brunswick, NJ: Rutgers University Press. pp. 153–178.

Gorelick, S. (2000) Facing the farm crisis, *The Ecologist*, 30 (4), 28–31.

Gould, A. and Keeble, D. (1984) New firms and rural industrialisation in East Anglia, *Regional Studies*, 18, 189–202.

Grant, W. (1983) The National Farmers Union: the classic case of incorporation?, in D. Marsh (ed.), *Pressure Politics*. London: Junction Books. pp. 129–143.

Grant, W. (2000) *Pressure Groups and British Politics*. Basingstoke, UK: Macmillan.

Gray, I. and Lawrence, G. (2001) *A Future for Regional Australia*. Cambridge, UK and Oakleigh, Australia: Cambridge University Press.

Green, B. (1996) *Countryside Conservation*. London: E & FN Spon.

Green, M.B. and Meyer, S.P. (1997a) An overview of commuting in Canada with special emphasis on rural commuting and employment, *Journal of Rural Studies*, 13, 163–175.

Green, M.B. and Meyer, S.P. (1997b) Occupational stratification of rural commuting, in R.D. Bollman and J.M. Bryden, *Rural Employment: An International Perspective*. Wallingford, UK: CAB International. pp. 225–238.

Gregory, D. (1994) Discourse, in R.J. Johnston, D. Gregory and D.M. Smith (eds), *The Dictionary of Human Geography*, Third Edition. Oxford, UK and Cambridge, MA: Blackwell. p. 136.

Hajesz, D. and Dawe, S.P. (1997) De-mythologizing rural youth exodus, in R.D. Bollman and J.M. Bryden (eds), *Rural Employment: An International Perspective*. Wallingford, UK: CAB International. pp. 114–135.

Halfacree, K. (1992) The Importance of Spatial Representations in Residential Migration to Rural England in the 1980s. Unpublished PhD thesis, Lancaster University.

Halfacree, K. (1993) Locality and social representation: space, discourse and alternative definitions of the rural, *Journal of Rural Studies*, 9, 23–37.

Halfacree, K. (1994) The importance of 'the rural' in the constitution of counterurbanization: evidence from England in the 1980s, *Sociologia Ruralis*, 34, 164–189.

Halfacree, K. (1995) Talking about rurality: social representations of the rural as expressed by residents of six English parishes, *Journal of Rural Studies*, 11, 1–20.

Halfacree, K. (1996) Out of place in the countryside: travellers and the 'rural idyll', *Antipode*, 29, 42–71.

Hall, A. and Mogyorody, V. (2001) Organic farmers in Ontario: an examination of the conventionalization argument, *Sociologia Ruralis*, 41, 399–422.

Hall, P. (2002) *Urban and Regional Planning*, 2nd edn. London and New York: Routledge.

Hall, R.J. (1987) Impact of pesticides on bird populations, in G.J. Marco, R.M. Hollingworth and W. Durham (eds), *Silent Spring Revisited*. Washington, DC: American Chemical Society. pp. 85–111.

Halliday, J. and Little, J. (2001) Amongst women: exploring the reality of rural childcare, *Sociologia Ruralis*, 41, 423–437.

Halseth, G. and Rosenberg, M. (1995) Complexity in the rural Canadian housing landscape, *The Canadian Geographer*, 39, 336–352.

Hanbury-Tenison, R. (1997) Life in the Countryside, *Geographical Magazine,* November, pp. 88–95 (sponsored feature).

Hannan, D.F. (1970) *Rural Exodus*. London: Chapman.

Hanson, S. (1992) Geography and feminism: worlds in collision?, *Annals of the Association of American Geographers*, 82, 569–586.

Harper, S. (1989) The British rural community: an overview of perspectives, *Journal of Rural Studies*, 5, 161–184.

Harper, S. (1991) People moving to the countryside, in T. Champion and C. Watkins (eds), *People in the Countryside*. London: Paul Chapman. pp. 22–37.

Harris, T. (1995) Sharecropping, in Davidson, C.N. and Wagner-Martin, L. (eds), *The Oxford Companion to Women's Writing in the United States*. New York: Oxford University Press.

Harrison, A. (2001) *Climate Change and Agriculture in NSW: The Challenge for Rural Communities*. Sydney, NSW: Nature Conservation Council of New South Wales.

Hart, J.F. (1975) *The Look of the Land*. Englewood Cliffs, CA: Prentice Hall.

Hart, J.F. (1998) *The Rural Landscape*. Baltimore, MD and London: Johns Hopkins University Press.

Harvey, G. (1998) *The Killing of the Countryside*. London: Vintage.

Heimlich, R.E. and Anderson, W.D. (2001) *Development at the Urban Fringe and Beyond*. ERS Agricultural Economic Report No. 803. Washington, DC: USDA Economic Research Service.

Held, D., McGrew, A., Goldblatt, D. and Perraton, J. (1999) *Global Transformations: Politics, Economics and Culture*. Cambridge, UK: Polity Press.

Henderson, G. (1998) *California and the Fictions of Capital*. New York: Oxford University Press.

Hendrickson, M. and Heffernan, W.D. (2002) Opening spaces through relocalization: locating potential resistance in the weaknesses of the global food system, *Sociologia Ruralis*, 42, 347–369.

Herbert-Cheshire, L. (2000) Contemporary strategies for rural community development in Australia: a governmentality perspective, *Journal of Rural Studies*, 16, 203–215.

Herbert-Cheshire, L. (2003) Translating policy: power and action in Australia's country towns, *Sociologia Ruralis*, 43, 454–473.

Hilchey, D. (1993) *Agritourism in New York State: Opportunities and Challenges in Farm-based Recreation and Hospitality*. Ithaca, NY: Department of Rural Sociology, Cornell University.

Hinrichs, C.C. (1996) Consuming images: making and marketing Vermont as a distinctive rural place, in E.M. DuPuis and P. Vandergeest (eds), *Creating the Countryside*. Philadelphia: Temple University Press. pp. 259–278.

Hodge, I. (1996) On penguins on icebergs: The Rural White Paper and the assumption of rural policy, *Journal of Rural Studies*, 12, 331–337.

Hodge, I., Dunn, J., Monk, S. and Fitzgerald, M. (2002) Barriers to participation in residual rural labour markets, *Work, Employment and Society*, 16, 457–476.

Hoggart, K. (1990) Let's do away with rural, *Journal of Rural Studies*, 6, 245–257.

Hoggart, K. (1995) The changing geography of council house sales in England and Wales, 1978–1990, *Tijdschrift voor Economische en Sociale Geografie*, 86, 137–149.

Hoggart, K. and Buller, H. (1995) Geographical differences in British property acquisitions in rural France, *Geographical Journal*, 161, 69–78.

Hoggart, K. and Mendoza, C. (1999) African immigrant workers in Spanish agriculture, *Sociologia Ruralis*, 39, 538–562.

Hoggart, K. and Paniagua, A. (2001) What rural restructuring?, *Journal of Rural Studies*, 17, 41–62.

Holloway, L. and Ilbery, B. (1997) Global warming and navy beans: decision making by farmers and food companies in the UK, *Journal of Rural Studies*, 13, 343–355.

Holloway, L. and Kneafsey, M. (2000) Reading the spaces of the farmer's market: a case study from the United Kingdom, *Sociologia Ruralis*, 40, 285–299.

Hopkins, J. (1998) Signs of the post-rural: marketing myths of a symbolic countryside, *Geografiska Annaler*, 80B, 65–81.

Horton, J. (2003) Different genres, different visions? The changing countryside in postwar British children's literature, in P. Cloke (ed.), *Country Visions*. Harlow, UK: Pearson. pp. 73–92.

Howkins, A. (1986) The discovery of rural England, in R. Colls and P. Dodd (eds), *Englishness: Politics and Culture, 1880–1920*. London: Croom Helm. pp. 62–88.

Hugo, G. (1994) The turnaround in Australia: some first observations from the 1991 Census, *Australian Geographer*, 25, 1–17.

Hugo, G. and Bell, M. (1998) The hypothesis of welfare-led migration to rural areas: the Australian case, in P. Boyle and K. Halfacree (eds), *Migration into Rural Areas*. Chichester, UK: Wiley.

Humphries, S. and Hopwood, B. (2000) *Green and Pleasant Land*. London: Channel 4 Books/ Macmillan.

Hunter, K. and Riney-Kehrberg, P. (2002) Rural daughters in Australia, New Zealand and the United States: an historical perspective, *Journal of Rural Studies*, 18, 135–144.

Huws, U., Korte, W.B. and Robinson, S. (1990) *Telework: Towards the Elusive Office*. Chichester, UK: Wiley.

Ilbery, B. (1985) *Agricultural Geography*. Oxford: Oxford University Press.

Ilbery, B. (1992) State-assisted farm diversification in the United Kingdom, in R. Bowler, C.R. Bryant and M.D. Nellis (eds), *Contemporary Rural Systems in Transition, Volume 1: Agriculture and Environment*. Wallingford, UK: CAB International. pp. 100–116.

Ilbery, B. and Bowler, I. (1998) From agricultural productivism to post-productivism, in B. Ilbery (ed.), *The Geography of Rural Change*. Harlow: Addison Wesley Longman. pp. 57–84.

INSEE (1993) *Les Agriculteurs*. Paris: INSEE.

INSEE (1995) *La Population de la France*. Paris: INSEE.

INSEE (1998) *Les Campagnes et leurs villes*. Paris: INSEE.

IPCC (Intergovernmental Panel on Climate Change) (2001) *Climate Change 2001: Impacts, Adaption and Vulnerability*. Contribution of Working Group II to the Third Assessment

Report of the Intergovernmental Panel on Climate Change. Cambridge, UK, and New York: Cambridge University Press.

Isserman, A.M. (2000) Creating new economic opportunities: the competitive advantages of rural America in the next century, in Center for the Study of Rural America (eds), *Beyond Agriculture: New Policies for Rural America*. Kansas City: The Federal Reserve Bank of Kansas City. pp. 123–142.

Jessop, B. (1995) The regulation approach, governance and post-Fordism: alternative perspectives on economic and political change?, *Economy and Society*, 24, 307–333.

Johnsen, S. (2003) Contingency revealed: New Zealand farmers' experiences of agricultural restructuring, *Sociologia Ruralis*, 43, 128–153.

Johnson, T.G. (2000) The rural economy in a new century, in Center for the Study of Rural America (eds), *Beyond Agriculture: New Policies for Rural America*. Kansas City: The Federal Reserve Bank of Kansas City. pp. 7–20.

Jones, G.E. (1973) *Rural Life*. London: Longman.

Jones, J. (2002) The cultural symbolism of disordered and deviant behaviour: young people's experiences in a Welsh rural market town, *Journal of Rural Studies*, 18, 213–218.

Jones, N. (1993) *Living in Rural Wales*. Llandysul, UK: Gomer.

Jones, O. (1995) Lay discourses of the rural: development and implications for rural studies, *Journal of Rural Studies*, 11, 35–49.

Jones, O. (1997) Little figures, big shadows: country childhood stories, in P. Cloke and J. Little (eds), *Contested Countryside Cultures*. London and New York: Routledge. pp. 158–179.

Jones, O. (2000) Melting geography: purity, disorder, childhood and space, in S.L. Holloway and G. Valentine (eds), *Children's Geographies: Playing, Living, Learning*. London and New York: Routledge. pp. 29–47.

Jones, O. and Little, J. (2000) Rural challenge(s): partnership and new rural governance, *Journal of Rural Studies*, 16, 171–183.

Jones, R. and Tonts, M. (2003) Transition and diversity in rural housing provision: the case of Narrogin, Western Australia, *Australian Geographer*, 34, 47–59.

Jones, R.E., Fly, J.M., Talley, J. and Cordell, H.K. (2003) Green migration into rural America: the new frontier of environmentalism?, *Society and Natural Resources*, 16, 221–238.

Juntti, M. and Potter, C. (2002) Interpreting and reinterpreting agri-environmental policy: communication, goals and knowledge in the implementation process, *Sociologia Ruralis*, 42, 215–232.

Kelly, R. and Shortall, S. (2002) 'Farmer's wives': women who are off-farm breadwinners and the implications for on-farm gender relations, *Journal of Sociology*, 38, 327–343.

Kennedy, J.C. (1997) At the crossroads: Newfoundland and Labrador communities in a changing international context, *Canadian Review of Sociology and Anthropology*, 34, 297–317.

Kenyon, P. and Black, A. (eds) (2001) *Small Town Renewal: Overview and Case studies*. Barton, Australia: Rural Industries Research and Development Corporation.

Kimmel, M. and Ferber, A.L. (2000) 'White men are this nation': right-wing militias and the restoration of rural American masculinity, *Rural Sociology*, 65, 582–604.

Kinsman, P. (1995) Landscape, race and national identity: the photography of Ingrid Pollard, *Area*, 27, 300–310.

Kneafsey, M., Ilbery, B. and Jenkins, T. (2001) Exploring the dimensions of culture economies in rural West Wales, *Sociologia Ruralis*, 41, 296–310.

Kontuly, T. (1998) Contrasting the counter-urbanisation experience in European nations, in P. Boyle and K. Halfacree (eds), *Migration Into Rural Areas*. Chichester, UK: Wiley. pp. 61–78.

Kramer, J.L. (1995) Bachelor farmers and spinsters: gay and lesbian identities and communities in rural North Dakota, in D. Bell and G. Valentine (eds), *Mapping Desire: Geographies of Sexualities*. London and New York: Routledge. pp. 200–213.

LaDuke, W. (2002) Klamath water, Klamath life, *Earth Island Journal*, 17.

Lapping, M.B., Daniels, T.L. and Keller, J.W. (1989) *Rural Planning and Development in the United States*. New York: Guilford.

Lash, S. and Urry, J. (1987) *The End of Organized Capitalism*. Cambridge, UK: Polity Press.

Lawrence, G. (1990) Agricultural restructuring and rural social change in Australia, in T. Marsden, P. Lowe and S. Whatmore (eds), *Rural Restructuring, Global Processes and their Responses*. London: David Fulton. pp. 101–128.

Lawrence, M. (1995) Rural homelessness: a geography without a geography, *Journal of Rural Studies*, 11, 297–307.

Laws, G. and Harper, S. (1992) Rural ageing: perspectives from the US and UK, in I.R. Bowler, C.R. Bryant and M.D. Nellis (eds), *Contemporary Rural Systems in Transition: Volume 2, Economy and Society*. Wallingford, UK: CAB International. pp. 96–109.

Leach, B. (1999) Transforming rural livelihoods: gender, work and restructuring in three Ontario communities, in S. Neysmith (ed.), *Restructuring Caring Labour.* New York: Oxford University Press.

Le Heron, R. (1993) *Globalized Agriculture.* London: Pergamon Press.

Le Heron, R. and Roche, M. (1999) Rapid reregulation, agricultural restructuring and the reimaging of agriculture in New Zealand, *Rural Sociology*, 64, 203–218.

Lehning, J. (1995) *Peasant and French: Cultural Contact in Rural France during the Nineteenth Century.* Cambridge: Cambridge University Press.

Lewis, G. (1998) Rural migration and demographic change, in B. Ilbery (ed.), *The Geography of Rural Change.* Harlow: Addison Wesley Longman. pp. 131–160.

Lichfield, J. (1998) The death of the French countryside, *Independent on Sunday Review*, 8 March, 12–15.

Liepins, R. (2000a) New energies for an old idea: reworking approaches to 'community' in contemporary rural studies, *Journal of Rural Studies*, 16, 23–35.

Liepins, R. (2000b) Exploring rurality through 'community': discourses, practices and spaces shaping Australian and New Zealand rural 'communities', *Journal of Rural Studies*, 16, 325–341.

Liepins, R. (2000c) Making men: the construction and representation of agriculture-based masculinities in Australia and New Zealand, *Rural Sociology*, 65, 605–620.

Little, J. (1991) Women in the rural labour market: a policy evaluation, in T. Champion and C. Watkins (eds), *People in the Countryside.* London: Paul Chapman. pp. 96–107.

Little, J. (1997) Employment marginality and women's self-identity, in P. Cloke and J. Little (eds), *Contested Countryside Cultures.* London and New York: Routledge. pp. 138–157.

Little, J. (1999) Otherness, representation and the cultural construction of rurality, *Progress in Human Geography*, 23, 437–442.

Little, J. (2002) *Gender and Rural Geography.* Harlow, UK: Prentice Hall.

Little, J. (2003) Riding the rural love train: heterosexuality and the rural community, *Sociologia Ruralis*, 43, 401–417.

Little, J. and Austin, P. (1996) Women and the rural idyll, *Journal of Rural Studies*, 12, 101–111.

Little, J. and Jones, O. (2000) Masculinity, gender and rural policy, *Rural Sociology*, 65, 621–639.

Little, J. and Leyshon, M. (2003) Embodied rural geographies: developing research agendas, *Progress in Human Geography*, 27, 257–272.

Little, J. and Panelli, R. (2003) Gender research in rural geography, *Gender, Place and Culture*, 10, 281–289.

Littlejohn, J. (1964) *Westrigg: The Sociology of a Cheviot Parish.* London: Routledge and Kegan Paul.

Lloyds TSB Agriculture (2001) *Focus on Farming: Survey Results 2001.* London: Lloyds TSB.

Lockie, S. (1999a) The state, rural environments and globalisation: 'action at a distance' via the Australian Landcare program, *Environment and Planning A*, 31, 597–611.

Lockie, S. (1999b) Community movements and corporate images: Landcare in Australia, *Rural Sociology*, 64, 219–233.

Looker, E.D. (1997) Rural–urban differences in youth transition to adulthood, in R.D. Bollman and J.M. Bryden, *Rural Employment: An International Perspective.* Wallingford, UK: CAB International. pp. 85–98.

Lowe, P., Buller, H. and Ward, N. (2002) Setting the next agenda? British and French approaches to the second pillar of the Common Agricultural Policy, *Journal of Rural Studies*, 18, 1–17.

Lowe, P., Clark, J., Seymour, S. and Ward, N. (1997) *Moralizing the Environment: Countryside Change, Farming and Pollution.* London: UCL Press.

Lowe, P., Cox, G., MacEwen, M., O'Riordan, T. and Winter, M. (1986) *Countryside Conflicts: The Politics of Farming, Forestry and Conservation.* London: Gower.

Lowe, R. and Shaw, W. (1993) *Travellers: Voices of the New Age Nomads.* London: Fourth Estate.

MacEwen, A. and MacEwen, M. (1982) *National Parks: Conservation or Cosmetics?* London: Allen & Unwin.

MacLaughlin, J. (1999) Nation-building, social closure and anti-traveller racism in Ireland, *Sociology*, 33, 129–151.

Macnaghten, P. and Urry, J. (1998) *Contested Natures.* London and Thousand Oaks, CA: Sage.

MAFF/DETR (2000) *Our Countryside: the future. A fair deal for rural England.* London: The Stationery Office.

Malik, S. (1992) Colours of the countryside – a whiter shade of pale, *Ecos*, 13, 33–40.

Manning, R. (1997) *Grassland: The History, Biology, Politics and Promise of the American Prairie.* New York: Penguin Books.

Markusen, A. (1985) *Profit Cycles, Oligopoly and Regional Development.* Cambridge, MA: MIT Press.

Marsden, T., Milbourne, P., Kitchen, L. and Bishop, K. (2003) Communities in nature: the construction and understanding of forest natures, *Sociologia Ruralis*, 43, 238–256.

Marsden, T., Murdoch, J., Lowe, P., Munton, R. and Flynn, A. (1993) *Constructing the Countryside.* London: UCL Press.

Marsh, D. and Rhodes, R. (eds) (1992) *Policy Networks in British Governance*. Oxford, UK: Oxford University Press.

Marshall, R. (2000) Rural policy in a new century, in Center for the Study of Rural America (eds), *Beyond Agriculture: New Policies for Rural America*. Kansas City: The Federal Reserve Bank of Kansas City. pp. 25–46.

Martin, R.C. (1956) *TVA: The First Twenty Years*. Tuscaloosa, AL: University of Alabama Press and Knoxville, TN: University of Tennessee Press.

Massey, D. (1984) *Spatial Divisions of Labour*. London: Macmillan.

Massey, D. (1994) *Space, Place and Gender*. Cambridge, UK: Polity Press.

Mather, A. (1998) The changing role of forests, in B. Ilbery (ed.), *The Geography of Rural Change*. Harlow, UK: Longman. pp. 106–127.

Matless, D. (1994) Doing the English village, 1945–90: an essay in imaginative geography, in P. Cloke, M. Doel, D. Matless, M. Phillips and N. Thrift, *Writing the Rural*. London: Paul Chapman. pp. 7–88.

Matthews, H., Taylor, M., Sherwood, K., Tucker, F. and Limb, M. (2000) Growing up in the countryside: children and the rural idyll, *Journal of Rural Studies*, 16, 141–153.

Mattson, G.A. (1997) Redefining the American small town: community governance, *Journal of Rural Studies*, 13, 121–130.

McCormick, J. (1988) America's third world, *Newsweek*, 8 August, pp. 20–24.

McCullagh, C. (1999) Rural crime in the Republic of Ireland, in G. Dingwall and S.R. Moody (eds), *Crime and Conflict in the Countryside*. Cardiff, UK: University of Wales Press. pp. 29–44.

McDonagh, J. (2001) *Renegotiating Rural Development in Ireland*. Aldershot, UK: Ashgate.

McKay, G. (1996) *Senseless Acts of Beauty*. London and New York: Verso.

McManus, P. (2002) The potential and limits of progressive neopluralism: a comparative study of forest politics in Coastal British Columbia and South East New South Wales during the 1990s, *Environment and Planning A*, 34, 845–865.

Meyer, F. and Baker, R. (1982) Problems of developing crime policy for rural areas, in W. Browne and D. Hadwinger (eds), *Rural Policy Problems: Changing Dimensions*. Lexington, KY: Lexington Books. pp. 171–179.

Michelsen, J. (2001) Organic farming in a regulatory perspective: the Danish case, *Sociologia Ruralis*, 41, 62–84.

Middleton, A. (1986) Marking boundaries: men's space and women's space in a Yorkshire village, in T. Bradley, P. Lowe and S. Wright (eds),

Deprivation and Welfare in Rural Areas. Norwich, UK: Geo Books.

Miele, M. and Murdoch, J. (2002) The practical aesthetics of traditional cuisines: slow food in Tuscany, *Sociologia Ruralis*, 42, 312–328.

Milbourne, P. (1997a) Introduction: challenging the rural: representation, power and identity in the British countryside, in P. Milbourne (ed.), *Revealing Rural 'Others': Representation, Power and Identity in the British Countryside*. London: Pinter. pp. 1–12.

Milbourne, P. (1997b) Hidden from view: poverty and marginalization in rural Britain, in P. Milbourne (ed.), *Revealing Rural 'Others': Representation, Power and Identity in the British Countryside*. London: Pinter. pp. 89–116.

Milbourne, P. (1998) Local responses to central state restructuring of social housing provision in rural areas, *Journal of Rural Studies*, 14, 167–184.

Milbourne, P. (2003a) The complexities of hunting in rural England and Wales, *Sociologia Ruralis*, 43, 289–308.

Milbourne, P. (2003b) Hunting ruralities: nature, society and culture in 'hunt countries' of England and Wales, *Journal of Rural Studies*, 19, 157–171.

Milbourne, P. (2003c) Nature-Society-Rurality: Making Critical Connections, *Sociologia Ruralis*, 43, 193–195.

Mingay, G. (ed.) (1989) *The Unquiet Countryside*. London: Routledge.

Mitchell, C.J.A. (2004) Making sense of counter-urbanization, *Journal of Rural Studies*, 20, 15–34.

Mitchell, D. (1996) *The Lie of the Land: Migrant Workers and the California Landscape*. Minneapolis, MN: University of Minnesota Press.

Monk, S., Dunn, J., Fitzgerald, M. and Hodge, I. (1999) *Finding Work in Rural Areas*. York, UK: York Publishing Services.

Mordue, T. (1999) Heartbeat country: conflicting values, coinciding visions, *Environment and Planning A*, 31, 629–646.

Mormont, M. (1987) The emergence of rural struggles and their ideological effects, *International Journal of Urban and Regional Research*, 7, 559–575.

Mormont, M. (1990) Who is rural? Or, How to be rural: Towards a sociology of the rural, in T. Marsden, P. Lowe and S. Whatmore (eds), *Rural Restructuring: Global Processes and Their Responses*. London: David Fulton. pp. 21–44.

Morris, C. and Potter, C. (1995) Recruiting the new conservationists: farmers' adoption of agri-environmental schemes in the UK, *Journal of Rural Studies*, 11, 51–63.

Morris, C. and Evans, N. (2001) Cheesemakers are always women: gendered representations of farm life in the agricultural press, *Gender, Place and Culture*, 8, 375–390.

Morris, C. and Evans, N. (2004) Agricultural turns, geographical turns: retrospect and prospect, *Journal of Rural Studies*, 20, 95–111.

Moseley, M. (1995) Policy and practice: the environmental component of LEADER, *Journal of Environmental Planning and Management*, 38, 245–252.

Moseley, M. (2003) *Rural Development*. London: Sage.

Murdoch, J. (1997) The shifting territory of government: some insights from the Rural White Paper, *Area*, 29, 109–118.

Murdoch, J. (2003) Co-constructing the countryside: hybrid networks and the extensive self, in P. Cloke (ed.), *Country Visions*. London: Pearson. pp. 263–282.

Murdoch, J. and Abram, S. (2002) *Rationalities of Planning*. Aldershot: Ashgate.

Murdoch, J. and Lowe, P. (2003) The preservation paradox: modernism, environmentalism and the politics of spatial division, *Transactions of the Institute of British Geographers*, 28, 318–332.

Murdoch, J. and Marsden, T. (1994) *Reconstituting Rurality*. London: UCL Press.

Murdoch, J. and Marsden, T. (1995) The spatialization of politics: local and national actor-spaces in environmental conflict, *Transactions of the Institute of British Geographers*, 20, 368–380.

Nash, R. (1980) *Schooling in Rural Societies*. London and New York: Methuen.

Naylor, E.L. (1994) Unionism, peasant protest and the reform of French agriculture, *Journal of Rural Studies*, 10, 263–273.

NCES (National Center for Education Statistics) (1997) *Statistical Analysis Report: Characteristics of Small and Rural School Districts*. Washington, DC: NCES.

Nelson, M.K. (1999) Between paid and unpaid work: gender patterns in supplemental economic activities among white, rural families, *Gender and Society*, 13, 518–539.

Newby, H. (1977) *The Deferential Worker*. London: Allen Lane.

Newby, H., Bell, C., Rose, D. and Saunders, P. (1978) *Property, Paternalism and Power: Class and Control in Rural England*. London: Hutchinson.

NFU (National Farmers' Union) (2002) *Farmers' Markets: A Business Survey*. London: NFU.

Ni Laoire, C. (2001) A matter of life and death? Men, masculinities and staying 'behind' in rural Ireland, *Sociologia Ruralis*, 41, 220–236.

Nord, M. (1999) Rural poverty remains unobserved, *Rural Conditions and Trends*, 8, 18–21.

Norris, D.A. and Johal, K. (1992) Social indicators from the General Social Survey: some urban–rural differences, in R.D. Bollman (ed.), *Rural and Small Town Canada*. Toronto: Thompson Educational Publishing. pp. 357–368.

North, D. (1998) Rural industrialization, in B. Ilbery (ed.), *The Geography of Rural Change*. Harlow: Addison Wesley Longman. pp. 161–188.

ODPM (2002) *A Review of Urban and Rural Area Definitions: Project Report*. London: Office of the Deputy Prime Minister.

O'Hagan, A. (2001) *The End of British Farming*. London: Profile Books.

Okihoro, N.P. (1997) *Mounties, Moose and Moonshine*. Toronto: University of Toronto Press.

Oliveira Baptista, F. (1995) Agriculture, rural society and the land question in Portugal, *Sociologia Ruralis*, 35, 309–325.

Pahl, R.E. (1968) The rural–urban continuum, in R.E. Pahl (ed.), *Readings in Urban Sociology*. Oxford, UK: Pergamon Press.

Panelli, R., Nairn, K. and McCormack, J. (2002) 'We make our own fun': reading the politics of youth with(in) community, *Sociologia Ruralis*, 42, 106–130.

Parker, G. (1999) Rights, symbolic violence and the micro-politics of the rural: the case of the Parish Paths Partnership Scheme, *Environment and Planning A*, 31, 1207–1222.

Parker, G. (2002) *Citizenships, Contingency and the Countryside: Rights, Culture, Land and the Environment*. London: Routledge.

Paxman, J. (1998) *The English: A Portrait of a People*. London: Michael Joseph.

Petersen, D. (2000) *Heartsblood: Hunting, Spirituality and Wildness in America*. Washington, DC: Island Press.

Phillips, D. and Williams, A. (1984) *Rural Britain: A Social Geography*. Oxford, UK: Blackwell.

Phillips, M. (1993) Rural gentrification and the process of class colonisation, *Journal of Rural Studies*, 9, 123–140.

Phillips, M. (2002) The production, symbolization and socialization of gentrification: impressions from two Berkshire villages, *Transactions of the Institute of British Geographers*, 27, 282–308.

Philo, C. (1992) Neglected rural geographies: a review, *Journal of Rural Studies*, 8, 193–207.

Philo, C. and Parr, H. (2003) Rural madness: a geographical reading and critique of the rural mental health literature, *Journal of Rural Studies*, 19, 259–281.

Pieterse, J. (1996) Globalisation as hybridization, in M. Featherstone, S. Lash, and R. Robertson (eds), *Global Modernities*. London: Sage. pp. 45–68.

Pirog, R., Van Pelt, T., Enshayan, K. and Cook, E. (2001) *Food, Fuel and Freeways: An Iowa Perspective on How Far Food Travels, Fuel*

Usage and Greenhouse Gas Emissions. Ames, IA: Leopold Center for Sustainable Agriculture.

Popper, D.E. and Popper, F. (1987) The Great Plains: from dust to dust, *Planning*, 53, 12–18.

Popper, D.E. and Popper, F. (1999) The Buffalo Commons: metaphor as method, *The Geographical Review*, 89, 491–510.

Porter, K. (1989) *Poverty in Rural America: A National Overview*. Washington, DC: Center on Budget and Policy Priorities.

Potter, C. (1998) Conserving nature: agri-environmental policy development and change, in B. Ilbery (ed.), *The Geography of Rural Change*. Harlow: Addison Wesley Longman. pp. 85–106.

Price, C.C. and Harris, J.M. (2000) *Increasing Food Recovery From Farmers' Markets: A Preliminary Analysis*. Report FANRR-4. Washington, DC: USDA Economic Research Service.

Radin, B., Agranoff, R., Bowman, A., Buntz, G., Ott, J.S., Romzek, B. and Wilson, R. (1996) *New Governance for Rural America*. Lawrence, KS: University of Kansas Press.

Ramet, S. (1996) Nationalism and the 'idiocy' of the countryside: the case of Serbia, *Ethnic and Racial Studies*, 19, 70–86.

Ray, C. (1997) Towards a theory of the dialectic of rural development, *Sociologia Ruralis*, 37, 345–362.

Ray, C. (2000) The EU LEADER programme: rural development laboratory, *Sociologia Ruralis*, 40, 163–171.

Rees, A.D. (1950) *Life in a Welsh Countryside*. Cardiff: University of Wales Press.

Reimer, B., Ricard, I. and Shaver, F.M. (1992) Rural deprivation: a preliminary analysis of census and tax family data, in R.D. Bollman (ed.), *Rural and Small Town Canada*. Toronto: Thompson Educational Publishing. pp. 319–336.

Reissman, L. (1964) *The Urban Process*. New York: Free Press.

Rhodes, R.A.W. (1996) The new governance: governing without government, *Political Studies*, 44, 652–667.

Ribchester, C. and Edwards, B. (1999) The centre and the local: policy and practice in rural education provision, *Journal of Rural Studies*, 15, 49–63.

Richardson, J. (2000) *Partnerships in Communities: Reweaving the Fabric of Rural America*. Washington, DC: Island Press.

Robinson, G. (1990) *Conflict and Change in the Countryside*. Chichester, UK: Wiley.

Robinson, G. (1992) The provision of rural housing: policies in the United Kingdom, in I.R. Bowler, C.R. Bryant and M.D. Nellis (eds), *Contemporary Rural Systems in Transition. Volume 2: Economy and Society*. Wallingford, UK: CAB International. pp. 110–126.

Rogers, A. (1987) Issues in English rural housing: an assessment and prospect, in D. MacGregor, D. Robertson and M. Shucksmith (eds), *Rural Housing in Scotland: Recent Research and Policy*. Aberdeen: Aberdeen University Press.

Rome, A. (2001) *The Bulldozer in the Countryside*. Cambridge, UK and New York: Cambridge University Press.

Rosenzweig, C. and Hillel, D. (1998) *Climate Change and the Global Harvest*. Oxford, UK and New York: Oxford University Press.

Rowles, G. (1983) Place and personal identity in old age: observations from Appalachia. *Journal of Environmental Psychology*, 3, 299–313.

Rowles, G. (1988) What's rural about rural aging? An Appalachian perspective, *Journal of Rural Studies*, 4, 115–124.

Rugg, J. and Jones, A. (1999) *Getting a Job, Finding a Home: Rural Youth Transitions*. Bristol, UK: Policy Press.

Runte, A. (1997) *National Parks: The American Experience*. Lincoln, NE: University of Nebraska Press.

Rural Policy Research Institute (2003) The rural in numbers, available at: www.rupri.org.

Sachs, C. (1983) *Invisible Farmers: Women's Work in Agricultural Production*. Totowa, NJ: Rhinehart Allenheld.

Sachs, C. (1991) Women's work and food: a comparative perspective, *Journal of Rural Studies*, 7, 49–56.

Sachs, C. (1994) Rural women's environmental activism in the USA, in S. Whatmore, T. Marsden and P. Lowe (eds), *Gender and Rurality*. London: David Fulton. pp. 117–135.

Saugeres, L. (2002) Of tractors and men: masculinity, technology and power in a French farming community, *Sociologia Ruralis*, 42, 143–159.

Saville, J. (1957) *Rural Depopulation in England and Wales, 1851–1951*. London: Routledge & Kegan Paul.

Schindegger, F. and Krajasits, C. (1997) Commuting: its importance for rural employment analysis, in R.D. Bollman and J.M. Bryden, *Rural Employment: An International Perspective*. Wallingford, UK: CAB International. pp. 164–176.

Schools Health Education Unit (1998) *Young People and Illegal Drugs in 1998*. Exeter, UK: Schools Health Education Unit.

Selby, E.F., Dixon, D.P. and Hapke, H.P. (2001) A woman's place in the crab processing industry of Eastern Carolina, *Gender, Place and Culture*, 8, 229–253.

Sellars, R.W. (1997) *Preserving Nature in the National Parks*. New Haven, CT: Yale University Press.

Senior, M., Williams, H. and Higgs, G. (2000) Urban–rural mortality differentials: controlling for material deprivation, *Social Science and Medicine*, 51, 289–305.

Serow, W. (1991) Recent trends and future prospects for urban–rural migration in Europe, *Sociologia Ruralis*, 31, 269–280.

Sharpe, T. (1946) *The Anatomy of a Village*. Harmondsworth, UK: Penguin.

Shaw, G. and Williams, A.M. (2002) *Critical Issues in Tourism: A Geographical Perspective*. Oxford, UK: Blackwell.

Sheppard, B.O. (1999) Black farmers and institutionalized racism, *The Black Business Journal*, available online at www.bbjonline.com

Shields, R. (1991) *Places on the Margin: Alternative Geographies of Modernity*. London: Routledge.

Short, J.R. (1991) *Imagined Country*. London: Routledge.

Sibley, D. (1997) Endangering the sacred: nomads, youth cultures and the English countryside, in P. Cloke and J. Little (eds), *Contested Countryside Cultures*. London and New York: Routledge. pp. 218–231.

Silvasti, T. (2003) Bending borders of gendered labour division on farms: the case of Finland, *Sociologia Ruralis*, 43, 154–166.

Simon, S. (2002) Iowa's tough stand against runoff is gaining support, *Los Angeles Times*, 19 March, p. A8.

Smith, A. (1998) The politics of economic development in a French rural area, in N. Walzer and B.D. Jacobs (eds), *Public– Private Partnership for Local Economic Development*. Westport, CT and London: Praeger. pp. 227–241.

Smith, F. and Barker, J. (2001) Commodifying the countryside: the impact of out-of-school care on rural landscapes of children's play, *Area*, 33, 169–176.

Smith, M.J. (1989) Changing policy agendas and policy communities: agricultural issues in the 1930s and 1980s, *Public Administration*, 67, 149–165.

Smith, M.J. (1992) The agricultural policy community: maintaining a closed relationship, in D. Marsh and R. Rhodes (eds), *Policy Networks in British Governance*. Oxford, UK: Oxford University Press. pp. 27–50.

Smith, M.J. (1993) *Pressure, Power and Policy*. Hemel Hempstead, UK: Harvester Wheatsheaf.

Snipp, C.M. (1996) Understanding race and ethnicity in rural America, *Rural Sociology*, 61, 125–142.

Snipp, C.M. and Sandefur, G.D. (1988) Earnings of American Indians and Alaskan Natives: the effects of residence and migration, *Social Forces*, 66, 994–1008.

Sobels, J., Curtis, A. and Lockie, S. (2001) The role of Landcare group networks in rural Australia: exploring the contribution of social capital, *Journal of Rural Studies*, 17, 265–276.

Sokolow, A.D. and Zurbrugg, A. (2003) *A National View of Agricultural Easement Programs: Profiles and Maps – Report 1*. Washington, DC: American Farmland Trust.

Sorokin, P. and Zimmerman, C. (1929) *Principles of Rural–Urban Sociology*. New York, NY: Henry Holt.

Soumagne, J. (1995) Deprise commerciale dans les zones rurales profondes et nouvelles polarisations, in R. Béteille and S. Montagné-Villette (eds), *Le 'Rural Profond' Français*. Paris: SEDES. pp. 31–44.

Spain, D. (1993) Been-heres versus come-heres: negotiating conflicting community identities. *Journal of the American Planning Association*, 59, 156–171.

Spencer, D. (1997) Counterurbanisation and rural depopulation revisited: landowners, planners and the rural development process, *Journal of Rural Studies*, 13, 75–92.

Squire, S.J. (1992) Ways of seeing, ways of being: literature, place and tourism in L.M. Montgomery's Prince Edward Island, in P. Simpson-Housley and G. Norcliffe (eds), *A Few Acres of Snow: Literary and Artistic Images of Canada*. Toronto: Dundurn Press. pp. 137–147.

Stabler, J. and Rounds, R.C. (1997) Commuting and rural employment on the Canadian Prairies, in R.D. Bollman and J.M. Bryden (eds), *Rural Employment: An International Perspective*. Wallingford, UK: CAB International. pp. 193–204.

Stacey, M. (1960) *Tradition and Change: a Study of Banbury*. Oxford: Oxford University Press.

Stebbing, S. (1984) Women's roles and rural society, in T. Bradley and P. Lowe (eds), *Locality and Rurality: Economy and Society in Rural Regions*. Norwich, UK: Geo Books.

Stenson, K. and Watt, P. (1999) Crime, risk and governance in a southern English village, in G. Dingwall and S.R. Moody (eds), *Crime and Conflict in the Countryside*. Cardiff, UK: University of Wales Press. pp. 76–93.

Stock, C.M. (1996) *Rural Radicals: Righteous Rage in the American Grain*. Ithaca, NY: Cornell University Press.

Stoker, G. (ed.) (2000) *The New Politics of British Local Governance*. London: Macmillan.

Storey, D. (1999) Issues of integration, participation and empowerment in rural development: the case of LEADER in the Republic of Ireland, *Journal of Rural Studies*, 15, 307–315.

Storey, P. and Brannen, J. (2000) *Young People and Transport in Rural Areas*. Leicester, UK: Youth Work Press/Joseph Rowntree Foundation.

Storper, M. and Walker, R. (1984) The spatial division of labour: labour and the location of industries, in L. Sawyers and W. Tabb (eds), *Sunbelt/Snowbelt: Urban Development and Regional Restructuring*. New York: Oxford University Press.

Strathern, M. (1981) *Kinship at the Core*. Cambridge: Cambridge University Press.

Sumner, D.A. (2003) Implications of the US Farm Bill of 2002 for agricultural trade and trade negotiations, *Australian Journal of Agricultural and Resource Economics*, 46, 99–122.

Swanson, L. (1993) Agro-environmentalism: the political economy of soil erosion in the USA, in S. Harper (ed.), *The Greening of Rural Policy*. London: Belhaven. pp. 99–118.

Swanson, L.E. (2001) Rural policy and direct local participation: democracy, inclusiveness, collective agency and locality-based policy, *Rural Sociology*, 66, 1–21.

Swarbrooke, J., Beard, C., Leckie, S. and Pomfret, G. (2003) *Adventure Tourism*. Oxford, UK and Boston, MA: Butterworth– Heinemann.

Thomson, M.L. and Mitchell, C.J.A. (1998) Residents of the urban field: a study of Wilmot township, Ontario, Canada, *Journal of Rural Studies*, 14, 185–202.

Thrift, N. (1987) Manufacturing rural geography, *Journal of Rural Studies*, 3, 77–81.

Thrift, N. (1989) Images of social change, in C. Hamnett, L. McDowell and P. Sarre (eds), *The Changing Social Structure*. London: Sage. pp. 12–42.

Tillberg Mattson, K. (2002) Children's (in)dependent mobility and parents' chauffeuring in the town and the countryside, *Tijdschrift voor Economische en Sociale Geografie*, 93, 443–453.

Tönnies, F. (1963) *Community and Society*. New York: Harper and Row.

Townsend, A. (1993) The urban–rural cycle in the Thatcher growth years, *Transactions of the Institute of British Geographers*, 18, 207–221.

Trant, M. and Brinkman, G. (1992) Products and competitiveness of rural Canada, in R.D. Bollman (ed.), *Rural and Small Town Canada*. Toronto: Thompson Educational Publishing. pp. 69–90.

Troughton, M., (1992) The restructuring of agriculture: the Canadian example, in I.R. Bowler, C.R. Bryant and M.D. Nellis (eds), *Contemporary Rural Systems in Transition, Volume 1:* *Agriculture and Environment*. Wallingford, UK: CAB International. pp. 29–42.

Tyler, P., Moore, B. and Rhodes, J. (1988) Geographical variation in industrial costs, *Scottish Journal of Political Economy*, 35, 22–50.

Urry, J. (1995) A middle-class countryside?, in T. Butler and M. Savage (eds), *Social Change and the Middle Classes*. London: UCL Press. pp. 205–219.

Urry, J. (2002) *The Tourist Gaze*, 2nd edn. London, UK and Thousand Oaks, CA: Sage.

USDA (United States Department of Agriculture) (1997) *America's Private Land: A Geography of Hope*. Washington, DC: USDA.

USDA (United States Department of Agriculture) (2000) *Agriculture Factbook 2000*. Washington, DC: United States Department of Agriculture.

Valentine, G. (1997a) A safe place to grow up? Parenting, perceptions of children's safety and the rural idyll, *Journal of Rural Studies*, 13, 137–148.

Valentine, G. (1997b) Making space: lesbian separatist communities in the United States, in P. Cloke and J. Little (eds), *Contested Countryside Cultures*. London and New York: Routledge. pp. 109–122.

Vias, A.C. (2004) Bigger stores, more stores, or no stores: paths of retail restructuring in rural America, *Journal of Rural Studies*, 20, 303–318.

Vining, D. and Kontuly, T. (1978) Population dispersal from major metropolitan regions: an international comparison, *International Regional Science Review*, 3, 49–73.

Vining, D. and Strauss, A. (1977) A demonstration that the current deconcentration of population in the United States is a clean break with the past, *Environment and Planning A*, 9, 751–758.

Vistnes, J. and Monheil, A. (1997) *Health Insurance Strategies of the Civilian Non-Institutionalised Population*. Medical Experts Panel Survey Research Report. Rockville, MD: Agency for Health Care Policy Research.

Von Meyer, H. (1997) Rural employment in OECD countries: structure and dynamics of regional labour markets, in R.D. Bollman and J.M. Bryden, *Rural Employment: An International Perspective*. Wallingford, UK: CAB International. pp. 3–21.

Wald, M.L. (1999) Tribe in Utah fights for nuclear waste dump, *New York Times*, 18 April, p. 16.

Walker, G. (1999) Contesting the countryside and changing social composition in the greater Toronto area, in O.J. Furuseth and M.B. Lapping (eds), *Contested Countryside: The Rural Urban Fringe in North America*. Aldershot, UK and Brookfield, VT: Ashgate. pp. 33–56.

Walker, R.A. (2001) California's golden road to riches: natural resources and regional capitalism, 1848–1940, *Annals of the Association of American Geographers*, 91, 167–199.

Walley, J.Z. (2000) Blueprint for the destruction of rural America? Available at www. paragonpowerhouse.org/blueprint_for_the_destruction_of.htm

Walmsley, D.J. (2003) Rural tourism: a case of lifestyle-led opportunities, *Australian Geographer*, 34, 61–72.

Walmsley, D.J., Epps, W.R. and Duncan, C.J. (1995) *The New South Wales North Coast, 1986–1991: Who Moved Where, Why and With What Effect?* Canberra: Australian Government Publishing Service.

Ward, C. (1990) *The Child in the Country*, 2nd edn. London: Bedford Square Press.

Ward, N. and McNicholas, K. (1998) Reconfiguring rural development in the UK: Objective 5b and the new rural governance, *Journal of Rural Studies*, 14, 27–40.

Ward, N. and Seymour, S. (1992) Pesticides, pollution and sustainability, in R. Bowler, C.R. Bryant and M.D. Nellis (eds), *Contemporary Rural Systems in Transition, Volume 1: Agriculture and Environment*. Wallingford, UK: CAB International.

Watts, J. (2001) Rural Japan braced for new riches, *Guardian*, 27 September, p. 19.

Weekley, I. (1988) Rural depopulation and counterurbanisation: a paradox, *Area*, 20, 127–134.

Weisheit, R. and Wells, L. (1996) Rural crime and justice: implications for theory and research, *Crime and Delinquency*, 42, 379–397.

Welch, R. (2002) Legitimacy of rural local government in the new governance environment, *Journal of Rural Studies*, 18, 443–459.

Westholm, E., Moseley, M. and Stenlås, N. (1999) *Local Partnerships and Rural Development in Europe*. Falun, Sweden: Darlana Research Institute.

Whatmore, S. (1990) *Farming Women: Gender, Work and Family Enterprise*. London: Macmillan.

Whatmore, S. (1991) Lifecycle or patriarchy? Gender divisions in family farming, *Journal of Rural Studies*, 7, 71–76.

Whatmore, S., Marsden, T. and Lowe, P. (1994) Feminist perspectives in rural studies, in S. Whatmore, T. Marsden and P. Lowe (eds), *Gender and Rurality*. London: David Fulton. pp. 1–30.

White, S.D., Guy, C.M. and Higgs, G. (1997) Changes in service provision in rural areas. Part 2: Changes in post office provision in Mid Wales: a GIS-based evaluation, *Journal of Rural Studies*, 13, 451–465.

Whitener, L. (1997) Rural housing conditions improve but affordability continues to be a problem, *Rural Conditions and Trends*, 8, 70–74.

Wilcox, S. (2003) *Can Work – Can't Buy*. York, UK: York Publishing Services.

Wilkins, R. (1992) Health of the rural population: selected indicators, in R.D. Bollman, (ed.), *Rural and Small Town Canada*. Toronto: Thompson Educational Publishing.

Williams, B. (1999) Rural victims of crime, in G. Dingwall and S.R. Moody (eds), *Crime and Conflict in the Countryside*. Cardiff, UK: University of Wales Press. pp. 160–183.

Williams, K., Johnstone, C. and Goodwin, M. (2000) CCTV surveillance in urban Britain: beyond the rhetoric of crime prevention, in J. Gold and G. Revill (eds), *Landscapes of Defence*. London: Prentice Hall. pp. 168–187.

Williams, M.V. (1985) National park policy 1942–1984, *Journal of Planning and Environmental Law*, 359–377.

Williams, R. (1973) *The Country and the City*. London: Chatto and Windus.

Williams, W.M. (1956) *The Sociology of an English Village: Gosforth*. London: Routledge and Kegan Paul.

Williams, W.M. (1963) *A West Country Village: Ashworthy*. London: Routledge and Kegan Paul.

Wilson, A. (1992) *The Culture of Nature: North American Landscape from Disney to the Exxon Valdez*. Cambridge, MA and Oxford, UK: Blackwell.

Wilson, B. (1981) *Beyond the Harvest: Canadian Grain at the Crossroads*. Saskatoon, Saskatchewan: Western Producer Prairie Books.

Wilson, G. (2001) From productivism to post-productivism ... and back again? Exploring the (un)changed natural and mental landscapes of European agriculture, *Transactions of the Institute of British Geographers*, 26, 77–102.

Wilson, G. and Hart, K. (2001) Farmer participation in agri-environmental schemes: towards conservation-oriented thinking?, *Sociologia Ruralis*, 41, 254–274.

Wilson, J. (1999) Green and pleasant land 'at risk' as meadows disappear, *Guardian*, 15 March, p. 4.

Winson, A. (1997) Does class consciousness exist in rural communities? The impact of restructuring and plant shutdowns in rural Canada, *Rural Sociology*, 62, 429–453.

Winter, M. (1996) *Rural Politics*. London and New York: Routledge.

Wirth, L. (1938) Urbanism as a way of life, *American Journal of Sociology*, 44, 1–24.

Woods, M. (1997) Discourses of power and rurality: local politics in Somerset in the 20th century, *Political Geography*, 16, 453–478.

Woods, M. (1998a) Mad cows and hounded deer: political representations of animals in the British countryside, *Environment and Planning A*, 30, 1219–1234.

Woods, M. (1998b) Advocating rurality? The repositioning of rural local government, *Journal of Rural Studies*, 14, 13–26.

Woods, M. (1998c) Researching rural conflicts: hunting, local politics and actor-networks, *Journal of Rural Studies*, 14, 321–340.

Woods, M. (2000) Fantastic Mr Fox? Representing animals in the hunting debate, in C. Philo and C. Wilbert (eds), *Animal Spaces, Beastly Places*. London: Routledge. pp. 182–202.

Woods, M. (2003a) Deconstructing rural protest: the emergence of a new social movement, *Journal of Rural Studies*, 19, 309–325.

Woods, M. (2003b) Conflicting environmental visions of the rural: windfarm development in Mid Wales, *Sociologia Ruralis*, 43, 271–288.

Woods, M. (2004a) Politics and protest in the contemporary countryside, in L. Holloway and M. Kneafsey (eds), *The Geographies of Rural Societies and Cultures*. Aldershot, UK: Ashgate.

Woods, M. (2004b) Political articulation: the modalities of new critical politics of rural citizenship, in P. Cloke, T. Marsden and P. Mooney (eds), *The Handbook of Rural Studies*. London and Thousand Oaks, CA: Sage.

Woods, M. and Goodwin, M. (2003) Applying the rural: governance and policy in rural areas, in P. Cloke (ed.), *Country Visions*. London: Pearson. pp. 245–262.

Woodward, R. (1996) 'Deprivation' and 'the rural': an investigation into contradictory discourses, *Journal of Rural Studies*, 12, 55–67.

Worster, D. (1979) *Dust Bowl: The Southern Plains in the 1930s*. New York: Oxford University Press.

Yarwood, R. (2001) Crime and policing in the British countryside: some agendas for contemporary geographical research, *Sociologia Ruralis*, 41, 201–219.

Yarwood, R. and Edwards, B. (1995) Voluntary action in rural areas: the case of Neighbourhood Watch, *Journal of Rural Studies*, 11, 447–461.

Yarwood, R. and Evans, N. (2000) Taking stock of farm animals and rurality, in C. Philo and C. Wilbert (eds), *Animal Spaces, Beastly Places*. London and New York: Routledge. pp. 98–114.

Yarwood, R. and Gardner, G. (2000) Fear of crime, cultural threat and the countryside, *Area*, 32, 403–412.

Young, M. and Willmott, P. (1957) *Family and Kinship in East London*. London: Routledge and Kegan Paul.

Index

Indexed by Caroline Eley